Lecture Notes in Mecha

Series Editors

Francisco Cavas-Martínez, Departamento de Estructuras, Universidad Politécnica de Cartagena, Cartagena, Murcia, Spain

Fakher Chaari, National School of Engineers, University of Sfax, Sfax, Tunisia

Francesco Gherardini, Dipartimento di Ingegneria, Università di Modena e Reggio Emilia, Modena, Italy

Mohamed Haddar, National School of Engineers of Sfax (ENIS), Sfax, Tunisia

Vitalii Ivanov, Department of Manufacturing Engineering Machine and Tools, Sumy State University, Sumy, Ukraine

Young W. Kwon, Department of Manufacturing Engineering and Aerospace Engineering, Graduate School of Engineering and Applied Science, Monterey, CA, USA

Justyna Trojanowska, Poznan University of Technology, Poznan, Poland

Francesca di Mare, Institute of Energy Technology, Ruhr-Universität Bochum, Bochum, Nordrhein-Westfalen, Germany

Lecture Notes in Mechanical Engineering (LNME) publishes the latest developments in Mechanical Engineering—quickly, informally and with high quality. Original research reported in proceedings and post-proceedings represents the core of LNME. Volumes published in LNME embrace all aspects, subfields and new challenges of mechanical engineering. Topics in the series include:

- Engineering Design
- Machinery and Machine Elements
- Mechanical Structures and Stress Analysis
- Automotive Engineering
- Engine Technology
- Aerospace Technology and Astronautics
- Nanotechnology and Microengineering
- Control, Robotics, Mechatronics
- MEMS
- Theoretical and Applied Mechanics
- Dynamical Systems, Control
- Fluid Mechanics
- Engineering Thermodynamics, Heat and Mass Transfer
- Manufacturing
- Precision Engineering, Instrumentation, Measurement
- Materials Engineering
- Tribology and Surface Technology

To submit a proposal or request further information, please contact the Springer Editor of your location:

China: Ms. Ella Zhang at ella.zhang@springer.com
India: Priya Vyas at priya.vyas@springer.com
Rest of Asia, Australia, New Zealand: Swati Meherishi at swati.meherishi@springer.com
All other countries: Dr. Leontina Di Cecco at Leontina.dicecco@springer.com

To submit a proposal for a monograph, please check our Springer Tracts in Mechanical Engineering at http://www.springer.com/series/11693 or contact Leontina.dicecco@springer.com

Indexed by SCOPUS. All books published in the series are submitted for consideration in Web of Science.

More information about this series at http://www.springer.com/series/11236

José Machado · Filomena Soares ·
Justyna Trojanowska · Sahin Yildirim
Editors

Innovations in Mechatronics Engineering

Editors
José Machado
Department of Mechanical Engineering
University of Minho
Guimarães, Portugal

Justyna Trojanowska
Poznan University of Technology
Poznan, Poland

Filomena Soares
Department of Industrial Electronics
University of Minho
Guimarães, Portugal

Sahin Yildirim
Department of Mechatronics Engineering,
Faculty of Engineering
Erciyes University
Kayseri, Turkey

ISSN 2195-4356   ISSN 2195-4364 (electronic)
Lecture Notes in Mechanical Engineering
ISBN 978-3-030-79167-4   ISBN 978-3-030-79168-1 (eBook)
https://doi.org/10.1007/978-3-030-79168-1

© The Editor(s) (if applicable) and The Author(s), under exclusive license
to Springer Nature Switzerland AG 2022
This work is subject to copyright. All rights are solely and exclusively licensed by the Publisher, whether the whole or part of the material is concerned, specifically the rights of translation, reprinting, reuse of illustrations, recitation, broadcasting, reproduction on microfilms or in any other physical way, and transmission or information storage and retrieval, electronic adaptation, computer software, or by similar or dissimilar methodology now known or hereafter developed.
The use of general descriptive names, registered names, trademarks, service marks, etc. in this publication does not imply, even in the absence of a specific statement, that such names are exempt from the relevant protective laws and regulations and therefore free for general use.
The publisher, the authors and the editors are safe to assume that the advice and information in this book are believed to be true and accurate at the date of publication. Neither the publisher nor the authors or the editors give a warranty, expressed or implied, with respect to the material contained herein or for any errors or omissions that may have been made. The publisher remains neutral with regard to jurisdictional claims in published maps and institutional affiliations.

This Springer imprint is published by the registered company Springer Nature Switzerland AG
The registered company address is: Gewerbestrasse 11, 6330 Cham, Switzerland

Preface

This volume of Lecture Notes in Mechanical Engineering gathers selected papers presented at the first International Scientific Conference ICIE'2021, held in Guimarães, Portugal, on June 28–30, 2021. The conference was organized by School of Engineering of University of Minho, throughout MEtRICs and ALGORITMI Research Centres.

The aim of the conference was to present the latest engineering achievements and innovations and to provide a chance for exchanging views and opinions concerning the creation of added value for the industry and for the society. The main conference topics include (but are not limited to):

- Innovation
- Industrial Engineering
- Mechanical Engineering
- Mechatronics Engineering
- Systems and Applications
- Societal Challenges
- Industrial Property

The organizers received 213 contributions from 24 countries around the world. After a thorough peer review process, the committee accepted 126 papers written by 412 authors from 18 countries for the conference proceedings (acceptance rate of 59%), which were organized in three volumes of Springer Lecture Notes in Mechanical Engineering.

This volume, with the title "Innovations in Mechatronics Engineering," specifically reports on innovative control and automation concepts for applications in a wide range of fields, including industrial production, medicine and rehabilitation, education and transport, with a special focus on cutting-edge control algorithms for mobile robots and robot manipulators, innovative industrial monitoring strategies for industrial process, improved production systems for smart manufacturing, and discusses important issues related to user experience, training and education, as well as national developments in the field of mechatronics. Last but not least, it provides a timely overview and extensive information on trends and technologies behind the future developments of mechatronics systems in the era of Industry 4.0. This book consists of 41 chapters, prepared by 147 authors from 10 countries.

Extended versions of selected best papers from the conference will be published in the following journals: Sensors, Applied Sciences, Machines, Management and Production Engineering Review, International Journal of Mechatronics and Applied Mechanics, SN Applied Sciences, Dirección y Organización, Smart Science, Business Systems Research, and International Journal of E-Services and Mobile Applications.

A special thank to the members of the International Scientific Committee for their hard work during the review process.

We acknowledge all that contributed to the staging of ICIE'2021: authors, committees, and sponsors. Their involvement and hard work were crucial to the success of ICIE'2021.

June 2021

José Machado
Filomena Soares
Justyna Trojanowska
Şahin Yildirim

Contents

Screwing Process Monitoring Using MSPC in Large Scale Smart Manufacturing 1
Humberto Nuno Teixeira, Isabel Lopes, Ana Cristina Braga, Pedro Delgado, and Cristina Martins

Comparison of Neural Networks Aiding Material Compatibility Assessment 14
Izabela Rojek, Ewa Dostatni, and Piotr Kotlarz

Parameterized State Feedback Control Applied to the 1st Degree of Freedom of a Cylindric Pneumatic Robot 25
Marcos G. Q. Rijo, Eduardo A. Perondi, Mário R. Sobczyk S., and Carlos A. C. Sarmanho Jr.

Building a Mobile Application by Combining Third Party Platforms in Order to Reduce Time, Costs and Improve Functionality 37
Marian Tanasie and Ionel Simion

Conveyor Belts Joints Remaining Life Time Forecasting with the Use of Monitoring Data and Mathematical Modelling 44
Edward Kozłowski, Anna Borucka, Yiliu Liu, and Dariusz Mazurkiewicz

Experience in Implementing Computer-Oriented Methodological Systems of Natural Science and Mathematics Research Learning in Ukrainian Educational Institutions 55
Olena Hrybiuk

Optimal Preventive Maintenance Frequency in Redundant Systems ... 69
Guilherme Kunz

Experimental Investigation of the Effect of Mass Load on Flight Performance of an Octorotor and Dodecarotor UAV 81
Şahin Yildirim, Nihat Çabuk, and Veli Bakircioğlu

Indoor GPS System for Autonomous Mobile Robots Used in Surveillance Applications 90
Philip Coandă, Mihai Avram, Victor Constantin, and Bogdan Grămescu

Incorporating Inteco's 3D Crane into Control Engineering Curriculum 99
Frantisek Gazdos and Lenka Sarmanova

Design of Laser Scanners Data Processing and Their Use in Visual Inspection System 112
Ivan Kuric, Matej Kandera, Jaromír Klarák, Miroslav Císar, and Ivan Zajačko

Monitoring System of Taekwondo Athletes' Movements: First Insights ... 119
Tudor Claudiu Tîrnovan, Pedro Cunha, Vítor Carvalho, Filomena Soares, Camelia Avram, and Adina Aştilean

Performance Evaluation of the BioBall Device for Wrist Rehabilitation in Adults and Young Adults 129
Bárbara Silva, Ana Rita Amorim, Valdemar Leiras, Eurico Seabra, Luís F. Silva, Ana Cristina Braga, and Rui Viana

Smart Packages Tracking System 141
Camelia Avram, Mihai Modoranu, Dan Radu, and Adina Aştilean

Automatic Warehouse for Workshop Tools 154
Marco Ferreira, Miguel Rodrigues, and Caetano Monteiro

New Refinement of an Intelligent System Design for Naval Operations 164
M. Filomena Teodoro, Mário J. Simões Marques, Isabel Nunes, Gabriel Calhamonas, and Marina A. P. Andrade

ICT4Silver: Design Guidelines for the Development of Digital Interfaces for Elderly Users 178
Nuno Martins, Sónia Ralha, and Ricardo Simoes

Dynamic Analysis of a Robot Locomotion for an External Pipe Inspection and Monitoring 189
Bogdan Grămescu, Adrian Cartal, Ahmed Sachit Hashim, and Constantin Nițu

The Choice of the Electric Energy Storage Device Type for the Hybrid Power Drive of Military Wheeled Vehicles 201
Dmitriy Volontsevich, Sergii Strimovskyi, Ievgenii Veretennikov, Dmytro Sivykh, and Vadym Karpov

Machinery Retrofiting for Industry 4.0 213
Pedro Torres, Rogério Dionísio, Sérgio Malhão, Luís Neto, and Gil Gonçalves

Conceptual Design of a Positioning System for Systematic Production of Needle Beds 221
Luis Freitas, Rui Oliveira, Teresa Malheiro, A. Manuela Gonçalves, José Vicente, Paula Monteiro, and Pedro Ribeiro

Selection and Development of Technologies for the Education of Engineers in the Context of Industry 4.0 236
Pedro José Gabriel Ferreira, Silvia Helena Bonilla, and José Benedito Sacomano

An Exploratory Approach with EEG – Electroencephalography in Design as a Research and Development Tool 245
Bernardo Providência and Rute Silva

Modelling IT Specialists Competency in the Era of Industry 4.0 257
Maciej Szafrański, Selma Gütmen, Magdalena Graczyk-Kucharska, and Gerhard Wilhelm Weber

Original Constructive Solutions for the Development of Industry 4.0 in Romania 270
Gheorghe Gheorghe, Badea Sorin-Ionut, Iulian Ilie, and Despa Veronica

Metrology Information in Cyber-Physical Systems 285
João Sousa, João Silva, and José Machado

Overview of Collaborative Robot YuMi in Education 293
Jiri Vojtesek and Lubos Spacek

Reliability of Replicated Distributed Control Systems Applications Based on IEC 61499 301
Adriano A. Santos, António Ferreira da Silva, António Magalhães, and Mário de Sousa

Inspection Robotic System: Design and Simulation for Indoor and Outdoor Surveys 313
Pierluigi Rea, Erika Ottaviano, Fernando J. Castillo-García, and Antonio Gonzalez-Rodríguez

Dutch Auction Based Approach for Task/Resource Allocation 322
Eliseu Pereira, João Reis, Gil Gonçalves, Luís Paulo Reis, and Ana Paula Rocha

MOBEYBOU 334
Hugo Baptista Lopes, Vítor Carvalho, and Cristina Sylla

Model of Acquiring Transversal Competences Among Students on the Example of the Analysis of Communication Competences 351
Marek Goliński, Małgorzata Spychała, and Marek Miądowicz

Mobile Applications in Engineering Based on the Technology of Augmented Reality 366
Tetiana Zhylenko, Vitalii Ivanov, Ivan Pavlenko, Nataliia Martynova, Yurii Zuban, and Dmytro Samokhvalov

A Review in the Use of Artificial Intelligence in Textile Industry 377
Filipe Pereira, Vítor Carvalho, Rosa Vasconcelos, and Filomena Soares

***HiZeca:* A Serious Game for Emotions Recognition** 393
Pedro Santos, Vinícius Silva, João Sena Esteves, Ana Paula Pereira, and Filomena Soares

Portable Bathing System for Bedridden People 406
M. Leonor Castro-Ribeiro, A. A. Vilaça, Mariana A. Pires, Karolina Bezerra, Cândida Vilarinho, and Ana Olival

Manufacture of Facial Orthosis in ABS by the Additive Manufacturing Process: A Customized Application in High Performance Sports 422
Anna Kellssya Leite Filgueira, Isabella Diniz Gallardo, Ketinlly Yasmyne Nascimento Martins, Rodolfo Ramos Castelo Branco, Karolina Celi Tavares Bezerra, and Misael Elias de Morais

E-Health in IDPs Health Projects in Pakistan 433
M. Irfanullah Arfeen, Adil Ali Shah, and Demetrios Sarantis

Environmental Parameters Monitoring System with an Application Interface for Smartphone 449
Jorge Ramos, André Teixeira, Carlos Arantes, Sérgio Lopes, and João Sena Esteves

Mechatronic Design of a Wall-Climbing Drone for the Inspection of Structures and Infrastructure 460
Erika Ottaviano, Pierluigi Rea, Massimo Cavacece, and Giorgio Figliolini

Influence of Magnitude of Interaction on Control in Decentralized Adaptive Control of Two Input Two Output Systems 468
Karel Perutka

Author Index 481

Screwing Process Monitoring Using MSPC in Large Scale Smart Manufacturing

Humberto Nuno Teixeira[1(✉)], Isabel Lopes[1], Ana Cristina Braga[1], Pedro Delgado[2], and Cristina Martins[2]

[1] ALGORITMI Research Centre, University of Minho, Guimarães, Portugal
`b6440@algoritmi.uminho.pt`
[2] Bosch Car Multimedia Portugal SA, Braga, Portugal

Abstract. The ability to obtain useful information to support decision-making from big data sets delivered by sensors can significantly contribute to enhance smart manufacturing initiatives. This paper presents the results of a study performed in an automotive electronics assembly line. An approach that uses Multivariate Statistical Process Control based on Principal Component Analysis (MSPC-PCA) was applied to early detect undesirable changes in the screwing processes performance by extracting relevant information from the torque-angle curve data. Since the data of different torque-angle curves are not aligned, the proposed approach includes the linear interpolation of the original data to enable Principal Component Analysis (PCA). PCA proved to be an appropriate technique to obtain significant information from the process variables, which consist of the successive value of the torque at constant angular intervals. Score plots and multivariate control charts were used to detect defective tightening and identify behaviors that represent inefficient tightening. This is a new approach that can be applied to effectively monitor screwing processes in the assembly of different products either periodically or in real-time.

Keywords: Multivariate Statistical Process Control (MSPC) · Principal Component Analysis (PCA) · Screwing process · Smart manufacturing

1 Introduction

Modern manufacturing systems need to provide high quality products developed in efficient, fast and cost reduced processes [1]. The increased diversity of products also implies less time to ensure that the process is capable of producing appropriate quality [2]. Problems that affect quality are often related to assembly errors [3]. There are two basic categories of assembly tasks: parts mating and parts joining. Fasteners are commonly used to join components together in industry. The screwdrivers allow tightening of threaded fasteners with a specified torque value [4]. The usage of modern equipment involves complex parameter settings of and assessment requirements [5].

Assembly process control is a complex problem whose solution must consider all the particularities and connection requirements of a given mating [6]. Therefore, a high level of process automation demands a high level of automated monitoring and control [2]. Errors detection should occur at an early stage of the process to avoid any damage and to ensure that the assembly is completed according to the requirements and specifications [7]. Thus, to prevent technical risks, it becomes important to develop methods for providing process related information to the operator [8].

Most automatic screw fastening processes are monitored using torque sensors and a target torque for a consistent assembly is usually established. However, monitoring torque alone does not ensure the required clamping force. The torque applied to provide the desired clamping force in bolted joints can differ significantly for the same bolt type due to the combined effect of several factors, such as thread and under head friction, thread deformations and variations in bolt diameter. Therefore, problems during the tightening operation must be detected to avoid wrong conclusions about the screwing process quality. To obtain more complete information about the tightening operation performance, both torque and rotation angle should be monitored [9]. The torque-angle curves provide relevant information to properly qualify the capability of tightening tools [10].

Smart manufacturing aims to support accurate and timely decision-making from real-time data delivered by sensors [11]. The quick detection of changes, particularly increases in process variability, is essential for quality control [12]. Statistical Process Control (SPC) is frequently employed to monitor and detect relevant changes in manufacturing processes [13]. When Principal Component Analysis (PCA) technique is implemented to monitor industrial processes, Hotelling's T^2 and Squared Prediction Error (SPE) statistics are used for the detection of process disturbances which can originate failures. However, although control charts allow to detect deviations from normal operating region, they do not indicate reasons for the deviations [14]. Once the fault is exposed, the contribution of each original variable for T^2 or SPE statistics can be determined based on contribution plots. The contributions plot shows the most affected variables, so that the causes can be identified and actions to bring the process to the statistical control region can be implemented [15].

The monitoring and control of screwing process parameters has been addressed by several studies [7–9, 16]. Many of the proposed methods were designed to control the screwing process of specific parts and their generalization ability has not been proven. It was also noticed that error detection effectiveness is often assessed based on simulated experiments. Furthermore, the multivariate nature of the screwing processes is not contemplated by the exiting monitoring approaches. In industrial context, the monitoring strategies for automated tightening processes are generally focused on torque, torque-angle, torque rate and variation of rate monitoring [17]. For this purpose, an individual analysis of each monitored variable is usually performed to verify their conformance to specifications during each tightening operation. The investigation of causes of variation is not an integral part of these strategies. Therefore, a more extended application in real scenarios is still needed in most cases. For this reason, methods that provide a broader understanding of the screwing processes to the operator should be developed to support the definition of more informed improvement initiatives and analyze their impact over a sequence of tightening operations. This is particularly relevant for processes with short operating cycles, since when a change in the process is verified a high number of tightening operations can be affected.

In this paper, an approach is proposed to monitor screwing processes' through a visual representation of their performance over time and to assist the identification of factors responsible for inefficient tightening based on torque-angle curves data. This approach was tested and refined based on its application to the screwing process of an assembly line of Bosch Car Multimedia Portugal. PCA was performed using the torque-angle curves data collected by the screwing machine controller. To enable PCA, the data of different torque-angle curves were aligned by performing linear interpolations. Then, the process was monitored in different periods using multivariate control charts. The structure of the paper is organized as follows. Section 2 presents the approach steps through its application on the assembly line screwing process and the respective results. In Sect. 3, the main results of the study are discussed. Section 4 summarizes the conclusions derived from the study and presents recommendations for future research.

2 Screwing Process Monitoring and Analysis

2.1 Sample Definition

The screwing process analysis was performed using a data set composed of 12327 observations (Table 1). All observations belong to the same product type and were executed in a unique workplace. The product is an automotive electronics system which is assembled by placing seven screws according to a predetermined sequence. The overall data set includes data concerning defective ("bad") and non-defective ("good") tightening operations. The first sample includes only 7 non-sequential cases classified as "bad". These data were included in the analysis to better test the accuracy of the PCA model by considering a higher diversity of unsuccessful cases. Samples 2 to 5 are composed of sequential cases and include all operations performed in a working day.

Table 1. Description of the screwing process data samples.

Sample		1	2	3	4	5
Date		22/09/2017	29/09/2017	02/10/2017	11/10/2017	12/10/2017
Result	Good	–	2273	3189	2959	3880
	Bad	7	5	7	2	5
Number of cases		7	2278	3196	2961	3885

2.2 Data Interpolation

A preliminary analysis of original data collected from a screwdriver at regular time intervals revealed significant fluctuations in the amplitude of the rotation angle intervals (Table 2), since the angular displacement of the screw is not constant. Therefore, linear interpolation was the method selected to align the data and generate a torque value to each rotation angle of the predefined scale. To provide a quick result, an algorithm in R code that performs linear interpolations was developed. The obtained information allows to represent the different torque-angle curves in the same scale.

Table 2. Extract from a screwing process data set.

Angle	4	6	11	15	17	21	22	28	30	35	38	39	44
Torque	10,9	11	11,1	11,2	11,5	11,6	11,7	12	12,1	12,2	12,3	12,4	12,4
Angle	51	55	58	62	72	73	83	89	93	96	99	103	106
Torque	12,3	12,4	12,6	12,8	12,9	12,8	13	13,3	13,4	13,5	13,6	13,8	13,9
Angle	1	3	4	9	14	16	20	24	28	34	37	42	45
Torque	10,3	10,4	10,4	10,4	10,6	10,4	10	9,4	9,2	9	9,1	9	9,3

In this study, the linear interpolation was performed to determine values of torque in a range between 0 and 2300° at intervals of 10°. Since the considered variables consist of the successive value of the torque at constant angular intervals (Fig. 1), a data set composed of 231 primary variables was obtained.

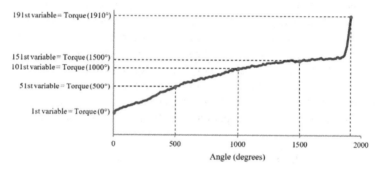

Fig. 1. Generic representation of a set of variables from a tightening operation.

2.3 Data Segmentation

The PCA model was formed by the values of 504 torque-angle curves (Fig. 2a). Only cases classified as "good" were included in the Normal Operating Conditions (NOC) data set and each screw, considering its position in the product, is represented by the same number of observations. Firstly, a set of 252 observations from a sequence of 36 consecutive products obtained in sample 3 was selected. Then, to endow the NOC data with a higher variety of behaviors, another group of 252 cases selected from samples 2 to 5 was added. Figure 2b shows three distinctive zones of a typical torque-angle curve, classified by Shoberg [17] as Rundown (R), Snugging (S) and Elastic Clamping (EC).

The torque-angle curves were represented with the Unscrambler® and the data analysis was performed using ProSensus MultiVariate software.

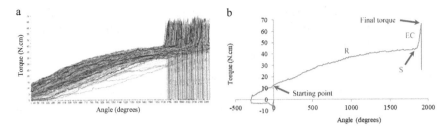

Fig. 2. NOC torque-angle curves (a) and typical torque-angle curve (b).

2.4 Principal Components Identification and Interpretation

The number of principal components extracted by the PCA model was defined based on the cumulative explained variation by the ordered components (Table 3). Since the fourth principal component has little relevance, only the first three were considered.

Table 3. Variation explained by the first four principal components.

Principal component	Variation explained	Cumulative variation explained
First	62.24%	62.24%
Second	20.40%	82.64%
Third	3.70%	86.34%
Fourth	2.06%	88.40%

The interpretation of the principal components was performed by statistics and process experts, based on the variables loadings' (Fig. 3) and considering technical knowledge about the process. In order to obtain a broader perspective of the tightening operation, the loadings of six secondary variables were also analyzed. These variables are maximum torque, absolute angle, total time, end torque, end angle and screwing energy.

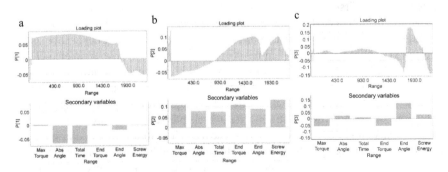

Fig. 3. Loading plots of the first (a), second (b) and third (c) principal components.

The loading plot of the first principal component (Fig. 3a) shows high loadings over the rundown zone and a significant decrease after 1740°, that represents the instant in which the snugging zone is reached. The higher loadings correspond to rotation angles wherein friction was higher. As at the elastic clamping zone, tightening is mainly affected by the elastic deformation of the screw, the corresponding variables' loadings tend to be substantially lower. After 1790°, the loadings become negative and start to assume a decreasing trend, as more tightening operations are completed. The negative loadings exhibited in this period are due to the decision of extending the tightening curves up to 0 after the final torque is achieved. The secondary variables' loadings are represented on the left side of the plot and shown with more detail below in Fig. 3a. Absolute angle, total time and end angle have negative loadings. All the other variables assume a positive correlation with the first component, since higher friction losses entail greater energy expenditure and higher torques to complete the tightening operation. However, the required clamping force is reached within a shorter period. Thus, the first principal component refers to the torque variation during the rundown zone. It allows to distinguish screws that exhibit high torques at the rundown zone, which is a behavior that can be caused by differences between screws or tighter threads.

In the second principal component loading plot (Fig. 3b), negative loadings until the angle 830 can be identified. The lowest loadings are exhibited at the beginning of the rundown zone. Afterwards, a growing trend is manifested. However, a decrease is verified in the approximation to the snugging zone and after 2090°. The secondary variables that show higher loadings are maximum torque, end torque and screw energy, since greater torques are needed to produce the required clamping force. The screwing energy loading is significantly higher in this principal component. Thus, the second principal component enables to differentiate screws that have low torques at the rundown zone, a relatively high final torque and whose tightening operation is concluded later. This behavior can be caused by more open threads due to rework, since the tightening operation must be repeated. Furthermore, this principal component also represents delays in engagement, manifested by the presence of positive loadings in the absolute angle, total time and end angle.

The third principal component loading plot exhibits several fluctuations over the entire scale of rotation angles (Fig. 3c). These shifts reflect slope variations in the torque-angle curves. Nevertheless, the loadings' value remains close to 0 along the angles that correspond to the first two-thirds of the tightening operation. Between 1260 and 1780°, the loadings become negative and reveal a decreasing trend until the snugging zone. At the snugging zone, the loadings are positive and significantly higher. Negative loadings are also obtained after 2130° when most operations were already completed. The end angle is the secondary variable which assumes the higher loading, and both end torque and maximum torque are the only variables with negative loadings. Thus, this principal component represents screws that manifest a gradual growth of the torque with some fluctuations at the rundown zone followed by an interval in which the increase of the torque ceases before the snugging zone is reached. This circumstance contributes to the delay of the tightening operation and leads to a lower end torque. In conclusion, the third principal component refers to differences between threads. The thread crests can often exhibit irregularities or slightly different formats (e.g., rounded, sharpened, flat) that can result in sudden variations in the torque evolution.

2.5 Control Limits Definition

To identify both defective and inefficient tightening, control limits defined by the confidence intervals of 99% and 99.73% were determined and represented using ProSensus MultiVariate. In order to test the reliability of the PCA model, the defective cases were represented with the NOC observations in the Hotelling's T^2 and SPE control charts defined based on the first three principal components (Fig. 4). Both charts show all the observations classified as "bad" above the control limits. However, in the SPE control chart the distance to the limits is shorter for a set of 8 cases that reached higher torques.

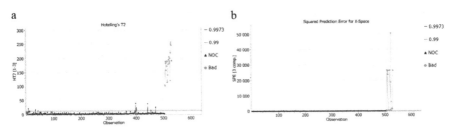

Fig. 4. Hotelling's T^2 (a) and SPE (b) control charts of the defective tightening cases.

2.6 Behavior Identification

The scatter score plot formed by the first two principal components (Fig. 5) was used to analyze the NOC observations. This analysis allowed to identify nine cases outside the limit defined by the confidence interval of 99.73%. Three correspond to significant delays at the beginning of the tightening operation (nearly 2 turns), five exhibit low torques during the entire tightening operation, and one case manifests a delay of nearly 1 turn in which the torque values are low at the rundown zone and high at the end. The first group is closer to the negative side of the first principal component axis. Whereas the second group is located at a similar distance from the negative side of both principal components' axes. These behaviors correspond to delays in engagement and well succeed reworks. The latter case is the observation that is further away from the limits. Although these behaviors resulted from special causes of variation, they cannot be completely removed. Thus, the cases that fall outside the limits were included in the NOC.

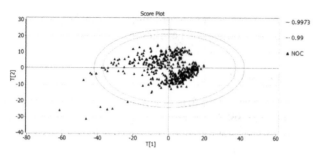

Fig. 5. Scatter score plot of the NOC observations.

2.7 Process Monitoring

The process was monitored with the data from the samples described in Table 1, except those included in the NOC. The sequential set of 11 823 observations was represented in scatter score plots, time series score plots and multivariate control charts.

In the scatter score plot formed by the first two principal components several outliers were identified (Fig. 6a). The observations that deviate most from the distribution mean are the cases classified as "bad" and delays in engagement. Although this plot allows to identify unusual behaviors in each sample, it does not reveal their distribution over time. Thus, the scores were also represented in time series score plots. These plots exhibit the scores distribution over time with respect to each principal component, allowing to determine if there are any trends or fluctuations and differences between samples.

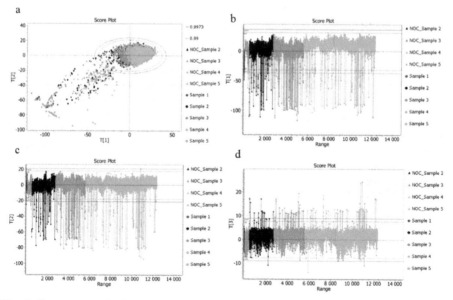

Fig. 6. Scatter score plot formed by the first two principal components (a); time series score plot of the first (b), second (c) and third (d) principal components.

In the first principal component time series score plot (Fig. 6b), several fluctuations are shown. Furthermore, the scores tend to align above the origin of the plot. However, this trend is more marked in the interval formed by the observations of samples 4 and 5. At this period, the duration of the tightening operations was shorter and higher torques at the rundown zone were attained. Since the first principal component represents the torque variation during the rundown zone, the scores of the cases with higher torques are located above the origin. In contrast, the scores that fall below the lower limit mainly correspond to defective tightening, delays and observations that exhibit low torques during the rundown zone. The scores distribution is more centered around the origin and shows lower dispersion in the second principal component time series score plot (Fig. 6c), which has lower accuracy in detecting delays. Thus, only observations that represent delays of approximately 3 turns or more comparing to the mean are outside the limits. Nevertheless, this plot is more effective in detecting observations with low torques during the entire tightening operation and shows all the defective cases below the lower limit. The existence of low torques during the rundown zone can be caused by product rework, since in most cases this behavior is verified in the torque-angle curves of the seven screws applied to the same product. The period formed by the observations of samples 4 and 5 includes a considerable number of cases with low torques at the rundown zone. This characteristic also occurs in a significant number of cases of sample 2 and is less frequent in sample 3. The third principal component time series score plot (Fig. 6d) manifests high dispersion in the scores' distribution in all the analyzed samples. However, at short periods the process becomes slightly more stable. The greater dispersion reflects a higher diversity of slope variations in the torque-angle curves. The observations identified above the upper limit comprise defective cases and the most extreme delays, whereas below the lower limit only a group of four observations that reveal very atypical variations in the torque-angle curve was detected.

The data sample in which greater instability is observed is sample 3. In the interval between observations 3435 and 3729, most of the scores lie above the origin in the plot of Fig. 6b. At the same time, a steep decrease of the scores related to the second principal component is verified and a higher number of observations below the origin is shown in the plot of Fig. 6d. Based on the analysis of the torque-angle curves that correspond to the observations of this interval (Fig. 7a) it was found that high torques are exhibited during the entire tightening operation and no delays occurred. A significant number of delays was verified thereafter in the period between observations 3730 and 4025 (Fig. 7b). This fact is reflected by the number of observations outside the limit defined by the confidence interval of 99.73% in the three time series score plots.

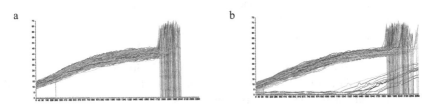

Fig. 7. Torque-angle curves: between 3435 and 3729 (a), and between 3730 and 4025 (b).

The Hotelling's T^2 control chart (Fig. 8a) shows several observations above the limits. The highest points correspond to the screwing cases classified as "bad". However, some observations which relate to delays and low torques at the rundown zone are also above the limits. In the SPE control chart (Fig. 8b), the observations which are further away from the control limits refer to defective tightening. Whereas observations that correspond to delays, although they are above the limits, reveal much lower deviations from the principal components' subspace. In addition, a small group of observations that were concluded considerably earlier are also outside the limits of this chart.

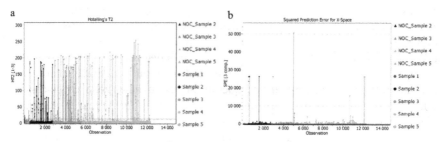

Fig. 8. Hotelling's T^2 control chart (a) and SPE control chart (b)

3 Discussion of Results

In this study, an approach for screwing processes monitoring using Multivariate Statistical Process Control based on Principal Component Analysis (MSPC-PCA) was developed and tested with data collected from a manual screwdriver of an automotive electronics assembly line. A PCA model that is deemed to represent the NOC of the process for a specific product type was defined and validated according to the proposed approach. The results show that 86.34% of the total variation can be explained only by three principal components. Since the analysis involves 231 primary variables, a univariate analysis would be extremely time consuming and would not consider correlations between variables. The interpretation of the principal components' physical meaning enabled to more accurately analyze the information provided by the monitoring tools. Moreover, it was possible to develop a detailed understanding of the process and to ascertain the causes of unusual behaviors. The structure of the principal components was partially influenced by the decision of assigning 0 to the variables that succeed the final torque to avoid missing values in the scale of rotation angles. Another option could be to maintain the value of the final torque until the end of the scale.

The time series score plots enabled to observe fluctuations in the scores' distribution, which represent sudden or gradual changes that affect the tightening behavior over time. These shifts are more visible in the first and second principal components time series score plots, since they are related to delays and increase of the number of reworks. The Hotelling's T^2 and SPE control charts show all the observations related to defective tightening above the control limits. In addition, cases that exhibit unusual behaviors, such as delays or low torques at the rundown zone, were also easily identified both in the score plots and control charts. Some of these observations formed distinct groups of outliers in the scatter score plot of the first two principal components. Comparing both control charts, it can be verified that delays are more easily detected in Hotelling's T^2, since those cases reflect more significant deviations from the distribution mean.

MSPC-PCA can be applied using an appropriate software available in the market. If the proposed approach is used to monitor screwing processes in real-time, this functionality must also be provided by the software package. The screwdriver operator and the responsible by the screwing process should be able to monitor the process using four main tools, such as the scatter score plots formed by pairs of principal components, time series score plots concerning each principal component, Hotelling's T^2 and SPE control charts. The uncommon behaviors depicted in the scatter score plots should be promptly identified and classified considering the principal components' physical meaning and making comparisons with the NOC observations scores, namely groups that are further away from the distribution mean. Furthermore, the knowledge acquired from the PCA model definition must also be used to interpret points outside the control limits, patterns and trends both in the time series score plots and multivariate control charts. Whenever significant anomalies are detected, their causes should be investigated considering the process variables which contributed most to the identified shifts or inferred based on experience. Subsequently, appropriate corrective actions must be planned and implemented.

4 Conclusion

This paper proposes an approach that provides the ability to monitor the overall screwing process behavior over time and considers the multivariate nature of each tightening operation. The defined approach uses PCA to extract relevant information from correlated variables that consist of the successive values of the torque at constant angular intervals. Afterwards, scatter score plots, time series score plots and multivariate control charts developed based on the defined PCA model are applied to enable the process monitoring in different perspectives. The application of the proposed approach to the study data showed the capacity of these tools to detect defective tightening. Screwing cases that reveal unusual behaviors are also identified and classified. Thus, performing linear interpolations to align the screwing process data proved to be an essential and appropriate decision for the successful application of PCA.

The case study enabled to define and validate a PCA model that represents the NOC of a workplace for a specific product type. Considering the obtained results, Bosch Car Multimedia Portugal decided to monitor its screwing processes in real-time based on the defined approach. The same procedure will be adopted to define NOC models to monitor other combinations of products and workplaces. It is expected that real-time monitoring will contribute to achieve significant improvements, particularly the reduction of operating cycles due to a higher awareness to delays and the identification of components out of specifications (e.g., tighter threads, irregularities in thread crests) which can often result in defective tightening. Each NOC model must be updated whenever adjustments are made in the process. This task requires considerable time and effort from the people involved. Therefore, the ability to perform self-adaptive model updating is highly desirable in this context. The PCA model used to enable the MSPC is based on data related to screws placed in seven different positions. In order to guarantee greater homogeneity, a PCA model for each screw joint can be defined. However, a more significant technical and human effort would be needed to coordinate the process monitoring due to the increase of the amount of data and tools involved.

Future work will address the development of an automatic classifier which performs feature recognition of the screwing process observations and their prompt association to specific behavior categories. Furthermore, it is intended to identify patterns or trends in the monitored data that can be associated with failure modes of the screwing equipment by analyzing the process evolution over time. Afterwards, this information will be used to prevent or early detect failures in the screwing equipment.

Acknowledgements. This work has been supported by FCT – Fundação para a Ciência e Tecnologia within the R&D Units Project Scope: UIDB/00319/2020.

References

1. Foehr, M., Lüder, A., Wagner, T., Jäger, T., Fay, A.: Development of a method to analyze the impact of manufacturing systems engineering on product quality. In: 2011 IEEE 16th Conference on Emerging Technologies & Factory Automation (ETFA), Toulouse, France, pp. 1–4. IEEE (2011)
2. Wolter, B., Dobmann, G., Boller, C.: NDT based process monitoring and control. Strojniški Vestn. – J. Mech. Eng. **57**, 218–226 (2011)
3. Tsarouchi, P., et al.: Robotized assembly process using dual arm robot. Procedia CIRP **23**, 47–52 (2014)
4. Longwic, R., Nieoczym, A.: Control of the process of screwing in the industrial screwdrivers. Adv. Sci. Technol. **10**, 202–206 (2016)
5. Yuliya, L., Ulrich, B.: Development of an information system with maturity degree management for automated screwing processes. In: Nunes, M.B., Isaías, P., Powell, P. (eds.) Proceedings of the IADIS International Conference Information Systems 2012, pp. 49–56. IADIS Press, Berlin (2012)
6. Chumakov, R.: Optimal control of screwing speed in assembly with thread-forming screws. Int. J. Adv. Manuf. Technol. **36**, 395–400 (2008)
7. Deters, C., Lam, H.K., Secco, E.L., Wurdemann, H.A., Seneviratne, L.D., Althoefer, K.: Accurate bolt tightening using model-free fuzzy control for wind turbine hub bearing assembly. IEEE Trans. Control Syst. Technol. **23**, 1–12 (2015)
8. Lebedynska, Y., Berger, U.: Intelligent knowledge-based system for the automated screwing process control. In: Proceedings of the 11th WSEAS International Conference on Artificial Intelligence, Knowledge Engineering and Data Bases (AIKED 2012), Cambridge, pp. 175–180 (2012)
9. Saygin, C., Mohan, D., Sarangapani, J.: Real-time detection of grip length during fastening of bolted joints: a Mahalanobis-Taguchi system (MTS) based approach. J. Intell. Manuf. **21**, 377–392 (2010)
10. Cristalli, C., et al.: Integration of process and quality control using multi-agent technology. In: 2013 IEEE International Symposium on Industrial Electronics, Taipei, Taiwan, pp. 1–6 (2013)
11. He, Q.P., Wang, J.: Statistical process monitoring as a big data analytics tool for smart manufacturing. J. Process Control. **67**, 35–43 (2018)

12. Yeh, A.B., Li, B., Wang, K.: Monitoring multivariate process variability with individual observations via penalised likelihood estimation. Int. J. Prod. Res. **50**, 6624–6638 (2012)
13. Siddiqui, Y.A., Saif, A.-W., Cheded, L., Elshafei, M., Rahim, A.: Integration of multivariate statistical process control and engineering process control: a novel framework. Int. J. Adv. Manuf. Technol. **78**(1–4), 259–268 (2014). https://doi.org/10.1007/s00170-014-6641-6
14. Lin, W., Qian, Y., Li, X.: Nonlinear dynamic principal component analysis for on-line process monitoring and diagnosis. Comput. Chem. Eng. **24**, 423–429 (2000)
15. Babamoradi, H., van den Berg, F., Rinnan, A.: Confidence limits for contribution plots in multivariate statistical process control using bootstrap estimates. Anal. Chim. Acta. **908**, 75–84 (2016)
16. Mura, M.D., Dini, G., Failli, F.: An integrated environment based on augmented reality and sensing device for manual assembly workstations. Procedia CIRP **41**, 340–345 (2016)
17. Shoberg, R.S.: Engineering Fundamentals of Threaded Fastener Design and Analysis. RS Technologies, a Division of PCB Load & Torque, Inc., Farmington Hills (2000)

Comparison of Neural Networks Aiding Material Compatibility Assessment

Izabela Rojek[1(✉)], Ewa Dostatni[2], and Piotr Kotlarz[1]

[1] Institute of Computer Science, Kazimierz Wielki University, Bydgoszcz, Poland
izarojek@ukw.edu.pl
[2] Faculty of Mechanical Engineering, Poznan University of Technology, Poznan, Poland

Abstract. A new method of selection of materials at the design step is presented in this paper. The method takes into recyclability of materials. The authors compare the effectiveness of neural networks (a multilayer perceptron, radial basis function networks, and self-organizing feature map - SOFM networks) as modelling tools aiding the selection of compatible materials in ecodesign. The best artificial neural networks were used in an expert system. The input data for the selection of materials was start point to initiate the study. The input data, specified in cooperation with designers, include both technological and environmental parameters which guarantee the desired compatibility of materials. Next, models were developed using the selected neural networks. The models were assessed and implemented into an expert system. The authors show which models best fit their purpose and why. Models aiding the compatible materials selection help boost the recycling properties of designed products. Neural networks are a very good tool to support the selection of materials in the ecodesign. This has been proven in the article.

Keywords: Compatibility · Neural networks · Classification models · Expert system · Materials selection

1 Introduction

Nowadays, fast development of environmental awareness is observed. Enterprises are increasingly focusing on environmentally friendly solutions due to legal regulations. Producers' attitudes are also shaped by marketing campaigns that promote eco-friendly products. Creating customer demand, they force manufacturers to supply eco-friendly products to the market. Some customer groups due to greater awareness are even willing to pay a higher price for an eco-friendly product. Manufacturers also sell products in ecological versions cheaper.

The study, exploring the materials selection in ecodesign, is a follow up on previous research into the application of artificial intelligence (AI) in the selection of materials in product design to provide for their recycling compatibility. The research has been described in [1–4]. Based on the decision tree induction methods and MLP artificial neural networks, the proposed tools automate the materials selection in the process of design, building upon the designer's knowledge gained through experience.

A new method supporting designers at the design stage in choosing materials that are compatible ensures that the product is recyclable and environmentally friendly. The aim of the authors study was to create expert system similar to human thinking in the concept of reasoning. Functions of the created expert system allows to solve certain tasks, in the similarly like a man who is an expert in the domain. The creation of such a system was possible to the use of artificial neural network methods.

2 Review of Literature

The ecodesign concept combines many aspects of traditional and environmental design. The main goal is to develop sustainable solutions that meet human needs [5]. Product recyclability is one of the basis of ecodesign. Recycling is one of the methods of environmental protection. It consists to reduce the amount of waste and the consumption of natural resources [6]. Therefore, at the initial stages of the product life cycle, you should consider what will happen to it after its end of life.

More and more ecodesign supporting tools use intelligent solutions, such as neural networks. The article [7] presents models of life cycle assessment (LCA) based on a BPNN, which allow estimating the amount of hazardous chemicals and the consumption of electronic product for the entire product lifecycle. The solution described in [8] uses the artificial neural network (ANN) for forecasting and performance of product lifecycle assessment (LCA). Any missing data required for the LCA is estimated using the ANN. Artificial intelligence is also applied in waste sorting is the deep learning based method, implemented by Refined Technologies of Sweden [9]. It recognizes products or product models with a high degree of similarity.

Many researchers have described the application of neural networks methodologies across different scientific and practical domains. Numerous research papers discuss Kohonen networks applied for multidimensional data visualization to evaluate classification possibilities of various coal types [10], collision free path planning and control of wheeled mobile robot [11], or parametric fault clustering in analog electronic circuits with the use of a self-organizing artificial neural network [12].

Radial basis function (RBF) networks are also widely discussed in the literature as a tool supporting, among others, rotor fault detection of the converter fed induction motor [13], local dynamic integration of ensemble in prediction of time series [14], predicting the corrections of the Polish time scale UTC(PL) (Universal Coordinated Time) [15], or accurate load forecasting in a power system [16]. However, solutions implementing the RBFN, Kohonen networks or MLP in ecodesign are scarce. Hence the authors' interest in the application of neural networks.

The method aiding the selection of materials in the ecodesign, which used neural networks was shown in the article. While developing the expert system, the authors used their experience from solutions supporting ecodesign [17–19] and creative of neural networks [4, 20–22].

3 Methods

Materials selection for product components is aided with the use of MLP, RBF and SOFM networks. Neural networks are very good suited to the obtain of knowledge and experience, in this case in the range of selection of compatible materials in the process of ecodesign.

The MLP network structure consists of many artificial neurons in a few connected layers. The artificial neural network is a very simplified model of the brain of a living organism, because the brain doesn't have a perfectly layered structure in which neurons are connected between the layers. The neural network model is very simplified. For simpler implementation, the neural network is built symmetrically. It consists in the fact that all neurons from one layer are connected to neurons of the next layer. The network learning process eliminates connection redundancy [23]. MLP networks have three basis features. The network is made up of layers in which neurons are located. These neurons in each layer may have a different number. The layers in the network connect to each other on the principle of: each neuron from one layer with each neuron of the previous layer. Data in the network flows from entry to exit in one direction. The network can be divided into input, hidden (or n hidden) and output layers.

MLP network is the most universal network commonly applied for resolving various problems, including technical ones [23]. But, RBF networks also have many advantages. Firstly, they are able to map any nonlinear function by a single hidden layer, unlike the MLP network where sometimes we need more than one hidden layer. Moreover, the RBF network typically has one hidden layer with radial neurons, each of which models the Gaussian process based response surface [23]. Radial networks are composed of neurons whose activation functions map (1). Their values change radially around center c.

$$x \rightarrow \varphi(\|x - c\|), \ x \in R^n \tag{1}$$

where ($\|\cdot\|$) is usually typically an Euclidean norm. Functions $\varphi(\|x - c\|)$ are referred to as radial basis functions.

Secondly, in the output layer we can performed to optimize simple linear transformation by means of traditional linear modelling techniques. The techniques are quick and free from such problems as local minima, which occur in the training of MLP networks. Therefore, we can train RBF networks in a very short time period (the difference in the training speed can reach orders of magnitude).

Kohonen networks are one of the basic types of self-organizing networks. They train in a way similar to the way human beings do, without defining any patterns – the patterns are created through the training process combined with normal functioning. Owing to their self-organizing capability, they open up new possibilities, such as adaptation to input data they have little knowledge of. SOFM networks represent an entire group of networks which learn by the self-organizing competitive method. On the network's inputs signals are set up to choose the winning neuron – the one which best corresponds to the input vector. SOFM's topology correct feature maps first choose the winning neuron (by means of the Euclidean distance), and subsequently determine the training coefficient of the winner's neighboring neurons [23].

Once the network is triggered by the input vector x, neurons compete among themselves. The winning neuron is the one whose weights are most similar to the respective components of that vector. A topological neighborhood G(i, x) is assumed around the ith neuron. In the standard Kohonen algorithm, the G(i, x) function is defined as follows (2):

$$G(i, x) = \begin{cases} 1 \text{ for } d(i, w) \leq R \\ 0 \text{ for others} \end{cases} \qquad (2)$$

where d(i, w) represents the Euclidean distance between the winning vector w_i and the ith neuron, and R – the neighborhood radius.

We identify three stages in Kohonen networks: construction, learning, and recognition.

4 Creation of Models Based on the Neural Networks

The selection of materials in the ecodesign is possible thanks to neural network models that were created in the following stages:

- analysis of input data for the selection of the materials (input data were developed on the basis of real data taken into account by the designers when selecting materials),
- creating training, testing and validation kits containing examples of the selection of materials to be used in creating neural networks and evaluation their effectiveness,
- creating models using neural networks (MLP, RBF, SOFM),
- model evaluation,
- selection of the most effective material selection models and their implementation in the expert system.

4.1 Data Preparation

Recycling oriented ecodesign relies primarily on the material selection and methods of connecting them. The main goal is to design a product made of the largest possible number of standardized and recyclable materials. This has a positive impact on the environment in the last stages of the product's lifecycle, such as maintenance or withdrawal from use. When selecting materials for a product, we are guided by their compliance in terms of recycling. Mainly the chemical composition of materials determines recycling. Special tables have been developed that show material compatibility due to recycling.

For a detailed analysis, selected properties of materials have been added upon consultation with designers. Parameters of material selection were chosen for AGD products. The files have been prepared based on an analysis of properties of materials, such as: MM - main material name (e.g. PVC), D - density in grams per cubic centimeter (e.g., 7.88), TS - tensile strength expressed in mega Pascal (e.g., 35.5), YPE - elongation at yield point (Re) expressed as a percentage value (e.g., 5.5), PT - processing temperature expressed in degrees centigrade (e.g., 20.8), DC - the dielectric constant (e.g., 2.0), ME - Young's modulus (elasticity) expressed in gigapascal (e.g., 4.61), WA - water absorbency expressed as a percentage value (e.g., 22.55), RC - the recycling cost expressed in PLN per kilogram (e.g., 4.25), where a positive value represents a profit from the sale of material, and a negative one – the disposal cost, and AM - the added material name (e.g. ABS). Properties of materials give inputs of MLP network and output is C - compatibility. The data set includes 980 examples.

4.2 Material Selection by Neural Networks

In experiments authors built many models of MLP neural networks. The input and output of the network was the same for all built models (10 inputs and 1 output) (see Fig. 1). In order to develop the best model of the neural network, various parameters were changed in the experiments. Changed: the number of neurons in the hidden layer (5–25), in the BFGS algorithm the number of training periods (10–120), the error function (SOS, Entropy), the activation function in the hidden and output layer (linear, logistic, exponential, Tanh, Softmax). Seven neural networks of varying efficacy (activity quality) were presented in Table 1.

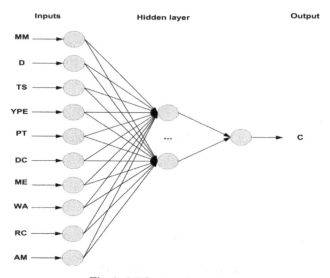

Fig. 1. MLP network structure.

MLP 10-23-1, MLP 10-16-1, and MLP 10-21-1 neural networks were the most effective neural networks (100%). MLP 10-25-1 (96.87%), MLP 10-9-1 (97.12%), MLP 10-16-1 (99.22%), and MLP 10-25-1 (96.34%) neural networks were a little less effective. MLP 10-12-1 turned out to be the least effective (76.48%). Table 2 shows the analysis of neural network errors. In the case of networks with lower efficiency, these networks classified some examples incorrectly.

The MLP 10-23-1 network had 0% errors (100%, best network). MLP 10-25-1 was slightly worse (35 errors) and MLP 10-12-1 showed the worst classification (220 errors).

Table 1. The neural networks (MLP) models for material selection.

ID	S	TQ	TEQ	VQ	TA	FE	FHA	FOA
1	10-25-1	96.06	97.96	96.60	38	Entropy	Logistic	Softmax
2	10-23-1	100.00	100.00	100.00	80	SOS	Tanh	Linear
3	10-16-1	100.00	100.00	100.00	30	Entropy	Tanh	Softmax
4	10-9-1	96.79	97.96	96.59	57	Entropy	Logistic	Softmax
5	10-16-1	95.19	96.59	95.24	35	Entropy	Logistic	Softmax
6	10-25-1	96.50	97.95	94.56	39	Entropy	Logistic	Softmax
7	10-21-1	100.00	100.00	100.00	60	SOS	Logistic	Tanh

where: ID – network of id, S – MLP network structure, TQ - quality of training, TEQ – quality of testing, VQ – quality of validation, TA - BFGS algorithm, FE - function of error, FHA – function of hidden activation, FOA – function of output activation

Table 2. Analysis of neural network errors.

Answer of network	Compatibility		
	Good	Incompatible	Limited
10-25-1-good	265	5	0
10-25-1-incompatible	10	170	15
10-25-1-limited	5	0	510
10-23-1-good	280	0	0
10-23-1-incompatible	0	175	0
10-23-1-limited	0	0	525
10-12-1-good	225	30	75
10-12-1-incompatible	10	115	30
10-12-1-limited	45	30	420

For RBF networks, the input parameters (10 inputs) for the construction of a neural network include material properties, including eco-friendliness, the added material for the assessment of compatibility with the main material, and one output – the decision class, which in this case is the compatibility of materials. Table 3 shows the most important data of the material selection models developed as RBF networks. The models have different numbers of neurons in the hidden layer (20–60). The activation function in the hidden layer is the Gaussian function, the activation function in the output layer is the Softmax function, and the training algorithm (RBFT).

The best RBF network (10-58-1) reached an efficiency of 94.91%. It featured 10 inputs, 58 neurons in the hidden layer, and one output. Measured with cross entropy (CE), the training error was 0.3956, the testing error – 0.2954, and the validation error – 0.3512.

Table 3. The neural networks (RBF) models for material selection.

ID	S	TQ	TEQ	VQ	TA	FE	FHA	FOA
1	10-20-1	76.03	75.09	74.08	RBFT	SOS	Gaussian	Softmax
2	10-28-1	80.66	78.93	76.65	RBFT	Entropy	Gaussian	Softmax
3	10-35-1	84.98	84.99	81.34	RBFT	Entropy	Gaussian	Softmax
4	10-50-1	93.05	92.11	90.01	RBFT	Entropy	Gaussian	Softmax
5	10-58-1	96.60	95.21	92.92	RBFT	Entropy	Gaussian	Softmax

where: ID – network id, S – RBF network structure, TQ – quality of training, TEQ – quality of testing, VQ - quality of validation, TA - RBFT algorithm, FE - function of error, FHA – function of hidden activation, FOA – function of output activation

For Kohonen networks, many neural network models were built in the course of the conducted experiments. They all featured the same input layer, whose size resulted from the amount of input data (11 inputs). The neural network models were parameterized by various values: the network topology, number of learning cycles and neighborhood were changed. The learning process consisted in the assignment of cluster centers to the radial neuron layer. The functioning of a self-organizing network during the learning process largely depends on the selection of the measure of distance between the winning neuron and the input vector. The learning coefficient represents the neighborhood radius, whose value decreases over time. The models of Kohonen networks differ between these the following parameters: error (learning), Kohonen's learning algorithm (Number of learning cycles – 100, 500, 1000), neighborhood (3, 5), and topology (6 × 10, 10 × 20, 15 × 25). Kohonen networks effectiveness was seen on Fig. 2.

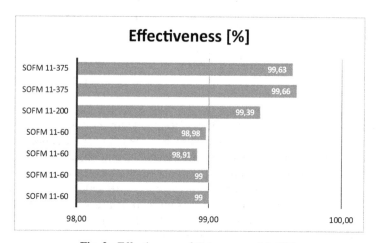

Fig. 2. Effectiveness of Kohonen models [%].

The most effective neural networks (100% effectiveness) were MLP 10-23-1, MLP 10-16-1, and MLP 10-21-1. Such good efficiency of neural networks was influenced by properly selected network parameters (number of neurons in the hidden layer, error function, activation functions, applied learning algorithm, and network learning time). Kohonen networks: 11-200 (99.39%) and 11-375 (99.66%) were a slightly less effective. RBF networks were the worst networks: RBF 10-50-1 (91,72%) and RBF 10-58-1 (94,91%) (see Table 4 and Fig. 3).

Table 4. Comparison of the best neural networks.

Structure	Effectiveness
MLP network	
10-23-1	100.00
10-16-1	100.00
10-21-1	100.00
RBF network	
10-50-1	91.72
10-58-1	94.91
Kohonen network	
11-200	99.39
11-375	99.66

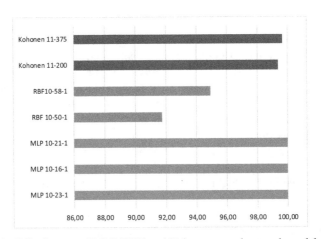

Fig. 3. Effectiveness of MLP, RBF, and Kohonen neural network models [%].

4.3 Example Implementation of a Neural Network into an Expert System

An expert system implementing a neural network model supporting the selection of materials and showing their compatibility (see Fig. 4) advises designers in the product development process. On the basis of the input data representing properties of the main and added materials, the system provides information on their compatibility. In the case under analysis, the response generated for PC and PA is: incompatible.

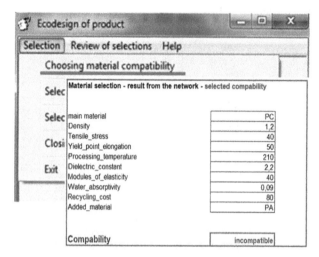

Fig. 4. Choosing material compatibility.

5 Summary

Considering large amounts of numerical input data used in ecodesign, application of neural networks as classification methods seems to be appropriate to support the process. Classification has been conducted using MLP, RBF, and Kohonen neural networks. All three methods are characterized by excellent classification abilities. It follows from the study that MLP networks are the best classifiers.

The selected classification methods have taken ecodesign to the next level. Owing to them, the knowledge which has so far been hidden in human minds or stored in databases, can be automatically acquired and used in the design process. The study has proven neural networks to be highly useful and effective in the aided of material selection in ecodesign.

The models developed and implemented into the decision support system aid in the design of new products. The system tells which materials are compatible and which are not, while the designer can take a decision on the materials to be used. What is important, the added material should meet both the environmental criteria and the technological requirements specified in the expert system, so as to maintain technological coherence of the product.

The more compatible the materials in the product are, the less time it takes to separate them. Therefore, materials with the highest degree of compatibility should be used to manufacture the product.

Acknowledgments. The work presented in the paper has been co-financed under 0613/SBAD/8727 and grants to maintain research potential Kazimierz Wielki University.

References

1. Rojek, I., Dostatni, E.: Artificial neural network-supported selection of materials in ecodesign. In: Trojanowska, J., Ciszak, O., Machado, J.M., Pavlenko, I. (eds.) MANUFACTURING 2019. LNME, vol. 1, pp. 422–431. Springer, Cham (2019). https://doi.org/10.1007/978-3-030-18715-6_35
2. Rojek, I., Dostatni, E., Hamrol, A.: Automation and digitization of the material selection process for ecodesign. In: Burduk, A., Chlebus, E., Nowakowski, T., Tubis, A. (eds.) ISPEM 2018. AISC, vol. 835, pp. 523–532. Springer, Cham (2019). https://doi.org/10.1007/978-3-319-97490-3_50
3. Rojek, I., Dostatni, E., Hamrol, A.: Ecodesign of technological processes with the use of decision trees method. In: Pérez García, H., Alfonso-Cendón, J., Sánchez González, L., Quintián, H., Corchado, E. (eds.) SOCO/CISIS/ICEUTE 2017. AISC, vol. 649, pp. 318–327. Springer, Cham (2018). https://doi.org/10.1007/978-3-319-67180-2_31
4. Dostatni, E., Rojek, I., Hamrol, A.: The use of machine learning method in concurrent ecodesign of products and technological processes. In: Hamrol, A., Ciszak, O., Legutko, S., Jurczyk, M. (eds.) Advances in Manufacturing. LNME, vol. 649, pp. 321–330. Springer, Cham (2018). https://doi.org/10.1007/978-3-319-68619-6_31
5. Karlsson, R., Luttropp, C.: EcoDesign: what's happening? J. Clean. Prod. **14**, 1291–1298 (2006)
6. Wikipedia. https://pl.wikipedia.org/wiki/Recykling. Accessed 21 Nov 2018
7. Chiang, T.A., Che, Z.H., Wang, T.T.: A design for environment methodology for evaluation and improvement of derivative consumer electronic product development. J. Syst. Sci. Syst. Eng. **20**(3), 260–274 (2011)
8. Li, J., Wu, Z., Zhang, H.-C.: Application of neural network on environmental impact assessment tools. Int. J. Sustain. Manuf. **1**(1/2), 100–121 (2008)
9. Artificial Intelligence put to use in Recycling. http://www.mistbreaker.com/sustainability/artificial-intelligence-put-use-recycling. Accessed 19 Apr 2019
10. Jamróz, D., Niedoba, T.: Application of multidimensional data visualization by means of self-organizing Kohonen maps to evaluate classification possibilities of various coal types. Arch. Min. Sci. **60**(1), 39–50 (2015)
11. Hendzel, Z.: Collision free path planning and control of wheeled mobile robot using Kohonen self-organising map. Bull. Pol. Acad. Sci. Tech. Sci. **53**(1), 39–47 (2005)
12. Grzecha, D.: Soft fault clustering in analog electronic circuits with the use of self-organizing neural network. Metrol. Meas. Syst. **XVIII**(4), 555–568 (2011)
13. Kamiński, M., Kowalski, C.T.: Rotor fault detector of the converter fed induction motor based on RBF neural network. Bull. Pol. Acad. Sci. Tech. Sci. **62**(1), 69–76 (2014)
14. Osowski, S., Siwek, K.: Local dynamic integration of ensemble in prediction of time series. Bull. Pol. Acad. Sci. Tech. Sci. **67**(3), 517–525 (2019)
15. Luzar, M., Sobolewski, Ł., Miczulski, W., Korbicz, J.: Prediction of corrections for the Polish time scale UTC(PL) using artificial neural networks. Bull. Pol. Acad. Sci. Tech. Sci. **61**(3), 589–594 (2013)
16. Osowski, S., Siwek, K., Szupiluk, R.: Ensemble neural network approach for accurate load forecasting in the power system. Int. J. Appl. Math. Comput. Sci. **19**(2), 303–315 (2009)

17. Dostatni, E., Diakun, J., Grajewski, D., Wichniarek, R., Karwasz, A.: Multi-agent system to support decision-making process in design for recycling. Soft. Comput. **20**(11), 4347–4361 (2016). https://doi.org/10.1007/s00500-016-2302-z
18. Dostatni, E., Diakun, J., Grajewski, D., Wichniarek, R., Karwasz, A.: Functionality assessment of ecodesign support system. Manag. Prod. Eng. Rev. **6**(1), 10–15 (2015)
19. Dostatni, E.: Recycling-oriented eco-design methodology based on decentralised artificial intelligence. Manag. Prod. Eng. Rev. **9**(3), 79–89 (2018)
20. Rojek, I.: Neural networks as prediction models for water intake in water supply system. In: Rutkowski, L., Tadeusiewicz, R., Zadeh, L.A., Zurada, J.M. (eds.) ICAISC 2008. LNCS (LNAI), vol. 5097, pp. 1109–1119. Springer, Heidelberg (2008). https://doi.org/10.1007/978-3-540-69731-2_104
21. Rojek, I.: Hybrid neural networks as prediction models. In: Rutkowski, L., Scherer, R., Tadeusiewicz, R., Zadeh, L.A., Zurada, J.M. (eds.) ICAISC 2010. LNCS (LNAI), vol. 6114, pp. 88–95. Springer, Heidelberg (2010). https://doi.org/10.1007/978-3-642-13232-2_12
22. Rojek, I.: Classifier models in intelligent CAPP systems. In: Cyran, K.A., Kozielski, S., Peters, J.F., Stańczyk, U., Wakulicz-Deja, A. (eds.) Man-Machine Interactions. AINSC, vol. 59, pp. 311–319. Springer, Heidelberg (2009). https://doi.org/10.1007/978-3-642-00563-3_32
23. Tadeusiewicz, R., Chaki, R., Chaki, N.: Exploring Neural Networks with C#. CRC Press Taylor & Francis Group, Boca Raton (2014)

Parameterized State Feedback Control Applied to the 1st Degree of Freedom of a Cylindric Pneumatic Robot

Marcos G. Q. Rijo[1(✉)], Eduardo A. Perondi[2], Mário R. Sobczyk S.[2], and Carlos A. C. Sarmanho Jr.[3]

[1] IFSUL Federal Institute, Sapiranga, RS, Brazil
mqrijo@gmail.com
[2] Mechanical Engineering Department, UFRGS University, Porto Alegre, RS, Brazil
[3] IFSUL Federal Institute, Charqueadas, RS, Brazil

Abstract. This paper addresses a gain-schedule trajectory controller applied to the first degree of freedom of a pneumatic five-degree cylindrical robot. The proposed control law is based on pole placement and state feedback techniques associated with a continuous gain-schedule scheme. Its gains are parameterized with respect to the trajectory-dependent mass moment of inertia of the manipulator with relation to its rotation axis. Therefore, the value of the equivalent translational inertia to be moved by the first degree of freedom actuator is calculated on line and used to update the gain set of the controller. As consequence, the poles of the closed-loop system remain unaltered, which results in small performance losses due to payload variations. Performance enhancement is verified by means of experimental results of position trajectory errors for the controlled system considering invariant and variable equivalent mass applied to the 1st DOF.

Keywords: Gain-schedule control · State feedback control · Pneumatic robotic manipulator

1 Introduction

Robotic manipulators are driven by electric motors or, less frequently, fluidic ones, according to several performance requirements such as precision, power/weight and power/volume ratios, maintainability, compliance, robustness, durability, reliability, response time, ease to control, energy source availability, energetic efficiency and cost. In this context, pneumatic actuators are attractive because they are fast, low-cost, easy to install (compressed air is common in industrial facilities), and durable, while presenting high power/volume and power/weight ratios. Moreover, air compressibility is useful in collaborative applications since it facilitates handling fragile objects and interacting with humans, thereby enhancing overall compliance. Nevertheless, these actuators are difficult to control accurately for a number of reasons associated with highly nonlinear phenomena such as air compressibility, pressure dynamics in the chambers, dead zones in the control valves, and dry friction [1, 2]. Thus, significant effort has been spent on improving control algorithms applied to pneumatic actuators, so their considerable operational advantages can be exploited with acceptable accuracy [3].

Due to their highly nonlinear nature, it is widely known that pneumatic actuators are not expected to achieve very good performance in precision applications when controlled by means of common linear feedback strategies, such as PID control [4–7]. On the other hand, nonlinear controllers tend to yield significant performance improvements, but at the cost of greatly increased complexity not only in theoretical terms, but also in the corresponding implementation hardware [8]. Thus, intermediate complexity algorithms such as State Feedback Control with position, velocity and acceleration feedback (the PVA controller) can often be regarded as an interesting "midway" approach, yielding acceptable performance at relatively reduced implementation costs when compared to more advanced algorithms such as Computed Torque Control [8].

When a linear model is provided, PVA control allows imposing the desired closed-loop dynamics to the system (at least for a limited range) with relatively little effort in design and implementation. If applied to pneumatic actuators, such controllers yield good responses when the mass is held constant, but its dynamic performance deteriorates with varying payloads [6], as demonstrated experimentally in [5]. When translated to the case of a pneumatically driven robot, this results in poor accuracy even if the payload remains constant, because pose variations will also affect the controller performance due to changes in the overall equivalent inertia perceived in each joint.

Here, we improve the PVA-controller robustness with respect to payload variations by parameterizing its gains in terms of the moment of inertia associated with the instantaneous robot pose, an approach that can be interpreted as a continuously updated gain-schedule control algorithm. It is applied to compensate for the variations in the moment of inertia perceived in the 1st degree of freedom (DOF) of a 5-DOF cylindrical robotic manipulator driven by a pneumatic power source. The performance of the proposed controller is evaluated experimentally, using the results obtained with a standard PVA-controller as reference for comparison.

Section 2 of this paper addresses the overall configuration of the pneumatic robot. Section 3 is dedicated to defining the equivalent mass perceived by the actuation system of the 1st DOF as a function of the manipulator pose. The synthesis of the control strategy is presented in Sect. 4, whereas experimental results are discussed in Sect. 5. Finally, the main conclusions are presented in Sect. 6.

2 The 5-DOF Pneumatic Robot

As shown in Fig. 1, the robot comprises one rotation joint related to the 1st DOF, followed by two prismatic orthogonal joints associated to the 2nd and 3rd DOFs, and two rotation joints associated to the 4th and 5th DOFs. The end effector is composed by a pneumatic gripper with two fingers. The first joint is driven by a rodless double action cylinder, connected with a 5/3 proportional valve to control the air mass flow rates that fill or exhaust the actuator chambers. Chamber are measured through pressure transducers. These components are mounted in a base constructed with aluminum profiles. Some of these hollow profiles are also used as accumulators for the compressed air coming from the pneumatic supply line, reducing its pressure oscillations. The robot base also houses a transmission system of three pulleys and a synchronizer belt, which convert the piston translation of the piston into the rotational movement of the 1st DOF of the robot.

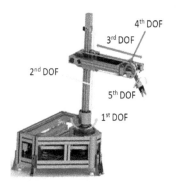 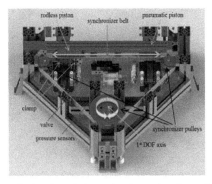

Fig. 1. Kinematic chain of the pneumatic cylindrical robot (left) and schematic view of the mechanical components of the 1st degree of freedom (right).

The main part of the transmission system of the 1st DOF is a timing belt pulley (primary pulley), assembled to a vertical rotational aluminum shaft that supports all other links of the robot. This pulley is connected to the other two by a timing belt, which ensures that all three rotations are synchronized. A metallic clamp is used to connect the extremities of the timing belt, which transmits the force from the pneumatic actuator to the pulley linked to the 1st joint. The timing belt is of HTD type, which, according to [9], is adequate for low speeds and high torque transfer operations.

A proprietary Microcontrolled Actuation and Control Unit (UCAM) was developed to communicate the control board with the measurement devices. Typical functions of UCAM include analogic signals sampling from pressure sensors, A/D and D/A conversions, and performing the serial communication (RS-485) with the magnetostrictive piston position sensor mounted inside of the pneumatic cylinder. A centralized robot control algorithm is programmed in Matlab-Simulink® and is processed in a PC hosted dSPACE® DS-1104 board with a hard real-time control cycle of 1.8 ms.

3 Parameterization of the Equivalent Mass

As the robot moves, the moment of inertia with respect to the rotation axis of the 1st DOF changes with the current robot joint coordinate values. The parametric control strategy is based on, by means the Steiner's law, calculate a so-called *equivalent lumped mass* that is instantaneously added to that of the moving parts of its linear pneumatic cylinder, resulting in the value of the total *equivalent* mass that have to be moved by the linear piston in result of the control driving. Taking into account this value, the gains of the state feedback controller are suitably updated by means a parametric algorithm. Therefore, when the arm changes its pose, the control gains are modified, seeking to keep constant the closed-loop poles associated with the 1st DOF. This strategy takes into account the positions of the 3rd and 4th DOFs of the robot, which alter the mass moment of inertia with respect the 1st DOF. Figure 2 presents the definition of the coordinate systems.

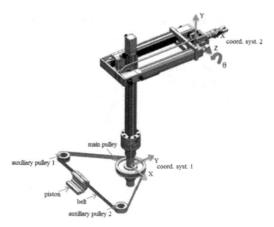

Fig. 2. Manipulator and coordinate systems for the moment of inertia parameterization

The moment of inertia I of a given robot link with respect to the 1st DOF Z-axis is:

$$I = I_G + md^2, \qquad (1)$$

where m is the link's mass, I_G is the link mass moment of inertia related to its mass center, and d is the distance from this mass center to the 1st DOF Z-axis.

To parameterize its equivalent lumped mass, the manipulator was divided into three subsets. The first one comprises the transmission shaft and the pulleys of the 1st DOF, the coupling part between the 1st and 2nd DOFs, the pneumatic actuator of the 2nd DOF, the coupling part between the 2nd and 3rd DOFs, and the piston of the 3rd DOF actuator. Since no one of these moving parts affects the mass moment of inertia with respect to the 1st DOF Z-axis, its equivalent lumped mass m_{E1} is constant.

The second subset is formed by the actuators of the 3rd and 4th DOFs. When these parts move, the mass moment of inertia relative to 1st DOF Z-axis varies. Thus, m_{E2} represents the inertia variations due to the displacements in the 3rd DOF, whereas the effects of the 4th DOF are represented by another variable equivalent mass m_{E3}. The equivalent lumped mass for this subsystem is given by $m_{E2} + m_{E3}$.

The third subset is formed by the wrist, whose mass moment of inertia I_G varies with the translation of the 3rd DOF and the rotation θ of the 4th DOF, for that, Steiner's law is used:

$$I_G = \left(\frac{I_{XG} + I_{YG}}{2}\right) - \frac{I_{XG} - I_{YG}}{2}\cos(2\theta) + I_{XYG}\sin(2\theta), \qquad (2)$$

where I_{XG}, I_{YG} and I_{XYG} are the wrist moment of inertia relative to the associated axes x, y and the rotation angle θ (Fig. 2). Thus, once obtaining the value of I_G, the equivalent mass m_{E4} is calculated through (1).

After calculating the equivalent masses of the three subsystems, the total equivalent mass is determining by summing all involved masses:

$$m_E = m_{CE} + m_C + 2m_{EP} + m_{E1} + m_{E2} + m_{E3} + m_{E4}, \qquad (3)$$

where m_{CE}, m_C and m_C are, respectively, the piston, clamp and belt masses, whereas $m_{EP} = I_P/R_P^2$ is the equivalent mass of each the auxiliary pulley in the transmission mechanism of the 1st DOF (Fig. 2), with I_P and R_P standing for their mass moment of inertia and effective radius.

4 State Feedback Controller

The modeling of a standard pneumatic servopositioning system is extensively discussed in several works [4–7, 10], where it is shown to be satisfactorily represented by means of the following transfer function:

$$G(s) = \frac{Y(s)}{I(s)} = \frac{b_0}{s(s^2 + a_1 s + a_2)}, \tag{4}$$

where I is the control signal voltage applied to the pneumatic servovalve, y is the piston position, s is the Laplace varialble, and b_0, a_1, a_2 are parameters that depend upon the characteristics of the system. These coefficients can be expressed as:

$$b_0 = \frac{4ArRTk_q}{m_E V}; \quad a_1 = \frac{4rA^2 P_0}{m_E V}; \quad a_2 = \frac{C_f}{m_E}, \tag{5}$$

where A is the cross-sectional area of the pneumatic piston, r is specific heat ratio of the air, R is the gas constant of compressed air, T is the average absolute temperature, V is the volume of the piston, k_q is the average mass flow rate gain of the valve orifices, and C_f is the is the viscous friction coefficient (linear Newton's friction coefficient). Table 1 presents the numerical values used in the present work for the controller design.

Table 1. Numerical values of the model parameters [8].

T	Average absolute temperature [K]	293,15
P_0	Initial pressure in the chambers [Pa]	3,85.10^5
R	Gas constant [J/kgK]	286,9
A	Cross-sectional area of the pneumatic piston actuator [m^2]	8,04.10^{-4}
r	Specific heat ratio of the air	1,4
k_q	Average mass flow rate gain of the servovalves [kg/s]	6,7.10^{-3}
V	Volume capacity of the cylinder [m^3]	3,62.10^{-4}
C_f	Viscous friction coefficient [Ns/m]	266,55

As the 3rd order model does not present zero-pole cancelation, the system is controllable. Therefore, within the actuating range of its control valve, it is possible to arbitrarily choose its closed-loop poles by using the following feedback law:

$$u(t) = k_p \tilde{y} + k_v \dot{\tilde{y}} + k_a \ddot{\tilde{y}}, \tag{6}$$

where $\tilde{y}$, $\dot{\tilde{y}}$ and $\ddot{\tilde{y}}$ are the errors between the desired values of position (y_d), velocity ($\dot{y}_d$), and acceleration ($\ddot{y}_d$), and their measured counterparts y, $\dot{y}$, and $\ddot{y}$, with respective gains

k_p, k_v and k_a. This law is schematically shown in Fig. 3. It is usually referred to as PVA because it uses position (P), velocity (V) and acceleration (A) as feedback states. Several authors applied PVA controllers to pneumatic positioners based on the 3rd order linear model given in (4) [7]. Velocity and acceleration signals were obtained by numerical differentiation of the position data.

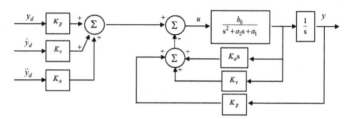

Fig. 3. Linear 3rd order system with state feedback controller.

The characteristic equation for this closed-loop system is

$$s^3 + (b_0 k_a + a_2)s^2 + (b_0 k_v + a_1)s + b_0 k_p = 0. \quad (7)$$

Since each gain affects one independent power of s in Eq. (7), the allocation of all three poles is a straightforward process. Defining P_{1d}, P_{2d} and P_{3d} as the desired poles, the correspondent characteristic coefficients are:

$$a_{0d} = -P_{1d} P_{2d} P_{3d}, \quad (8)$$

$$a_{1d} = P_{1d} P_{2d} + P_{2d} P_{3d} + P_{1d} P_{3d}, \quad (9)$$

$$a_{3d} = -(P_{1d} + P_{2d} + P_{3d}). \quad (10)$$

Thus, the desired dynamics' characteristic equation is

$$s^3 + a_{2d} s^2 + a_{1d} s + a_{0d} = 0. \quad (11)$$

Finally, matching Eqs. (7) and (11), the controller gains are determined as:

$$k_p = \frac{a_{0d}}{b_0}; \; k_v = \frac{a_{1d} - a_1}{b_0}; \; k_a = \frac{a_{2d} - a_2}{b_0}. \quad (12)$$

The desired dynamics is defined as a dominant pair of poles for the 1st DOF (approaching its dynamics of that of a 2nd order system), with the third pole located in a position ten times farthest than the real part of the dominant ones. The maximum overshoot is chosen as 20% and the settling time is 1.4 s (with 2% tolerance), implying a damping $\zeta = 0,5$ and a natural frequency $\omega_n = 5,71$ rad/s. The corresponding complex conjugate poles are $P_{12d} = -2,86 \pm 4,95i$, whereas the third pole is $P_{3d} = -28,6$. Once the desired dynamics is established, the complete proposed control strategy consists of performing the following three steps on each control cycle:

1. Compute the equivalent total lumped mass value as described in (3);
2. Update the open loop coefficients (b_0, a_1 and a_2) using (5);
3. Calculate the control gains k_p, k_v and k_a (12) and obtain the corresponding control signal to be applied to the servovalve (6).

5 Experimental Results

Two cases were analyzed through experimental tests. In the first, only the 1st DOF was moved, whereas the 3rd and 4th arm joints were fixed so as to generate three conditions for the equivalent moment of inertia: maximum, medium, and lowest. In the second one, all these three joints were made to track their respective desired trajectories, as described in Table 2, so that the equivalent inertia with respect to the 1st DOF was continuously varying.

Table 2. Desired trajectories

Trajectory	Position	Time
	$0{,}13$	$t < 10$
	$0{,}57\left(\frac{t-10}{3}\right)^5 - 1{,}425\left(\frac{t-10}{3}\right)^4 + 0{,}95\left(\frac{t-10}{3}\right)^3 + 0{,}13$	$10 \leq t < 13$
	$0{,}225$	$13 \leq t < 23$
	$0{,}57\left(\frac{t-23}{3}\right)^5 - 1{,}425\left(\frac{t-23}{3}\right)^4 + 0{,}95\left(\frac{t-23}{3}\right)^3 + 0{,}225$	$23 \leq t < 26$
$y_{d\_1st}(t)[m] =$	$0{,}32$	$26 \leq t < 36$
	$-0{,}57\left(\frac{t-36}{3}\right)^5 + 1{,}425\left(\frac{t-36}{3}\right)^4 - 0{,}95\left(\frac{t-36}{3}\right)^3 + 0{,}32$	$36 \leq t < 39$
	$0{,}225$	$39 \leq t < 49$
	$-0{,}57\left(\frac{t-49}{3}\right)^5 + 1{,}425\left(\frac{t-49}{3}\right)^4 - 0{,}95\left(\frac{t-49}{3}\right)^3 + 0{,}225$	$49 \leq t < 52$
	$0{,}13$	$52 \leq t \leq 60$
	0	$t < 10$
	$1{,}5\left(\frac{t-10}{3}\right)^5 - 3{,}75\left(\frac{t-10}{3}\right)^4 + 2{,}5\left(\frac{t-10}{3}\right)^3$	$10 \leq t < 13$
	$0{,}25$	$13 \leq t < 23$
	$0{,}3\left(\frac{t-23}{3}\right)^5 - 0{,}75\left(\frac{t-23}{3}\right)^4 + 0{,}5\left(\frac{t-23}{3}\right)^3 + 0{,}25$	$23 \leq t < 26$
$y_{d\_3th}(t)[m] =$	$0{,}3$	$26 \leq t < 36$
	$-0{,}3\left(\frac{t-36}{3}\right)^5 + 0{,}75\left(\frac{t-36}{3}\right)^4 - 0{,}5\left(\frac{t-36}{3}\right)^3 + 0{,}3$	$36 \leq t < 39$
	$0{,}25$	$39 \leq t < 49$
	$-1{,}5\left(\frac{t-49}{3}\right)^5 + 3{,}75\left(\frac{t-49}{3}\right)^4 - 2{,}5\left(\frac{t-49}{3}\right)^3 + 0{,}225$	$49 \leq t < 52$
	0	$52 \leq t \leq 60$
	$2{,}817$	$t < 23$
	$-16{,}191\left(\frac{t-23}{3}\right)^5 + 40{,}477\left(\frac{t-23}{3}\right)^4 - 26{,}985\left(\frac{t-23}{3}\right)^3 + 2{,}817$	$23 \leq t < 26$
$y_{d\_4th}(t)[rad] =$	$0{,}1185$	$26 \leq t < 36$
	$16{,}191\left(\frac{t-23}{3}\right)^5 - 40{,}477\left(\frac{t-23}{3}\right)^4 + 26{,}985\left(\frac{t-23}{3}\right)^3 + 0{,}1185$	$36 \leq t < 39$
	$2{,}817$	$39 \leq t \leq 60$

In both cases, two PVA controllers were used: a *standard* (SPVA) one, based on the nominal values of the linear model for the 1st DOF, and the proposed *parameterized* (PPVA) one, with continuously updated gains.

Due to the geometric configuration, the 2nd (vertical displacement) and the 5th DOF do not significantly influence the mass moment of inertia of the entire manipulator with respect to the 1st DOF. Therefore, only the 3rd and the 4th DOF displacement were taken into account in the variable mass case tests, in which the desired trajectory consists of a combination of a 7th order polynomial curve with constant values in the initial and final positions, as presented on Table 2. These kinds of trajectory have already been used as a test pattern in other studies of tracking control [11] and was prescribed in a way that the manufacturer pneumatic cylinder's limits of velocity and acceleration are always observed. Figure 4 presents the desired trajectory for the 1st DOF.

5.1 Case 1 – Invariant Equivalent Mass Applied to the 1st DOF

In this case, the tests were performed with three different geometric configurations of the robot, where the 3rd and 4th DOF were fixed so the equivalent mass perceived in the 1st DOF presented minimum (86.5 kg), intermediate (134 kg) and maximum values (180.5 kg). For simplicity, the results shown in this section refer only to the minimum inertia case. The trajectory-tracking errors in all cases are discussed in Sect. 5.3.

A typical trajectory tracking position for both controllers when the system operates with the minimum inertia is presented in Fig. 4.

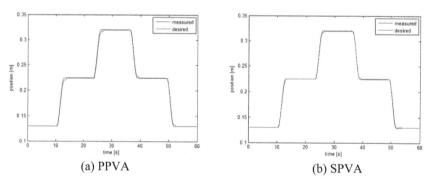

Fig. 4. Position trajectory tracking response (smaller inertia case: 86,5 kg).

Figure 5 presents the control signals for the minimum inertia case. It is clear that the control action required for the SPVA control law is much more severe than the one given in the PPVA case. Although no saturation occurred (the valve operates in the range of −5 V to +5 V), the chattering in control action for the PPVA is undesirable because it leads to premature wear of the control valve and to the risk of exciting unmodeled dynamics in the robot arm. In fact, when this law was used, it was possible to observe vibrations along the arm structure.

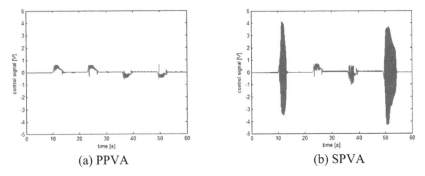

(a) PPVA (b) SPVA

Fig. 5. Control signals for minimum inertia case.

5.2 Case 2 – Variable Equivalent Mass Applied to the 1st DOF

Aiming at reproducing actual operation conditions, each DOF tracks its respective desired trajectory as defined in Table 2, so the equivalent mass with respect to the 1st DOF is time-varying, as depicted in Fig. 6.

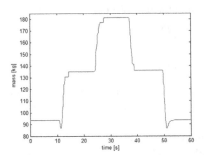

Fig. 6. Time evolution of the equivalent mass calculated through (4).

Figure 7 present the position tracking and position tracking error for the two control algorithms operating with all the joints moving concomitantly.

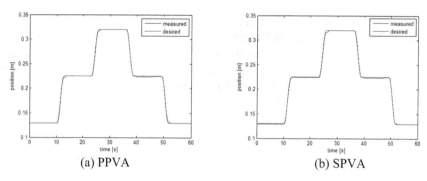

Fig. 7. Position trajectory tracking response (variable inertia).

Figure 8 shows the control actions applied to the valve for both controllers.

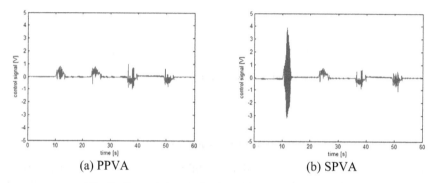

Fig. 8. Control signals for the time varying inertia case.

Once again, it is clear that the use of the proposed PPVA scheme leads to reductions in the chattering of the valve opening, which is desirable.

5.3 Position Trajectory-Tracking Errors

The results plotted in Fig. 9 presents the position trajectory-tracking errors in the closed-loop system for all performed tests. These results are also summarized numerically in Table 3.

Overall, error amplitudes are similar for both controllers, suggesting no clear advantage for the proposed scheme. However, closer inspection shows that PPVA control leads to significant reductions in steady-state errors, especially when the inertia in the actual system is larger than the estimate used in designing the SPVA. Coupled with the chattering reduction observed in the previous sections, this indicates that the proposed controller tends to improve the precision of the robot arm as a whole, since it reduces steady-state errors *and* structure vibrations as the overall trajectory is performed.

Still with respect to steady-state errors, this result is in apparent contradiction with the corresponding open-loop linear system model given in (4): since there is a pole on the

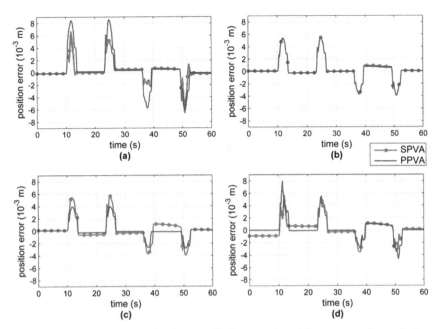

Fig.9. Trajectory-tracking errors for the controlled system: (a) minimum mass, (b) nominal mass, (c) maximum mass, (d) time-varying mass.

Table 3. Experimental results summary

Mass [kg]	Steady State Error [mm]		Tracking error [mm]		Gains
	Max.	Average	Max.	Average	
Variable	0,73	0,21	7,95	5,25	Parameterized
	0,63	0,38	5,72	4,52	Fixed
86,5	0,47	0,26	8,58	7,20	Parameterized
	0,53	0,32	6,72	5,23	Fixed
134	0,57	0,24	5,42	4,61	Parameterized
	0,69	0,26	5,52	4,63	Fixed
180,5	0,24	0,15	3,98	3,37	Parameterized
	0,57	0,40	5,72	4,73	Fixed

origin, the corresponding integral action should force these errors to zero. However, we must stress that the *real* system is highly nonlinear, with at least two major hindrances to such "ideal" integral action: (i) dry-friction forces, which cause the actuating piston to stick unless pneumatic forces are high enough to cause movement [12]; (ii) a dead zone in the control valve, which, for the employed model, is typically about 5% of its opening stroke. Thus, small but nonzero steady-state errors are to be expected when PVA-control is applied to this type of system.

6 Conclusions

Even though there is no significant reduction in error amplitudes when compared to a standard, fixed-gain PVA controller, the proposed parameterized control scheme leads to important reductions in the steady state error values, especially when the real inertia in the system is greater than expected during the design of the standard controller. Moreover, the proposed algorithm leads to significantly reduced chattering in the corresponding control signals applied to the servovalve. This particular result is highly desirable, since it leads to extended lifespan of the control valves and reduced vibrations in the overall structure of the robotic arm. For these reasons, the proposed control scheme can be considered an attractive one.

Further work would include stability and robustness analysis of the proposed parameterized PVA control scheme, as well as the extension of this technique to compensate for other undesirable effects in the system.

References

1. Saravanakumar, D., Mohan, B., Muthuramalingam, T.: A review on recent research trends in servo pneumatic positioning systems. Precis. Eng. **49**, 481–492 (2017). https://doi.org/10.1016/j.precisioneng.2017.01.014
2. Najafi, F., Fathi, M., Saadat, M.: Dynamic modelling of servo pneumatic actuators with cushioning. Int. J. Adv. Manuf. Technol. **42**(7–8), 757–765 (2009). https://doi.org/10.1007/s00170-008-1635-x
3. Rouzbeh, B., Bone, G.M., Ashby, G.: High-accuracy position control of a rotary pneumatic actuator. IEEE/ASME Trans. Mechatron. **23**(6), 2774–2781 (2018). https://doi.org/10.1109/TMECH.2018.2870177
4. Bobrow, J.E., McDonell, B.W.: Modeling, identification, and control of a pneumatically actuated, force controllable robot. IEEE Trans. Robot. Autom. **14**(5), 732–742 (1998)
5. Ning, S., Bone, G.M.: High steady-state accuracy pneumatic servo positioning system with PVA/PV control and friction compensation. In: Proceedings of the IEEE International Conference on Robotics and Automation, Washington, DC, pp. 2824–2829 (2002)
6. Ren, H.P., Fan, J.T.: Adaptive backstepping slide mode control of pneumatic position servo system. Chin. J. Mech. Eng. **29**(5), 1003–1009 (2016). https://doi.org/10.3901/CJME.2016.0412.050
7. Sobczyk, M.R., Perondi, E.A., Suzuki, R.: Feedback linearization control with friction compensation applied to a pneumatic positioning. In: ABCM Symposium Series in Mechatronics, Section II – Control Systems, vol. 5, pp 252–261 (2012)
8. Sarmanho Jr., C.A.C.: Desenvolvimento de um robô pneumático de 5 graus de Liberdade com controlador não linear com compensação de atrito. Thesis (Ph.D. in Mechanical Engineering), Universidade Federal do Rio Grande do Sul (2014)
9. Urethane Timing Belts and Pulleys GATES CATALOG. www.http://misbelt.com/wp-content/uploads/2018/01/Mectrol-Belt-Pulley-Catalog_5_11.pdf. Accessed 01 Jan 2018
10. Virvalo, T.: Designing a pneumatic position servo system. Power Int. **35**, 141–147 (1989)
11. Borges, F.A.P., Perondi, E.A., Sobczyk, M.R., Cunha, M.A.B.: A hydraulic actuator model using feedforward neural networks. In: 25th ABCM International Congress of Mechanical Engineering, Uberlândia, MG, Brazil (2019)
12. Sobczyk, M.R., Gervini, V.I., Perondi, E.A., Cunha, M.A.B.: A continuous version of the LuGre friction model applied to the adaptive control of a pneumatic servo system. J. Franklin Inst. **353**(13), 3021–3039 (2016)

Building a Mobile Application by Combining Third Party Platforms in Order to Reduce Time, Costs and Improve Functionality

Marian Tanasie[✉] and Ionel Simion

University Politehnica of Bucharest, 060042 Bucharest, Romania

Abstract. The article explains the collaboration between two different programs like structure (web design and android programming) in order to shorten the time process in design, reduce costs and improve the graphical results. Exemplifying will be explained using Wordpress and Android studio demonstrating the effectiveness of the programs in their achievement.

Keywords: Graphics · Design · Programming

1 Introduction

1.1 Goal

In this article we will demonstrate how we can develop a mobile app using Wordpress and Android Studio. Normally building a native mobile application would take weeks, months. But we will prove that less than a day it's enough to know how to build a website and then transform it into a mobile application. So in order to develop a mobile app we would need to learn java and Android Studio. In our case we only need to know few knowledge about how to make a website because we will use wordpress to create it. Of course this method works on any custom or other platforms made websites, but the point is to improve time, quality and design so we chose Wordpress for it's flexibility and large community.

1.2 Related Work

There are some plugins for Wordpress [1], whom authors sustain that can transform your Wordpress Website into a mobile application, but the reviews and reality shows us that they are spam, most of them don't work and can crash your website or some of them that work charges you for hundreds of dollars per month. Furthermore these kind of plugins can affect your website's integrity making it vulnerable for attackers and they provide some tools that forces you to use certain templates that may not bring the results you want.

2 How it Works

2.1 Creating an Online Store with Wordpress

We are going to demonstrate our work on an existing online store (www.feminista.ro) made on Wordpress with Woocommerce. But first we are going to explain how simple is to create a website with wordpress.

First we are going to install the wordpress on localhost machine or, in our case, on a server using FTP connection or cPanel. For that we need to create a database in phpmyadmin and a user for the database.

The next step is uploading wordpress files into the root directory through FTP or File Manager in cPanel. After uploading Wordpress we will access the file install.php in a browser and the installation will begin. We will need to complete the setup steps where we will fill in the name of the database created, the user and password (see Fig. 1).

Fig. 1. Setting up Wordpress

If we set up the database correctly the installation should be a success and we can start designing our website. In our case we want to make a online market where people can buy things. So the first thing we will do is install a theme in wordpress by going to Appearance – Themes.

After installing a theme we need to install woocommerce from Plugins menu so we can sell stuff on our website. Then we will have to setup the woocommerce settings (currency, country, shipping methods, payment gateways) (see Fig. 2).

After setting up woocommerce we will start creating our pages: Home, About, Shop, Terms and other pages that we like to have. All this is done from the WP admin menu and no programming is needed.

If we want to customize the theme we can access the files through FTP and start coding or we can add custom CSS in the admin custom code section. The platform is easy to learn and even a beginner can make a simple website. If we don't have knowledge at all about web programming we can search for plugins in the Plugin section in admin panel where we can find different plugins for what we need (e.g.: Image sliders, contact forms, social share, etc.). After the design is finished we can start adding products into

Fig. 2. Setting up Woocommerce

the shop. We can add products from products section in wordpress where we will set the name, price, category, attributes, images and description of the product (see Fig. 3).

Fig. 3. Adding products into the shop

2.2 Advantages of Using Wordpress for Creating the Website in Order to Easily Develop the Mobile Application

For developing a native application in Android Studio from scratch it requires lots of knowledges, libraries, modules, etc. [4], knowledges that should need years of learning. Using Wordpress to create the online store gives us many advantages for the developing of the application.

First of all we need to know few about html and css in order to adjust the wordpress theme if we don't like it. And that's all the coding we need to know. The most important thing that we need to achieve is a mobile responsive website because practically what the android app will to is to load the website mobile version. So as long as the design is responsive we won't have any problem. All Wordpress themes comes with fully responsive design but in order to be sure we just need to know a little html and css if we need to arrange some elements if they don't show properly.

So one of the important advantage is that we need few knowledge of programming. But related to this fact, another important advantage is time. Because wordpress themes gives us almost everything we need and we don't need programming, the time of implementation is very little.

Another advantage is money. For a company that has android developers as employees, there are a lot of costs in salary and resources. On the other hand a team of web designers is less expensive and if the company already has that there is no need for full stack programmers or Android/IOS programmers.

And of course time is very important for a project deadline. To make a native mobile app that would be connected to the website would take weeks, months depends on the project. So using our method will decrease time significantly.

The design of the application and user experience is another very important advantage. Making a mobile application from scratch would require front-end and back-end programmers because the quality of the design is very important. Most mobile application are not so eye-catching [2, 5] because it's hard to achieve a good looking interface in java programming.

Wordpress comes with a huge directory of themes and templates, always updated and improved. Also they are responsive so they look very good and have an interactive design, always keeping the user entertained (see Fig. 4).

Fig. 4. Adding products into the shop

One of the most important advantage is the functionalities of the app such as: ajax live search suggestions, membership features, login custom fields, advanced discount types and much more. This kind of features is needed for example in our store to give clients different benefits and user experience so they can become permanent clients.

These kind of features are predefined in Wordpress or can be find in different modules. In Android such features would require weeks or months of programming and testing.

2.3 Converting the Website into an Android Application

After we set up the website now we can head up in Android Studio and start baking the app. We will use WebView library [3] in Android studio and we will create an app that will practically load the website.

The first thing we are going to do next is adding the main code in MainActivity. So we are going to make the app to access the website's link. But that it's not enough because if we do only that the app will close and will open a browser to access the link, practically we've just give the OS a command. So we will set the app to open the website inside it using WebViewClient (see Fig. 5).

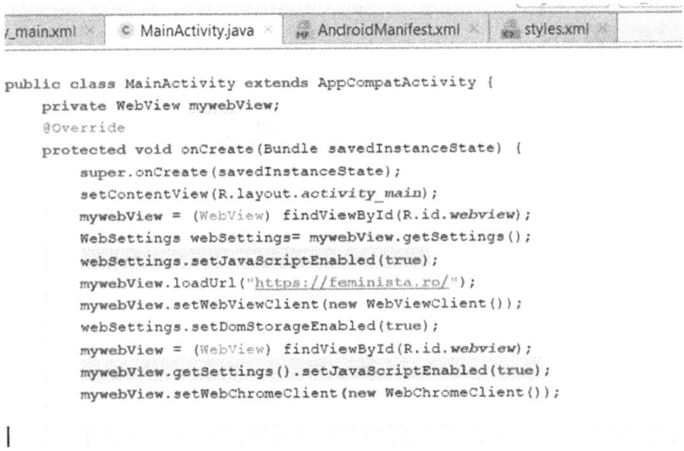

```
/_main.xml    c MainActivity.java    AndroidManifest.xml    styles.xml

public class MainActivity extends AppCompatActivity {
    private WebView mywebView;
    @Override
    protected void onCreate(Bundle savedInstanceState) {
        super.onCreate(savedInstanceState);
        setContentView(R.layout.activity_main);
        mywebView = (WebView) findViewById(R.id.webview);
        WebSettings webSettings= mywebView.getSettings();
        webSettings.setJavaScriptEnabled(true);
        mywebView.loadUrl("https://feminista.ro/");
        mywebView.setWebViewClient(new WebViewClient());
        webSettings.setDomStorageEnabled(true);
        mywebView = (WebView) findViewById(R.id.webview);
        mywebView.getSettings().setJavaScriptEnabled(true);
        mywebView.setWebChromeClient(new WebChromeClient());
```

Fig. 5. Main code for accessing the website

The next code we are going to add is in the AndroidManifest.xml where we are going to add permission to Android to let the app access the internet.

After we added the permissions we will go to activity_main.xml where we will set the width and height of the application. We will set them to fill the whole screen of the Phone so it will look good on every resolution. We can test the app to see if it works and check if it needs some modifications. In order to do that we will use AVD Manager. We can set up our own emulator with the resolution we want or we can connect our phone to the computer and it will appear on the list and test on it. After we select the emulator a virtual device will show up on our screen where we can see and test the app. We will go to Run in the menu and select Run App. The emulator will load the app and we can test it (see Fig. 6).

We notice that our app is working but we need to set up one more thing before we can export the app. When we access the pages on the app we notice that when we try press the back button the result is closing the app not going to the previous page. So in order to fix that we need to add some code to the MainActivity where we tell Android that back button will work in app too not globally.

After we finished the app we can head up and prepare the settings for export. To export the app we go to "Build" in the menu and select "Build APK". But before we do

Fig. 6. Testing the app

that we have to set the app an icon. In order to do that we right click on the project and select "New" and then "Image Asset" (see Fig. 7).

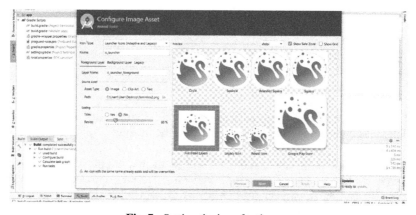

Fig. 7. Setting the icon for the app

3 Conclusions

In conclusion, mobile app's are needed more and more because most of people use their mobile phones to surf the internet or to buy their food, clothes or to rent a cab [2] but they are expensive and take a lot of time and resources to develop and can have poor quality design. Combining multiple free open source applications such as Wordpress and Android Studio, we significantly shorten time and reduced resources to create an app which can have great impact in the future.

We have shown that using multiple programs to achieve some projects (in our case an online store) it's far more better: advanced graphics quality, faster rendering, reduced time, low costs, low human resources, easier to administrate and modify and more.

Acknowledgement. This work has been funded by the European Social Fund from the Sectoral Operational Programme Human Capital 2014–2020, through the Financial Agreement with the title "Scholarships for entrepreneurial education among doctoral students and postdoctoral researchers (Be Antreprenor!)", Contract no. 51680/09.07.2019 - SMIS code: 124539.

References

1. Wordpress Plugins. https://www.wpbeginner.com/showcase/best-plugins-to-convert-wordpress-into-mobile-app/. Accessed 21 Feb 2020
2. Verma, N., Kansal, S., Malvi, H.: Development of native mobile application using android studio for cabs and some glimpse of cross platform apps. Int. J. Appl. Eng. Res. **13**, 12527–12530 (2018)
3. Documentation Page. https://developer.android.com/docs/. Accessed 16 Jan 2020
4. Jose, D.V., Lakshmi Priya, C., Priyadarshini, G., Singh, M.: Challenges and issues in android app development - an overview. Int. J. Adv. Res. Comput. Sci. Softw. Eng. IJARCSSE 811–814 (2015)
5. Kathuria, A., Gupta, A.: Challenges in Android application development: a case study. Int. J. Comput. Sci. Mob. Comput. IJCSMC **4**(5), 294–299 (2015)

Conveyor Belts Joints Remaining Life Time Forecasting with the Use of Monitoring Data and Mathematical Modelling

Edward Kozłowski[1], Anna Borucka[2(✉)], Yiliu Liu[3], and Dariusz Mazurkiewicz[4]

[1] Faculty of Management, Lublin University of Technology, ul. Nadbystrzycka 38, 20-618 Lublin, Poland
e.kozlovski@pollub.pl
[2] Faculty of Security, Logistics and Management, Military University of Technology, ul. gen. Sylwestra Kaliskiego 2, 00-908 Warsaw, Poland
anna.borucka@wat.edu.pl
[3] Faculty of Engineering, Norwegian University of Science and Technology, 7491 Trondheim, Norway
yiliu.liu@ntnu.no
[4] Mechanical Engineering Faculty, Lublin University of Technology, ul. Nadbystrzycka 36, 20-618 Lublin, Poland
d.mazurkiewicz@pollub.pl

Abstract. Monitoring the condition of conveyor belt joints by measuring and continuously analysing changes in their length raises a number of research challenges. One of them is the need for a more advanced method of assessment of the condition of each individual joint. Suitable mathematical modelling that would allow to identify the change in the length of such joints over its life time could be considered as a potential solution. Therefore, the aim of this study was to indicate that on the basis of the operation history of such elements, it is possible to estimate the parameters of the time series model, allowing to identify the joints which should be observed more carefully due to the unfavourable changes occurring in them. As a results of this research aim - the article presents the methodology allowing to identify the proper model and the function estimated for the selected joint.

Keywords: Real-time monitoring · Conveyor belts joints · Remaining useful time · Mathematical modelling

1 Introduction

Belt conveyor systems are usually used to transport large quantities of rock materials over long distances. High cost of manufacturing highly damage-resistant belts, e.g. fabric and rubber ones, the sections of which are usually joined by vulcanized or cold bond splicing, makes them particularly important for maintenance services as well. From the point of view of maintenance of the elements of, usually complex, transport system, cold bond joints pose the greatest challenge in terms of ensuring the expected reliability, as their breaking strength is noticeably lower than that of the belt itself. A typical single belt conveyor, which constitutes the basic component of a belt conveyor system, usually consists of several belts spliced together. These joints are the most vulnerable element of the entire structure, as breaking one of them results in downtime of the entire conveyor belt along with the other associated elements of the conveyor system, which operates in series. Therefore, monitoring of the condition of individual joints of the belt is vital [1]. Monitoring the condition of conveyor belt joints by measuring and continuously analysing changes in their length raises a number of research challenges [1–3]. One of them is the need for a more advanced method of assessment of the condition of individual joints. Suitable mathematical modelling that would allow to identify the change in the length of such joints over time could be considered as a potential solution. Momentary change in the length of the cold bond joint is a process that is influenced by several factors, such as the momentary load on the belt, time of operation, significant load at start-up or when stopping the conveyor, the amount of material transported and the method of its distribution on the belt, quality of joints, etc., many of which are difficult or even impossible to assess or measure precisely. Therefore, the aim of the presented study was to indicate that on the basis of the operation history of such elements, it is possible to estimate the parameters of the time series model, allowing to identify the joints which should be observed more carefully due to the unfavourable changes occurring in them. The article presents the research methodology allowing to identify the proper model and the function estimated for the selected joint.

2 Methodology

The article examines the problem of identification of the time series, describing the change in joint length over time, which is an important element influencing the reliable operation of a conveyor belt. Exceeding the limit values may result in rupture and significant losses in the mining process. The undesirable phenomenon identified in such a time series is a trend that is described with a linear function. If the gradient determining the occurrence of the trend in the model is not significantly different from zero, particular attention should be paid to the residuals. If they do not meet the condition of homoscedasticity, this means that the variance is unstable, which results from e.g. uneven load distribution on the conveyor belt. This is also an adverse phenomenon, heterogeneous changes in joint length (especially stretching) lead to faster wear rate and increase the possibility of rupture.

Let $(\Omega, \mathcal{F}, P)$ be the probability space, R – a set of real numbers and N – a set of natural numbers, and for any $t \in N$ random variable $x_t : \Omega \to R$.

Definition 1. Time series $\{x_t\}_{t \in N}$ is defined as a stochastic process in discrete time [4, 5].

The development trend model (a model describing the development of the phenomenon over time) is used to identify the joint length. At each moment (cycle) the $t \in N$ joint length is presented in the following form:

$$x_t = \alpha_0 + \alpha_1 t + \varepsilon_t, \tag{1}$$

where $\{\varepsilon_t\}_{t \in N}$ it is a sequence of independent random variables with a normal distribution $N(0, \sigma^2)$. The following assumptions are made for regression models (1):

- the variance of the random component is constant (does not depend on the moment $t = 1, 2, \ldots, n$);
- there is no external factor influence on the realization of the dependent variable;
- there is no autocorrelation of random components.

Failure to meet the above assumptions indicate the existing dependencies in the residuals, which were not explained by the model (1). The solution to this problem may be to additionally identify the series of residuals $\{\varepsilon_t\}_{t \in N}$ using models that take into account the dynamics of external disturbances. In order to identify the structural parameters in the model (1), the least squares method is used [6]. For the realization of $\{x_t\}_{1 \le t \le n}$, the values of structural parameters are determined using the formula:

$$\begin{cases} \widehat{\alpha_1} = \dfrac{n \sum_{t=1}^{n} t x_t - \sum_{t=1}^{n} x_t \sum_{t=1}^{n} t}{n \sum_{t=1}^{N} t^2 - \left(\sum_{t=1}^{N} t\right)^2}, \\ \widehat{\alpha_0} = \dfrac{1}{n} \left(\sum_{t=1}^{n} x_t - \alpha_1 \sum_{t=1}^{n} t\right). \end{cases} \tag{2}$$

Using the formulas $\sum_{t=1}^{n} t = \frac{n(n+1)}{2}$ and $\sum_{t=1}^{n} t^2 = \frac{n(n+1)(2n+1)}{6}$ the system (2) can be presented in the following form:

$$\begin{cases} \widehat{\alpha_1} = \dfrac{12 \sum_{t=1}^{N} t x_t - \frac{N(N+1)}{2} \bar{x}}{N(N^2-1)}, \\ \widehat{\alpha_0} = \bar{x} - \alpha_1 \dfrac{N+1}{2}, \end{cases} \tag{3}$$

where $\bar{x} = \frac{1}{n} \sum_{t=1}^{n} x_t$. In the linear model (1), the unbiased estimator of the variance of the random σ^2 component is:

$$S_\varepsilon^2 = \frac{\sum_{t=1}^{n} \varepsilon_t^2}{n-2}, \tag{4}$$

where $\{\varepsilon_t\}_{1 \le t \le n}$ means the realization of residuals, $\varepsilon_t = x_t - \widehat{\alpha_0} - \widehat{\alpha_1} t$ for $1 \le t \le n$. Variance of parameters α_0 and α_1 is determined as:

$$\begin{cases} S_{\alpha_0}^2 = S_\varepsilon^2 \dfrac{\sum_{t=1}^{n} t^2}{n \sum_{t=1}^{N} t^2 - \left(\sum_{t=1}^{N} t\right)^2} = \dfrac{S_\varepsilon^2}{n - \frac{3}{2} \frac{n+1}{2n+1}}, \\ S_{\alpha_1}^2 = S_\varepsilon^2 \dfrac{n}{n \sum_{t=1}^{N} t^2 - \left(\sum_{t=1}^{N} t\right)^2} = \dfrac{12 S_\varepsilon^2}{n(n+1)^2}. \end{cases}$$

The analysis of trend significance in the linear model (1) allows to determine whether the time factor (the number of cycles) affects the joint length significantly. To this end, the working hypothesis is verified at the significance level $0 < \alpha < 1$:

H_0: $\alpha_1 = 0$ (slope in the linear model (1) is not significantly different from zero) against the alternative hypothesis:
H_1: $\alpha_1 \neq 0$ (slope in the linear model (1) is significantly different from zero).

Test statistics $T = \frac{|\widehat{\alpha_1}|}{S_{\alpha_1}}$ is characterized by Student's t-distribution with $n-2$ degrees of freedom. If $T \leq t\left(1 - \frac{\alpha}{2}, n - 2\right)$, then at the significance level α there are no grounds to reject the working hypothesis, therefore the α_1 coefficient is not significantly different from zero and the time factor (cycles) does not have a significant impact on the joint length. If $T > t\left(1 - \frac{\alpha}{2}, n - 2\right)$, then at the significance level α the working hypothesis is rejected, therefore the α_1 coefficient is significantly different from zero and the time factor (cycles) has a significant impact on the joint length.

In addition to the determination of the trend in the series $\{x_t\}_{1 \leq t \leq n}$, the identification of the residual series $\{\varepsilon_t\}_{1 \leq t \leq n}$ should also be performed. Both its stationarity and homogeneity (homoscedasticity) are verified.

Definition 2. Time series $\{\varepsilon_t\}_{t \in N}$ is strictly stationary, if for each $m \in N$, any $t_1 < t_2 < \ldots < t_m$ and each $\tau \in N$ the total probability distributions m of elementary random sequences $\varepsilon_{t_1}, \ldots, \varepsilon_{t_m}$ and $\varepsilon_{t_1+\tau}, \ldots, \varepsilon_{t_m+\tau}$ are identical.

Definition 2 shows that for stationary time series, for any time shift, the static and dynamic properties remain unchanged, therefore the mean value and the variance of the elements of the time series $\{\varepsilon_t\}_{t \in N}$ are constant [4, 5].

Definition 3. Time series $\{\varepsilon_t\}_{t \in N}$ is homogeneously (homoscedasticly) non-stationary if by separating a non-random component from the time series a stationary series is obtained.

The most commonly used tests for determining time series stationarity are unit root tests: the Augmented Dickey–Fuller test (ADF) test, the Kwiatkowski-Phillips-Schmidt-Shin (KPSS) test [7, 8]. For the purpose of this study the ADF test was used. Time series $\{\varepsilon_t\}_{t \in N}$ of the following form is considered:

$$\Delta \varepsilon_t = \theta \varepsilon_{t-1} + \sum_{i=1}^{k} \gamma_i \Delta \varepsilon_{t-i} + \epsilon_t \tag{5}$$

where $\{\epsilon_t\}_{t \in N}$ is a sequence of independent random variables with a normal distribution $N(0, \sigma^2)$. The $k \in N$ order of autoregression is selected with the aim of eliminating correlations between the elements of the series $\{\epsilon_t\}_{t \in N}$. A working hypothesis is created at the significance level $0 < \alpha < 1$:

H_0: time series $\{\varepsilon_t\}_{t \in N}$ is non-stationary; and the following alternative hypothesis is assumed:
H_1: time series $\{\varepsilon_t\}_{t \in N}$ is stationary.

Test statistic:

$$DF = \frac{\hat{\theta}}{S(\theta)} \tag{6}$$

is characterized by the Dickey-Fuller distribution, where $\hat{\theta}$ is an estimator of the parameter θ and $S(\theta)$ is the standard deviation of this parameter. The value of the estimator of the parameter θ and the standard deviation are determined using the least squares method. For the Dickey-Fuller distribution the critical value DF^* is determined. If $DF^* \leq DF$, then at the significance level α there are no grounds for rejecting the working hypothesis, therefore it is assumed that the series $\{\varepsilon_t\}_{t \in N}$ is non-stationary. If $DF < DF^*$, then at the significance level α the working hypothesis is rejected in favor of the alternative hypothesis and it is assumed that the series $\{\varepsilon_t\}_{t \in N}$ is stationary. In the case where the elements of the series $\{\varepsilon_t\}_{t \in N}$ are correlated and meet the condition of stationarity, the series is identified using the $ARMA(p, q)$, $p, q \in N$ class models (*AutoRegressive Moving Average*). In this case, the residuals are presented in the following form:

$$\varepsilon_t = \alpha_0 + \alpha_1 \varepsilon_{t-1} + \cdots + \alpha_p \varepsilon_{t-p} + \epsilon_t - \theta_1 \epsilon_{t-1} - \cdots - \theta_q \epsilon_{t-q} \quad (7)$$

where $\{\epsilon_t\}_{t \in N}$ is a sequence of independent random variables with distribution $N(0, \sigma^2)$.

To verify the homoscedasticity of the series of residuals $\{\varepsilon_t\}_{t \in N}$ the Breusch-Pagan test is used [9, 10]. A working hypothesis is created at the significance level $0 < \alpha < 1$:

H_0: $Var(\varepsilon_t) = \sigma^2$ for $t = 1 \ldots n$ (the variance of residuals is constant) and an alternative hypothesis:
H_1: $Var(\varepsilon_t) = \sigma_t^2 \neq const$ (the random component of the model (1) is heteroscedastic, the variance of residuals is a function dependent on time t).

To verify the working hypothesis using the Breusch-Pagan test, the linear regression equation of the following form is considered:

$$\varepsilon_t^2 = \gamma_0 + \gamma_1 t + \vartheta_t \quad (8)$$

and the coefficient of determination R^2 for the model (8) is determined. The test statistic

$$LM = nR^2 \quad (9)$$

has the χ^2 distribution with 1 degree of freedom. If the value of the statistics $LM < \chi^2(1 - \alpha, 1)$ (the value of $\chi^2(1 - \alpha, 1)$ is a quantile of order $1 - \alpha$ for distribution χ^2 with 1 degree of freedom), then at the significance level α there are no grounds to reject the working hypothesis H_0 and it is assumed that the series $\{\varepsilon_t\}_{t \in N}$ is homoscedastic. If $LM \geq \chi^2(1 - \alpha, 1)$, then at the significance level α the working hypothesis H_0 is rejected in favour of the alternative hypothesis H_1 (series $\{\varepsilon_t\}_{t \in N}$ is heteroscedastic).

Residual series for which the variance changes over time are identified using ARCH *(Autoregressive Conditional Heteroscedasticity)* class models or their extension GARCH *(Generalized Autoregressive Conditional Heteroscedasticity)* models [11, 12]. The ARCH model [3, 14, 15] was proposed by Robert Engle in 1982. For the ARCH(p) model, the $p \in N$ elements of the time series $\{\varepsilon_t\}_{t \in N}$ are determined using the following equation:

$$\varepsilon_t = \sigma_t \epsilon_t, \tag{10}$$

where the conditional variance σ_t^2 is described by the following formula:

$$\sigma_t^2 = \alpha_0 + \alpha_1 \varepsilon_{t-1}^2 + \cdots + \alpha_p \varepsilon_{t-p}^2 \tag{11}$$

and $\{\epsilon_t\}_{t \in N}$ is a sequence of independent random variables with distribution $N(0, 1)$ and $\alpha_0 > 0, \alpha_1, \ldots, \alpha_p \geq 0$. From formula (10), the conditional variance of the elements of a series of residuals $\{\varepsilon_t\}_{t \in N}$ is a linear function of the squares of past values of these residuals. In the case of strong disturbances in the time series having a long-term effect on the variability of the series, it is necessary to adopt a high value of autoregression order p. It is a disadvantage of the $ARCH(p)$ model.

A generalization of the $ARCH(p)$ model is $GARCH(p, q), p, q \in N$ [11, 12], which was proposed by T. Bollerslev in 1986. The model takes into account the dependence of the conditional variance not only on the previous values of the series of residuals, but also on the values of previous conditional variances. For the $GARCH(p, q)$ model, the $p, q \in N$ elements of the time series $\{\varepsilon_t\}_{t \in N}$ are determined using Eq. (10), while the conditional variance σ_t^2 is described by the following formula:

$$\sigma_t^2 = \alpha_0 + \alpha_1 \varepsilon_{t-1}^2 + \cdots + \alpha_p \varepsilon_{t-p}^2 + \beta_1 \sigma_{t-1}^2 + \cdots + \beta_q \sigma_{t-q}^2 \tag{12}$$

and $\alpha_0 > 0, \alpha_1, \ldots, \alpha_p, \beta_1, \ldots, \beta_q \geq 0$. In formula (11), the conditional variance of the elements of a series of residuals $\{\varepsilon_t\}_{t \in N}$ is a linear combination of past squares of residuals and past conditional variances.

The problem of reliability of technical systems is the subject of a number of analyses [16, 17]. The best models for predicting and preventing process disturbances under conditions of uncertainty are sought [18, 19]. Often a good solution is to use several methods of mathematical modelling, which allows to achieve higher quality of identification and prediction [19, 20].

3 Case Study

The collected data concerns several joints connecting individual elements of the conveyor belt. Three were selected for the analysis: joint A, joint B, joint C. The graph of change in their length for the first 200 test cycles is presented in Fig. 1.

The values of the length of individual joints differ, as shown in the frame graph presented in Fig. 2.

Due to clear differences between the momentary values of the length of individual joints, these values cannot be considered as one variable in the model [21]. It is therefore

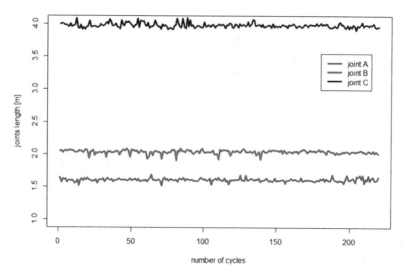

Fig. 1. Line graph of selected observations (joint: A, B, C).

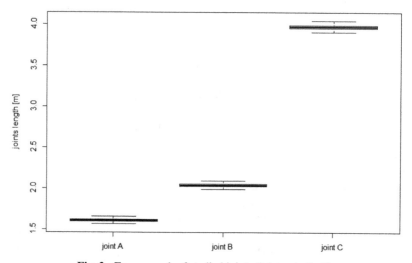

Fig. 2. Frame graph of studied joints (joints: A, B, C).

necessary to analyse each of them independently. A test aimed at identifying the existence of a trend first, and then at assessing the variance of the random component was conducted for all joints. The existence of a positive trend in the linear model means that, as the number of cycles of operation increases, a gradual stretching of the joint occurs, which is the result of the stresses occurring in the belt. This situation requires constant monitoring and assessment of potential exceeding of the limit values.

Special attention should also be paid to those joints in which no trend was identified, however statistical tests confirmed the heteroscedasticity of the variance of external disturbances. The variability of the variance could mean a significant impact of the conveyor belt momentary loads on the joint tension and, consequently, affect its reliability. The results for selected joints are presented in Table 1.

Table 1. Diagnostics of time series for joints A, B and C.

Joint	Trend assessment	Trend factor	Variance of the residual component	Residual component model
Joint A	A significant trend was found	0.000066	Homoscedasticity of variance	ARMA(1,1)
Joint B	No significant trend was found	–	Homoscedasticity of variance	–
Joint C	No significant trend was found	–	Heteroscedasticity of variance	GARCH(1,1)

For joint A, the analysis of the trend significance in the linear model allowed to conclude that the number of cycles has a significant impact on the change of the joint length. The linear model has the following form:

$$y_t = 0{,}000066 \cdot t + 1{,}6 + \varepsilon_t \qquad (13)$$

The joint stretches gradually over time, which means that it is necessary to constantly monitor and check for exceeding of the limit value. The graph of the function of the trend and empirical values is presented in Fig. 3.

Then the residuals of the linear model were diagnosed by testing the homoscedasticity of their variance. Test statistics $BP = 0.8313$ and p-value $= 0.3619$, therefore, there are no grounds to reject the working hypothesis of homoscedasticity of the random component. Therefore, the external disturbances of the linear model are described using the ARMA model. Results of estimation of its parameters are presented in Table 2.

For joint B, no trend was identified, the condition of homoscedasticity of variance is met (test statistic BP $= 1.445$ and $p - value = 0.2293$), therefore there are no adverse changes in its tension.

No trend was identified for joint C either, however the heteroscedasticity of variance was found (test statistics $BP = 0.1155$ and $p-value = 0.00068$). External disturbances occurring in the linear model were identified using the GARCH model, the structural parameters are presented in Table 3.

The model has the following form:

$$\varepsilon_t = \sqrt{0{,}000025 + 0{,}063\varepsilon_{t-1}^2 + 0{,}91\ \sigma_{t-1}^2}\ \epsilon_t \qquad (14)$$

The analysis of changes in the length of joint A indicated the existence of a significant trend. This means that consecutive work cycles cause it to elongate. Such a joint should be subject to thorough inspections, as an element conducive to belt rupture, which can be identified by the linear model proposed by the authors. Additionally, heteroscedasticity

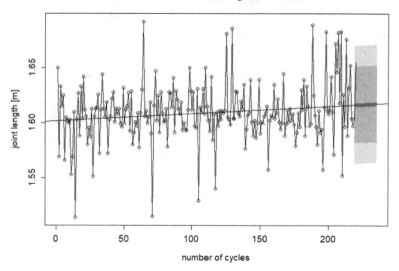

Fig. 3. Empirical observations and linear model of joint A.

Table 2. Parameters of the ARMA(1,1) model for joint A.

ARMA(1,1)	AR (p)	MA(q)
Coefficients	0.7867	0.7079
SE	0.1507	0.1674

Table 3. Estimated parameters of the GARCH model.

GARCH (1,1)	α_0	α_1	β_1
Coefficients	0.000025	0.063	0.91
SE	$2.764376\ 10^{-5}$	$2.567176\ 10^{-2}$	$4.002943\ 10^{-2}$

of the external disturbances was not found, which means no uncontrolled influence on the elongation of the tested element. The ARMA(1,1) model was proposed to describe the residuals of the model. In the case of joint C, heteroscedasticity of the component of the residuals was found. The GARCH(1,1) model was used to identify them. The heterogeneity of the variance of residuals indicates adverse changes in the joint length.

4 Final Thoughts

The aim of this article is to evaluate and present a method of mathematical modelling of the change in the length of joints of a conveyor belt, the reliability of which is determined by the quality of joints of all its individual components.

The study presented the possibility of using a linear model to identify the trend, the occurrence of which means unfavourable changes in the joint during operation of a conveyor belt. Moreover, the importance of the characteristics of the variance of external disturbances for making the assessment was demonstrated. Therefore, in order to assess the possibility of joint rupture more effectively, an analysis of the residual component was carried out and, depending on the result, an appropriate model was proposed.

The proposed method can assist in the management of the conveyor belt operation process. The existence of the trend and the diagnosed heteroscedasticity of the variance indicate the necessity of monitoring such joints, because the increase in the number of work cycles or excessive loads resulting from the increase in weight carried on the conveyor belt may cause the joints to rupture.

Further research should consider the prediction of the change in the length of such joints over time, understood as forecasting the moment of occurrence of a potential damage in the joint, which will allow to estimate the time of correct operation or the best moment to perform the necessary maintenance service activities.

References

1. Mazurkiewicz, D.: Computer-aided maintenance and reliability management systems for conveyor belts. Eksploatacja i Niezawodnosc – Maintenance Reliab. **16**, 377–382 (2014)
2. Vališ, D., Mazurkiewicz, D.: Application of selected Levy processes for degradation modelling of long range mine belt using real-time data. Arch. Civil Mech. Eng. **18**(4), 1430–1440 (2018). https://doi.org/10.1016/j.acme.2018.05.006
3. Valis, D., Mazurkiewicz, D., Forbelska, M.: Modelling of a transport belt degradation using state space model. In: 2017 IEEE International Conference on Industrial Engineering and Engineering Management (IEEM), Singapore, pp. 949–953. IEEE (2017). https://doi.org/10.1109/IEEM.2017.8290032.
4. Box, G.E.P., Jenkins, G.M.: Time Series Analysis: Forecasting and Control. Holden-Day, San Francisco (1970)
5. Kozłowski, E.: Analiza i identyfikacja szeregów czasowych. Politechnika Lubelska, Lublin (2015)
6. Chow, G.C.: Ekonometria. PWN, Warszawa (1995)
7. Kokoszka, P., Young, G.: KPSS test for functional time series. J. Theor. Appl. Stat. **50**(5), 957–973 (2016)
8. Kozłowski, E., Mazurkiewicz, D, Żabiński, T., Prucnal, S., Sęp, J.: Assessment model of cutting tool condition for real-time supervision system. Eksploatacja i Niezawodnosc – Maintenance Reliab. **21**(4), 679–685 (2019). https://doi.org/10.17531/ein.2019.4.18
9. Breusch, T.S., Pagan, A.R.: A simple test for heteroscedasticity and random coefficient variation. Econometrica **47**, 1287–1294 (1979)
10. Maddala, G.S., Lahiri, K.: Introduction to Econometrics, 4th edn. Wiley, Chichester (2009)
11. Bollerslev, T., Chou, R., Kroner, K.: ARCH modeling in finance. J. Econometrics **52**(1–2), 5–59 (1992). https://doi.org/10.1016/0304-4076(92)90064-X

12. Bollerslev, T.: Generalized autoregressive conditional heteroskedasticity. J. Econometrics **31**, 307–327 (1986)
13. Bera, A.K., Higgins, M.: ARCH Models: properties, estimation and testing. J. Econ. Surv. **7**(4), 305–336 (1993)
14. Engle, R.F.: Autoregressive conditional heteroscedasticity with estimates of the variance of United Kingdom inflation. Econometrica **50**, 987–1007 (1982)
15. Shimizu, K.: Parametric AR(p)-ARCH(q) models. In: Bootstrapping Stationary ARMA-GARCH Models, pp. 19–64, Vieweg+Teubner (2010)
16. Iscioglu, F., Kocak, A.: Dynamic reliability analysis of a multi-state manufacturing system. Eksploatacja i Niezawodnosc – Maintenance Reliab. **21**(3), 451–459 (2019). https://doi.org/10.17531/ein.2019.3.11
17. Selech, J., Andrzejczak, K.: An aggregate criterion for selecting a distribution for times to failure of components of rail vehicles. Eksploatacja i Niezawodnosc – Maintenance Reliab. **22**(1), 102–111 (2020). https://doi.org/10.17531/ein.2020.1.12
18. Li, J., Wang, Z., Ren, Y., Yang, D., Lv, X.: A novel reliability estimation method of multi-state system based on structure learning algorithm. Eksploatacja i Niezawodnosc – Maintenance Reliab. **22**(1), 170–178 (2020). https://doi.org/10.17531/ein.2020.1.20
19. Li, J., Wang, Z., Liu, C., Qiu, M.: Accelerated degradation analysis based on a random-effect Wiener process with one-order autoregressive errors. Eksploatacja i Niezawodnosc – Maintenance Reliab. **21**(2), 246–255 (2019). https://doi.org/10.17531/ein.2019.2.8
20. Alkali, M.A., Sipan, I.M., Razali, N.: Assessing the forecasting performance of ARIMA and ARIMAX models of residential prices in Abuja Nigeria. Asia Proc. Soc. Sci. **4**(1), 4–6 (2019)
21. Kissi, E., Adjei-Kumi, T., Amoah, P., Gyimah, J.: Forecasting construction tender price index in Ghana using autoregressive integrated moving average with exogenous variables model. Constr. Econ. Build. **18**(1), 70–82 (2018)
22. Borucka, A., Wiśniowski, P., Mazurkiewicz, D., Świderski, A.: Laboratory measurements of vehicle exhaust emissions in conditions reproducing real traffic. Measurement **174**, 108998 (2021). https://doi.org/10.1016/j.measurement.2021.108998

Experience in Implementing Computer-Oriented Methodological Systems of Natural Science and Mathematics Research Learning in Ukrainian Educational Institutions

Olena Hrybiuk

Institute of Information Technologies and Learning Tools NAES of Ukraine, Kiev, Ukraine

Abstract. Taking into account the specifics of the existing dissonance regarding the technical characteristics of computer equipment used in schools in the process of designing computer-oriented methodological systems of natural science and mathematics research learning (COMSRL), the solutions which meet all the necessary requirements, contribute to the effective organization of research activities during research studies at school and are a budget alternative to expensive software are offered. The experience is analyzed. With the aim to increase the motivation and effectiveness of students' learning the ways of solving the problems of designing variable models of research learning using the components of the COMSRL are considered. The results of the experimental study confirm that due to the use of COMSRL the optimal concentration of educational resources is ensured integrating training courses of robotics tested in the educational process and extracurricular activities; orientation of the content and technologies of students' preparation for research, manufacturing integration in terms of educational process. To organize the students' cognitive activity in the field of robotics, a number of constructors have been developed that help the students' to assemble the structure, create the programme and run the robot model. The results of a wide experiment allowed us to determine the most effective psychopedagogical factors that influence the efficiency of mastering knowledge. It has been shown that in determining the feasibility of applying and selecting the COMSRL it is necessary to give preference to those ones that will help students to develop conceptual structures of knowledge.

Keywords: Modelling · Engineering · Computer oriented methodological systems of research learning · Variational models · Manufacturing · Educational robotics

1 Introduction

Programming and robotics training in the technological 21st century promotes the development of communicative skills of young people, develops interaction, autonomy in decision-making, reveals students' creativity. To achieve this goal, within the all-Ukrainian experimental work "Variable models of computer-oriented methodological systems of natural science and mathematics research learning in a comprehensive educational institution" [1] in the context of continuity of education and strategic initiative "Variable models of systems of general and extra education for kids and youth" [2] the first 27 experimental sites were created in 2016.

In the conditions of low motivation of kids and youth as for the cognitive activity sand scientific creativity the task to improve extra educational programmes, the creation of computer oriented methodological systems of research learning with the use of appropriate forms for kids' intellectual development, their preparation for clarifying educational engineering programs are of significant importance.

Introduction of the elements of robotics in school subjects has helped to increase the motivation and interest of students, has diversified educational and research activities [3]. To avoid the scholasticism of knowledge, demo experiments and laboratory works are used in the process of research learning of the subjects of the natural-mathematical cycle [4]. But often the appliances that make up the foundation of the laboratory equipment have a high error, which allows to evaluate the results of the experiment qualitatively, not quantitatively. To solve this problem, robotic platforms, constructors, handmade devices, etc. are used. With the help of these tools, with the participation of children, various models are created and programmed, such as household appliances powered by electricity, the solar system, mobile robots with temperature sensors, magnetic field sensors, and others.

The theoretical and methodological basis of scientific research is the system of developmental learning [5], the theory of educational activity [6], the theory of the gradual formation of mental actions [7]), the psychological research [8], the theory of teaching methods [9], the theory of pedagogical systems [10], problems of variability in education and vocational training [11–13], Principles of Convergent Natural Sciences and Engineering Education [14–16]. However, there are a number of problems that prevent penetration of progressive methods in the engineering and science teaching process as well as the spreading of an interdisciplinary approach to teaching, including the international experience [17, 18]. The problem of pedagogical design with the use of variable models is fragmented in science, that is why the research was conducted to create a comprehensive scientific understanding of the features of computer-oriented design of the educational environment and the construction of variable models for the implementation of research learning in natural and mathematical disciplines. However, the problem of exploring characteristics necessary and sufficient for the organization of research learning at school, using COMSRL and taking into account the psychopedagogical features of the students in the process of designing COMSRL remains a priority. Determining the relationship between the theoretical foundations of research learning and the possibility of developing the teaching practice using the COMSRL in accordance with the goals of students' development, became one of the objectives of the research.

2 Methodology

In the experimental study in the process of information resources selection during the research learning we took into account the psychophysiological and psychopedagogical factors, among which the stages of intellectual progress (SIP) of the students were of great importance [3, 19].

In the process of determining the factors that most influence the learning outcome, research data from 7.7 thousand respondents (students, teachers) from 27 educational institutions located in different regions of Ukraine were analyzed. Significant differences between the learning and teaching styles of students and teachers have been experimentally proved. It is shown that the differences depend primarily on the SIP and the preparation. It has been identified that students who have different combinations of learning styles experience different cognitive pressure during research learning with the use of COMSRL [20].

Descriptive statistics, Pearson's correlation analysis with correlation coefficients, nonparametric tests for calculation the W Kendall coefficient of correlation were used for statistics analysis of the results in the study. A significance level of 0.05 was used to test the hypotheses, and appropriate bilateral significance values were provided as appropriate. All components of the COMSRL used in learning natural science and mathematics were organised into 17 groups according to their functions.

Indicators of students' preference for the use of information resources are considered as characteristics of the popularity of a separate COMSRL. Two parameters were identified regarding the need for certain restrictions on the practical use of information resources and the popularity of their use: the value of the average score obtained during the survey of respondents and the number of significant correlations.

The calculated correlations between the students' preference indicators towards the use of separate COMSRL and SIP information resources for individual groups of information resources are used to adjust the research teaching methodology to optimize the selection of learning resources to minimize contradictions considering students' SIP (Table 1), specific to the group.

The study developed and tested an algorithm for selecting the optimal set of resources of the COMSRL, which involves the use of quantitative criteria to determine the suitability of resources to the requirements of students and teachers. The first criterion is the average score of the preference for e-resource selection by students with different SIPs and learning styles, which is calculated for a typical group (class) profile. The second criterion is the difference between the expert rating of resources for each topic and the score given by the students.

An analysis of the experimental data obtained and the scientific literature made it possible to formulate such recommendations that would help teachers to prevent students from working ineffectively [3].

1. The formulation of the cognitive style of the individual in the process of professional training actually leads to the leveling of the difference in individual styles, taking into account SIP. In the process of research learning, it is necessary to develop a mechanism for integrating different types of student style behaviour.
2. Preliminary diagnosis of student behaviours in the group is desirable in order to organize effective research learning. Understanding the composition of the group helps the teacher create the conditions for the prevention of phenomena which help avoid the "conflict of learning styles" regarding the SIP of teachers and students (for example, ineffective design of teaching material, poor performance, high levels of discomfort, etc.); actualization and enrichment of the whole system of mechanisms

Table 1. Coefficients of correlation between students' preferences towards the use of separate information resources of COMSRL and students' SIP for the groups of resources "static visualizations" and "dynamic visualizations".

Resource group	Resource name	Stages of students' intellectual progress (SIP)			
		I	II	III	IV
Static visualisation	Drawings	−0,406	−0,489	0,014	−0,116
	Charts	−0,627	−0,427	−0,300	−0,359
	Diagrams	−0,461	−0,471	−0,221	−0,461
	Tables	−0,113	−0,556	−0,060	−0,441
	Schemes	−0,428	−0,380	−0,080	−0,493
Dynamic visualisation	3D Model	−0,316	−0,239	0,067	−0,365
	Animation	−0,415	−0,101	0,467	−0,198
	Video playback of the experiment	−0.221	−0,09	0,042	−0,032
	Video playback of natural processes	−0,110	−0,198	0,124	−0,423
	Video playback of life examples	−0,070	−0,135	0,417	−0,327
	Video Excursions in manufacturing plants	−0,026	−0,202	0,571	−0,087

of students' stylistic behaviour, which determines the productivity of intellectual actions.
3. The task of organizing effective research learning involves the creation of didactic materials with the possibility of selecting a specific line of research learning and prerequisites for the gradual formation of individual cognitive styles.
4. In order to organize effective research learning in natural science and mathematics disciplines, COMSRL are required, which have adaptive tools and provide the opportunity to present educational material in various forms.

3 Collaborative Manufacturing and Education in the Context of Industry 4.0 (COMSRL)

The results of the study confirm that in the process of research learning, each student develops the vital qualities in modern realities (communicativeness, critical thinking, creativity, cooperation, individual responsibility) under conditions of effective and positive interaction [3, 13]. Theoretical analysis of scientific works, the study of experience in the use of information and communication technologies in the educational process indicates the existence of contradictions between: the development of modern information and communication technologies and the degree of their introduction in the educational process; variety of computer equipment and mobility of participants in the educational process; the presence of a new type of teacher in schools capable of organizing effective interaction with the use of COMSRL in natural science and mathematics, and the absence of scientifically based technologies of its organization; the growing demands on the management and organization of the educational process by society.

The urgency of the research was determined by the need to develop a new field of applied research: the use of variable models of COMSRL of natural science and mathematics in the educational process, management and dissemination of the author's methodology of research learning.

The hypothesis of the study was that the use of COMSRL natural sciences and mathematics would have a positive impact on the organization of learning, would create conditions for the development of new methods and technologies of students' research learning, would increase their motivation for learning, would ensure the development of information, research, foreign languages and professional teachers' competences, that would lead to positive qualitative changes in organization of the participants of educational process.

The computer-based system of teaching students and developing their technological competence is represented as a set of innovative research practices, which are implemented using the technosphere of educational organizations on the principles of variability and their usage contributes to the activation of cognitive activity [1]. The students' motivation to learn and choose the engineering professions is achieved through their interest in research and innovation practices, including the realization of design and graphic tasks. Research learning with COMSRL is ensured through the use of appropriate programmes, expositions, laboratory and demonstration equipment, appropriate software and content, active forms of organization of the educational process, research activities of students.

The uniqueness of the author's method of the research learning is, for example, the possibility of programming on examples made for a particular performer, designed by the students themselves [3]. The achievement of this goal is ensured by the fulfillment of the following tasks: promoting the formation of elementary knowledge in IT, physics, mathematics; ability to design models of robots and their construction, creation and software algorithms realisation; operating system and graphical programming language skills; developing the ability of students to use information and communication technologies to effectively solve the atypical tasks of obtaining and submitting information through physical devices, processing this data by the processing unit, saving for further processing; formation of the students' scientific worldview as an integral part of the general human culture, as the necessary condition for a fulfilling life in modern society; sustainable motivation to study; intellectual development of personality, development of students' logical thinking, algorithmic, informational and graphic culture, memory, attention, intuition; physical, ecological, aesthetic, civic education and formation of positive personality traits.

Among the main educational areas it is necessary to distinguish the following: interesting and/or in-depth study of mathematics, chemistry, biology, physics, ecology and other subjects of the natural science cycle, engineering graphics, information and communication technologies, design and construction, robotics, fundamentals of electrical mechatronic engineering, 3D-visualization, prototyping, nano-technologies and others (depending on the specific tasks of the educational institutions).

A computer-based learning system is presented as logistically and scientifically methodological support, which contains logically and logistically related objects and services. The core of a computer-based learning system is an innovative science complex which components give students the opportunity to learn the basics of mathematics, biology, physics, chemistry, electronics, geography, ecology, history, and other sciences in an exciting (non-standard) form. The innovative scientific complex consists of "permanent (static)" and "variable (dynamic)" components (blocks, exposures). "Static" components are intended for the basic sciences and may not change for a long time; "variable (dynamic)" components change periodically, the achievements of applied sciences and technological sphere are presented with their help, too. In different premises of the educational institution, due to the use of modern technologies, it is possible to hold both scientific conferences and creative events for preschoolers, workshops, seminars, mini-conferences, lectures and other events.

Here is a fragment of the contents of the elementary school physics program using educational robotics. For example, when studying the section "Physics and Physical Methods for the Study of Nature" such topics are considered "Physics – the science of nature", "Observation and description of physical phenomena", "Physical appliances", "Physical quantities and their measurements", "Physical experiment", "The role of mathematics in the development of physics", "Physics and Technology". Such research papers as "Calibration of measuring devices", "Timer" and others are recommended to implement.

In the process of studying the section "Thermal phenomena", when the environmental problems of the use of heat engine are investigated, the work "Electric car with a solar battery" is offered to perform. Studying the section "Electromagnetic oscillations and waves" it is advisable to pay attention to the implementation of research projects "Electrical measurements", "Lego-capacitor - energy storage", "Electricity storage", "Calculation of energy costs for lighting", "Connection of the generator to the engine", "Optimization of energy conversion process", "Solar battery power", "Electric power generation using a water wheel", "Electricity generation using a wind engine", etc.

4 The Use of Digital Laboratories in the Process of Teaching Natural Science and Mathematics

Author's Digital Labs are sets of equipment and software for conducting demonstration and laboratory experiments, a wide range of studies, laboratory workshops including the accumulation and analysis of data from natural experiments using the project and research work of students. With the use of digital equipment it is possible to carry out both laboratory work within the school programme and new researches, synchronizing the data with the PC with the possibility of their further processing, conversion.

Digital laboratories for physics are used to perform various laboratory work, including studying the movement on an inclined plane; simple oscillatory movements; current-voltage characteristics of resistance, heating lamps and diodes; magnetic fields; speed of sound; diffraction and interference of light.

Digital laboratories for biology and chemistry are used in the study of the effect of exercises on human body temperature and heart rate; study of water evaporation by terrestrial plants; the influence of vegetation on the microclimate of the area; acid-base titration.

The use of digital laboratories and individual sensors (humidity sensors, oxygen concentration, heart rate, temperature, acidity, etc.) significantly improves visualization both during the direct performing of the work and in the process of processing the results of the study.

The equipment the physics laboratory includes sensors: *Current; Distances; Forces; Humidity; Lighting; Induction of magnetic field; Pressure; Sound sensor (microphone); Thermocouple; Voltages.*

The Digital Laboratory of Chemistry and Biology includes sensors: Breathing; Heart rate; Humidity; Lighting; Oxygen; pH meter; Temperature.

The equipment of the digital laboratory is universal, thus, students can use it in different experimental installations, to make measurements in the "field conditions", to save the time for students and teacher, to encourage students to creativity, allowing changing the parameters of measurements. Using video and analytics software, you can retrieve video snippets. All this contributes to the exploration of real life situations recorded on video by students and to the exploration of fragments of educational and popular science videos.

Using digital labs, students participate in scientific experimentation with the support of teachers and scientists being not limited to the topic of the lesson, gaining experience in analyzing and refining the results of the study.

For example, in the process of studying the acidity of various substances, students can independently make the conclusion that popular drinks are harmful to the digestive system. Accordingly, gloves should be used when using some detergents (containing chemical reagents).

Using digital laboratories, new approaches to learning are used that helps students to develop the ability to autonomous search, processing, and analyzing messages, they also help to reveal students' creativity.

In biology lessons, students perform laboratory works using the COMSTR and digital labs. Below there are some examples.

I. Middle school programme: *Cardiovascular system response to loading (Grade 8); Effect of enzymes on substrate (example of catalysis) (8, 9, 10 grades); Study of blood circulation (8 grade); Functional tests. Tests for the evaluation of the breathing system (8, 9, 10 grades); Relationship between loading and energy exchange (8–10 grades).*

II. High School Program: *Catalytic activity of enzymes in living tissues (9–10 grades); Adaptation of organisms to the habitat (9–10 grades).*

Undoubtedly, the motivation of students to study extracurricular researches using experimental tasks increases significantly.

For example, when studying topics in the section "Plant biology": *Water absorption by the roots. Root pressure (6 grade); CO_2 absorption and release of O_2 under the action of light (6 grade); Breathing of leaf; Breath of leaf; Evaporation of water by plants. Transpiration; Seed breathing; Cold-resistant and thermophilic plants; Seed germination conditions.*

Zoology Section: *Aquatic Animals (7 grade); Cold-blooded and warm-blooded animals (7 grade).*

Section "Human Biology" – "Man and his health" (8 grade): *Responses of the cardiovascular system of a man to exercise; Circulatory problems when you press your finger; Gas exchange in the lungs; Donder's model. Breath. Inhalation and exhalation mechanism; Vital capacity of the lungs. Breathing system response to exercise; Skin structure and functions. Skin selection. Respiratory and thermoregulatory functions of the skin.*

General Biology Section (9–11 grades): *Effect of enzymes on substrate (catalysis). Decomposition of hydrogen peroxide H_2O_2; Influence of pH on enzyme activity; The influence of external factors on photosynthesis. The rate of photosynthesis.*

In chemistry teaching, students can perform laboratory works using sensors and digital laboratories in the classroom and during extra-curricular activity.

For example, *Studying the Laws of Electrolysis and Application of Electrolysis in Practice (Current and Voltage Sensors); Water analysis. Chemical analysis of drinking water. Studying acidity of different water samples: from tap and drinking bottled water, drinking and mineral water (acidity sensor); Experimental verification of gas laws; Investigation of exothermic (interaction of copper chloride with aluminum) and endothermic (interaction of baking soda with citric acid, kefir, sauerkraut) reactions; Investigation of the thermal effect of fuel burning; Study of the chemical catalysis of the decomposition of hydrogen peroxide H_2O_2 in the presence of a catalyst (MnO_2).*

Demonstrative experiments can be performed during biology learning, including: *Gas exchange in the lungs. Breathing retention samples; Changes in blood circulation during clamping; Pressure change in the aquatic environment; Vein valves. The structure and functions of the venous system; Skin selection thermo control function; Response of the cardiovascular system to metered loading.*

When using digital labs in a demonstration experiment, the results are so obvious that students can not only quickly understand and memorize a learning topic, but also using concrete examples from their own life answer specific questions which are given by the teacher. For example, during a finger-pressing experiment, students will understand why their feet get frozen in tight shoes; why in order to stop bleeding in winter a braid cannot be applied for the same time interval as in summer. Finally, students will understand why warm-blooded animals can live in cold climates and cold-blooded animals do not; why cold-blooded animals can live without food for a long time, etc.

Demo experiments can be conducted in chemistry lessons. In all curricular, sufficient attention is paid to environmental issues. Laboratory workshops *(including the use of oxygen sensors, pH and light, pressure, temperature, humidity, etc.)* and excursions are required for a thorough study of this area of knowledge.

It is recommended to carry out ecological researches on the following topics during extra-curricular activities:

Research on the effectiveness of lighting in the school (Measurement of lighting in the school premises); Determination of acidity of different drinks; Influence of physical parameters of air on the health of the pupil in the school premises; Influence of ventilation on the microclimate of educational premises; Abiotic environmental factors; Ecology of urban areas; The oxygen content in the air of various accommodations of the city; Influence of change of soil acidity on the plants species; Determination of oxygen concentration in "flowering" reservoirs.

A new area of experimental work at school has been the study of physical phenomena with computer-processed results and the use of robots. The existence of the *LabVIEW* graphical programming environment, which is demanded and used in modern industry to control the production of a set of precision and convenient sensors attached to the *NXT* robot, transforms the process of experimentation into a cognitive, dynamic and exciting study.

Laboratory workshops with *Vernier* sensors were held at Summer School, where gifted young people were invited. The simplest and most understandable of all the temperature sensor was used to conduct experiments to determine the temperature of water in several glasses, finding their location. For this purpose, a voice thermometer was designed. The robot moved along a straight line, looking for a glass of water with different temperatures using a light sensor. When the object of study was found, the robot stopped and lowered the *Vernier* temperature sensor into the glass. Then the water temperature was being measured for 10 s. *LegoMindstorms* light sensor was used for searching of the glasses.

Software written in *LabVIEW* has a user-friendly interface. It was also interesting for the students to conduct a chemical experiment using robots. The idea was to manage a chemical reaction using *LabVIEW*. In the process of polymer formation, the temperature of the solution increased. After the polymer was formed, the temperature decreased and the polymer cooled and solidified. The experimenters observed changes in the optical density of the object and changes in temperature. The characteristics were measured using *Vernier* surface light and temperature sensors and were displayed on the screen (online) in graphical form.

All operations were performed automatically by the robot: *The process of releasing the reagent from one test-tube; Adding it to a test- tube for research; Immersion of a probe into a test tube for forming a polymer on it; Releasing polymer from the probe.*

Undoubtedly, using a *Lego NXT* microcomputer for a robot equipped with accurate Vernier sensors, working in a *LabVIEW* graphical programming environment to automatically process data will help students test their hypotheses, do their own cross-curricular experiments, and learn how to make scientific researches.

5 Perspectives of Introduction of Variational Models of Computer-Oriented Environment of Studying Subjects of Natural Sciences and Mathematical Cycle

The prospect of introducing variable models of computer-oriented systems of natural science and mathematics (COMSRL) research learning in schools is justified by the following alternative solutions.

1. Taking into account the specifics of the existing *dissonance regarding the technical characteristics of computer technology used in schools,* it is recommended that *xUbuntu and Ubuntu* distributions should be used in designing individual components of computer-based systems for the study of subjects in science and mathematics. The proposed solutions are a budget alternative to expensive software, meet all necessary requirements and contribute to the effective organization of the

educational process at school. So far, it is relevant to formulate and find ways to solve problems in the design of learning environment and the construction of variable models of the COMSRL in science and mathematics. As a result of the analysis, visual environments for *C++, C #* programming languages were selected; control of students' computers by the teacher; a system of management and training on-line, including group training, research and testing; creation of schedules of classes and distribution of teachers' training load; a set of administrative programmes for an educational institution; programmes for teaching algebra, geometry, physics, chemistry, geography, astronomy, etc. Detailed analysis was carried out and programmes for hearing, memory, students' eyesight, etc. were identified; work with electronic maps (GIS); systems of computer mathematics, including systems of dynamic mathematics *(GeoGebra), Cantor.* To construct graphs of mathematical functions we use *KAlgebra, OpenSource analogue MatLab, wxMaxima* - graphical interface for system of computer mathematics *Maxima,* system of computer algebra *Mathematic,* programme *Cabri, The Geometer's Sketchpad, GEONExT, Cinderella, TracenPoche,* etc.; a package of applied mathematical programmes for engineering and scientific calculations; creation and calculation of chemical structures, viewing macromolecules and preparation for publication of their images; editors of chemical schemes and reactions; schematic creation of electronic circuits, graphic stimuli of computer networks and electrical circuits; digital oscilloscopes, design of electronic devices and necessary boards.

2. *The construction of the curriculum as an individual educational project* is possible due to thoroughly comprehended *paths of students' individual learning, taking into account the possibilities of deductive design of the educational process of natural and mathematical disciplines.* The main emphasis is placed on the *project research activities,* and the corresponding goals of the subject research projects are creating the conditions for self-realization of the students' authorial position through their active participation in the subject research project; refinement and differentiation of the concept apparatus, systematization of students' knowledge, establishment *of cross-curricular connections* and preparation of Olympiad tasks, preparation for entry into higher educational institutions; correction of the level of formation of different learning actions using project-research activities.

3. In the study the variable models of COMSRL based on the *competence approach* in modern education, taking into account the basic stages of design *(target, methodological, factor, structural, functional, resource, deficient, procedural, prognostic and effective are represented)* [8]. In the process of designing a computer-oriented learning environment, the participants perform the following functions: *formation of relevant competences, including research ones, assessment of impact factors, determining the strategy of the institution, risk and educational resources assessment, selection of educational route, selection of educational content process; development of technologies and techniques for assimilation of necessary competences, examination of teaching programmes and assessment of competency result.*

4. To increase the motivation of students and the level of training, *FlashQard utilities, GNU it Flashcard Trainer, tkgate Circuit Simulator stimulators, KTurtle programming environment, KmPlot mathematical functions, K3DSurf mathematical tools, Euclidean geometry albums KSeg Geometry Sketchpad, the Euler interactive math*

programming environment are used in educational process. *Geography learners use KGeography, the Quantum GIS geographic information system,* and the *Emerillon Map Viewer map editor.*

5. It's appropriate to use the graphical simulators of computer networks *gns3, Kumir Language Implementation, Quite* universal circuit simulator, *Oregano electrical engineering tool* in the research work. To analyze the data visualisation *QtiPilot, MathGL*-based data rendering, *Qosmic's* recursive fractal editor, *Veusz* science graphics package, and *View Your Mind* link visualization tool are used. The development of mathematical thinking and relevant skills is accomplished through *GCompris, MathWar Tux, of Math Command*!; memory is developed with the help of the *Mnemosyne, PySyCache, Gamine, LMemory, Linux Letters, Childsplay, Blinken* software, and the hearing is trained via the *GNU Solfege* and others. *Tux !* and *KLettres* programmes are used in the educational process to develop imaginative thinking and perspective. Students like to explore the Universe using the *Celestia Space Simulator. Step* is also used for simulation of physical experiments. Using the *Kstars Desktop Planetarium, the Marble Globe,* and *the Pauker Cards, the Stellarium Planet, the Merkaator* electronic mapping programs are used to enhance students' cognitive activity and enhance the performance of the learning process. The periodic table *gElemental, GPeriodic, Kalzium,* molecular design and *Avogado* simulation systems *Chemtool, BKchem and xdrawchem Chemistry Editor* are used in chemistry lessons to represent and calculate chemical formulas and reactions. The *Unipro UGENE* package is recommended for conducting research in biology teaching, using *RasMol* and the *Xoscope* digital oscilloscope to view macro molecules and prepare their images for publication.

6. In the process of a research learning project work will be of a developmental character if only there weren't the *reproductive forms of activity, which will encourage students to search creatively for answers to problematic questions from mentors or literary sources.* When planning teaching process, it is advisable to take into account the control reflection in order to adjust the timing of work at each stage and improve the outcome of the project activity. The classification of software for research teaching of natural science and mathematics is developed [3].

7. It is important to take into account the system-conceptual approach to the design and operation of the COMSRL. The basic requirements for the creating of the COMSRL include: the use of licensed and freely distributed software; the existence of effective methods of protecting and determining the performance of the COMSRL; periodic analysis of safety and performance of the COMSRL, effective use of server equipment and workstations of training classes, multimedia equipment; availability of a file storage intended for creation of backup-files of the COMSRL server. Examples of architectures of the COMSRL using freely distributed software products have been developed [3].

One of the options of complex development of robotics is studying the functioning of numerically controlled machines. Undoubtedly, in addition to the basic lessons in robotics, there are various extracurricular activities, including those for the promotion of engineering, technical, technological specialties. For this purpose, competitions on educational robotics, round tables for students, quizzes, master classes "Designing and programming of robots" and the Olympiads are held. The possibilities and forms of studying robotics are not run out of those mentioned above. Students understand the learning material better when they create or invent something on their own. This fact is widely used in robotics classes, taking into account numerous prospects for further development. Using robots and robotic platforms, students create models of automated devices. Until now, the technology of research learning is used to solve applied problems, which inspires students for inventions in the sphere of science, technology, mathematics encouraging them to think creatively, to analyze a situation, to speculate, to apply their skills for solving environmental problems [3].

6 Conclusion

Within the experimental work we make sure that the study of educational robotics contributes to the formation of cross-curricular and meta-subject competences including the ability to do researches in a team, the decision-making autonomy in specific life situations, paying attention to the peculiarities of the environment and availability of additional materials [3]. Due to the implementation of robotics into educational process students anxiously look for relationships between different branches of knowledge on the basis of self-modeled prototypes of mechanical devices, while clarifying the specifics of the work of mechanical structures, clarifying for themselves the physical concepts while working in their own rhythm and taking into account the individual paths of the student's development.

Students' interest in research using educational robotics, a synergistic combination of engineering knowledge with interdisciplinarity, the development of new scientific and technical ideas will contribute to creating necessary conditions for increasing motivation including the use of COMSRL in the educational process and pedagogical approaches of research learning [2]. In the process of studying the course of educational robotics the basics of modern robotics and cognitive technologies as a whole are provided. Accordingly, students are shown the techniques of step-by-step robots' creation from constructor elements with the ability to connect programmable processors and sensor elements to build automated installations and anthropomorphic components on the open architecture platforms (both code and hardware).

The study offers the expediency of using a system-conceptual approach to the design and operation of COMSRL. In the context of the basic directions and principles of the use of systems of research learning of science and mathematics with the use of COMSRL blended learning approaches are recommended to use [1, 3, 4]. The analysis of the results of the conducted studies confirms the prospect and necessity of using variable models of COMSRL at school. Further work is underway to improve the scientific, methodological and didactic support of the components.

The research guidelines that teachers need to use to integrate the choice of components of COMSRL to the research learning methods including the style of teaching and SIP in real life conditions are formulated. For working out experimental techniques, taxonomy of choice of independent COMSRL and teaching techniques for each type of students has been drawn up. Taxonomy is based on the analysis of students' surveys and opinions of Ukrainian teachers as well as on the analysis of the results of scientific research of scientists from different countries and on our own experience. To choose the methods of teaching there were used classifications which reflect the peculiarities of the research learning at school and the specifics of the profile preparation. The identified methods and tools that are most suitable for the organization of research learning with the use of COMSRL gave the opportunity to substantiate pedagogical approaches and formulate methodological recommendations. It has been shown that the greatest effectiveness from the use of COMSRL may be expected with use of the research learning approach. The following problems also need to be addressed urgently: It is necessary to implement innovative activities of created educational and scientific clusters in order to carry out research cooperation with higher educational institutions and established regional STREAM centers for the possible collective use of unique and valuable scientific equipment.

References

1. Hrybiuk, O.: Improvement of the educational process by the creation of centers for intellectual development and scientific and technical creativity. In: Hamrol, A., Kujawińska, A., Barraza, M.F.S. (eds.) MANUFACTURING 2019. LNME, pp. 370–382. Springer, Cham (2019). https://doi.org/10.1007/978-3-030-18789-7_31
2. Hrybiuk, O.O.: Perspectives of Introduction of Variational Models of Computer-oriented Environment of Studying Subjects of Natural Sciences and Mathematical Cycle in General Educational Institutions of Ukraine. Collection of scientific works of the KPNU. Pedagogical series, vol. 22, pp. 184–190. KPNU (2016)
3. Hrybiuk, O.O.: Research learning of the natural science and mathematics cycle using computer-oriented methodological systems. In: Monograph, Kyiv, Drahomanov NPU, pp. 307–349 (2019)
4. Hrybiuk, O.O., Lukavyi, P.M., Kulish, N.Yu.: Implementation of STEM-STEAM-STREAM-education within the framework of the research-experimental work of the all-Ukrainian level "Variable models of computer-oriented environment for teaching subjects of the natural and mathematical cycle in a secondary school", Kyiv, pp. 127–134 (2019)
5. Vygotsky, L.: Thought and Language (A. Kozulin, Trans.). (Orig. 1934). MIT Press, Cambridge (1989)
6. Leontyev, A.N.: Activity, Consciousness, and Personality (M. J. Hall, Trans.). Prentice Hall, Englewood Cliffs (1978)
7. Halperin, P.Ya., Zaporozhets, A.V., Karpova, S.N.: Actual problems of developmental psychology, Moscow (1978)
8. Rubinstein, S.L.: The Principles of General Psychology, Moscow (1940)
9. Davydov, V.: Theory of developmental education, Moscow (1996)
10. Elkonin, D.: Problems of activity in theory of individuality, Moscow (1989)
11. Tomlinson, C.A., et al.: The Parallel Curriculum: A Design to Develop Learner Potential and Challenge Advanced Learners, 2nd edn. Corwin Press, Thousand Oaks (2009)
12. Hrybiuk, O,O.: Psychological and pedagogical requirements for computer-oriented systems of teaching mathematics in the context of improving the quality of education. Human. Bulletin of the SHEI, Kyiv, Issue 31, vol. IV, no. 46, pp. 110–123 (2013)
13. Hrybiuk, O.O.: Pedagogical design of a computer-oriented environment for the teaching of disciplines of the natural-mathematical cycle. Sci. Notes **7**(3), 38–50 (2015)
14. Renzulli, J.S., Koehler, J.L., Fogarty, E.A.: Operation Houndstooth intervention: social capital in today's schools. Gift. Child Today **29**(1), 14–24 (2006)

15. Zinchenko, V.P.: Psychological foundations of pedagogy: psychological and pedagogical basis of constructive development, Moscow, Gardariky, pp. 79–87 (2002)
16. Tejaswini, D., Wendell, K.: Community-based engineering. Sci. Child. **53**(1), 67–73 (2015)
17. Mann, E.L., Mann, R.L., Strutz, M., Duncan, D., Yoon, S.Y.: Integrating engineering into K–6 curriculum: developing talent in the STEM disciplines. J. Adv. Acad. **22**, 639–658 (2011)
18. Banchi, H., Bell, R.: The many levels of inquiry. Sci. Child. **46**(2), 26–29 (2008)
19. Hrybiuk, O.O.: The influence of information and communication technologies on psychophysiological development of the young generation. "Science". Publishing Center of the EAPP "Science", Prague, vol. 1, pp. 190–207 (2014)
20. Hrybiuk, O.O.: Cognitive theory of a computer-oriented system for learning natural sciences and mathematics and the interrelation of verbal and visual component. Humanitarian Bulletin "Higher education of Ukraine in the Context of Integration into the European Educational Space", Kiev, 36, vol. IV, no. 64, pp. 158–175 (2015)

Optimal Preventive Maintenance Frequency in Redundant Systems

Guilherme Kunz(✉)

Western Parana State University, Foz do Iguazu, Brazil
guilherme.kunz@unioeste.br

Abstract. The right frequency of preventive maintenance is essential for production availability with adequate safety and economic levels. The optimal preventive maintenance intervals are difficult to identify as production systems are becoming complex by combining electro-electronic and mechanical systems with large quantities. High rates of preventive maintenance boost quality costs, low availability, and high possibility of maintenance failures. Otherwise, it could happen an increase in unscheduled downtime and high costs by losing the production. The development of algorithms to evaluate the maintenance program performance becomes a challenge to validate when it includes continuous, discrete, and stochastic models. This paper proposes an approach to stochastic model checking for identifying the optimum frequency of preventive maintenance by mechanical equipment, simulated and verified through a network of timed automata. A case study was adopted to illustrate the effectiveness of the solution. It is useful to evaluate the optimal preventive maintenance frequency in similar circumstances.

Keywords: Reliability · Preventive maintenance · Stochastic model checker

1 Introduction

A good maintenance program is essential to keep industrial activities operational. There are three distinct strategies: corrective, preventive, and predictive. Corrective maintenance is useful when a failure happened. Preventive maintenance is a time-based replacement or fixed based on machine history failure. Predictive maintenance uses sensors and prediction models to recognize when they will happen.

Corrective maintenance is often used in redundant systems and usually is expansive. Predictive maintenance is the right choice because the machine will only be repaired when they indicate conditions to failure or abnormal performance. In the time of Industry 4.0, a lot of work has been doing to explore predictive maintenance [1–3]. However, in some cases, it is not very easy to create prediction models, and the machines do not show anomalous signals earlier enough to the anomalous detection algorithm reacts. In these situations, a proper maintenance strategy is preventive maintenance or time-based maintenance.

High frequency of time-based maintenance can increase the machine fails by the addition of human error during maintenance time and increase the equipment downtime. On the other hand, the lack of time-based maintenance results in unscheduled downtime and expensive maintenance. Unscheduled downtime increases the probability of manufacturing suspend by multiple failures, and it reflects in a loss of production and more expensive maintenance to the system restart in some cases. Scheduled shutdowns are significantly more efficient than unscheduled shutdowns because the time required for maintenance management (components purchase and transportation) is significantly smaller.

To identify the optimal time-based maintenance frequency is necessary to make models that reproduce the equipment statistical fail behavior and verify the frequency that coincides with the least amount of preventive maintenance that results in the least amount of corrective maintenance.

Analytical solutions are relatively easy to solve in simple systems. However, with the improvement of electronics in the production process, were received new monitoring capabilities [4–6] but also increased the complexity of the systems to obtain the maintenance frequencies of productive systems hard without the use of computational resources. The increase in complexity is showed when comparing equipment from different ages. One car, for example, in 1910, included about 2,000 pieces, and today it has about 20,000 pieces. Commercial aircraft of the 1930s had about 100,000 pieces, and today, it has about 1,000,000 pieces [7].

Different techniques could be used to analyze time-based maintenance [8–10], but in this paper, the main difference is the process of modeling, simulation, and verification techniques, applied to reliability. These techniques increase the knowledge about multiple fails probabilities to create strategies to avoid critical fail situations at connected mechanical, electrical, and electronic systems. This paper, when joined to papers [11,12], allows a detailed investigation of the mechanical, their controllers, communication protocols, and fail behavior of the operation.

This paper proposes a method to identify the optimal frequency of preventive maintenance of systems through the modeling, validation, and verification techniques modeled as networks of timed automata. By the nature of the problem is recommended to use stochastic timed automata to analyze the probability of failures. This paper will be limited to the scope of redundant mechanical systems.

2 Statistical Model Checking

Classical model checkers, based on their mathematical approach, can guarantee the absence of defects in system development since the models are considered acceptable according to the project specifications [13].

Initially, formal methods verification used to evaluate discrete hardware and software behavior. To address critical real-time systems was developed timed automata to analyze the behavior of clock-dependent systems. Uppaal is a toolbox that works directly on the real-time system checking by timed automata and applied to case studies such as communication protocols [11,14,15].

The Uppaal SMC (Statistical Model Checking) [16,17], was adopted as an alternative to avoid state explosion in highly complex systems, it allows an analysis of systems where their behavior depends on stochastic and nonlinear dynamics. Consequently, it allows the modeling of continuous (ordinary differential equations), discrete and stochastic systems behavior [18].

To identify the optimal preventive maintenance frequency in redundant systems through the stochastic model was chosen the UPPAAL Stratego version 4.1.20 [19]. UPPAAL Stratego incorporate techniques developed in Uppaal [14], Uppaal Tiga [20], Uppaal SMC [13] and Statistical learning methods [21].

3 Models

The goal is to determine the preventive maintenance interval that get higher availability and minimum maintenance. The system requirement needs models that are possible to represent clocks and state changes based on randomly generated values according to the distribution function that describes the behavior by the equipment under analysis.

To make electronic system failures models were needed random values of an exponential distribution function (constant failure rate). Mechanical systems failures are adequately described by Weibull distribution enabling setup the failure rates during the infant-mortality period, useful life period, and wear-out period [22].

The Log-Normal distribution describes the time required for maintenance where the average time to perform preventive maintenance, as well as its dispersion, are shorter than the time intervals needed for corrective maintenance, the effect of planned maintenance [23].

The UPPAAL Stratego toolbox provides builtin exponential and uniform distribution function. Were implemented in UPPAAL the Log-Normal distribution according to equation $X = e^{\mu + \sigma Z}$, where $Z = \mathcal{N}(\mu = 0, \sigma^2 = 1)$ and Weibull distribution according to equation $X = \eta \cdot -ln(Z)^{(1/\beta)}$, where $Z = U(0,1)$. It permits describe the equipment time to failure and support team behavior to corrective and preventive maintenance.

3.1 Case Study

This case study was inspired by the examples introduced in [24]. In this paper, the case study was changed to identify the optimal interval of preventive maintenance. It is supposedly impossible to observe the equipment degradation and the maintenance procedures turn the pump to the original condition. This case study demands to keep an operational process that demands four out of six cooling pumps operational.

Concerning maintenance teams, both preventive and corrective maintenance, restore the pump to its original condition, and there is only one team to perform the preventive maintenance activity. The preventive maintenance team chooses the pump with the longest maintenance-free time. There is always a team ready for corrective maintenance. Time preventive maintenance (planned maintenance) is faster then unscheduled maintenance [23].

The time to failure (TTF) depends on the pump's parts: bearing, rotor, seal, and shaft. All failure times were defined as Weibull distributions with parameters based on [25]. In real industrial conditions, the estimation of the Weibull parameters are based on maintenance records.

Concerning maintenance teams, both preventive and corrective maintenance, restore the pump to its original condition, and there is only one team to perform the preventive maintenance activity. The preventive maintenance team chooses the pump with the longest maintenance-free time. There is always a team ready for corrective maintenance. Time preventive maintenance (strongly planned maintenance) is faster then unscheduled maintenance [23]. Maintenance times was based on activity records and in [24]:

1. *Time to Preventive Maintenance* (T_{PM}). The time required to execute preventive maintenance, based on the performance records of the maintenance team. It is represented by a *Log Normal distribution* with $\mu = 80$ and $\sigma = 3$ hours.
2. *Time to Corrective Maintenance* (T_{CM}). The time required to perform corrective maintenance, based on the performance records of the maintenance team. It is represented by a *Log Normal distribution* with $\mu = 600$ and $\sigma = 25$ hours.

To obtain the optimal preventive maintenance interval implies to look at the highest availability and the lowest cost of maintenance. The costs values used was:

1. *Fail System Cost* (C_{fs}). Net revenues from the product sales are $250000/day.
2. *Preventive Maintenance Cost* (C_{pm}). The average preventive maintenance cost to pump is $480/day.
3. *Corrective Maintenance Cost* (C_{cm}). The average repair cost to pump is $3600/day.

3.2 Automata Models

To this case, automata models represent the time-to-failure probabilities in line with the block diagram to estimate the preventive maintenance intervals that reduce costs. Figure 1 shows one of the six pumps under analysis where $W1$ symbolizes the equipment working under reasonable condition, corrective maintenance, or preventive maintenance according to the following logic:

1. If the equipment is operational (SD[ID]!=F), the automata receive the next time for failure according to the elements (of the pump) that produce, in the simulation, the shortest time to failure (Weibull distribution).
2. If the equipment is under failure (SD[ID]==F), the automata receive the time needed for corrective maintenance (log-normal distribution).

3. If the equipment is operational, the automata can receive a preventive maintenance request (MP[ID]?) and receive the time required for preventive maintenance (log-normal distribution).

The model in Fig. 1 represents one of the N system equipment and replicates to represent six system pumps under analysis.

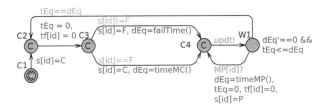

Fig. 1. Pumps model.

Two distinct models represent preventive maintenance. The first one (Fig. 2) controls the intervals and the conditions of starting preventive maintenance. Considering that there is only one (1) maintenance team, preventive maintenance only happens when all machinery is operational. It assures that the maintenance team is available and exist a reduced probability of a manufacturing failure by keeping more pumps working together during preventive maintenance.

Both maintenances move together if one maintenance is happening and another pump goes into fault condition. If complete the time required for preventive maintenance, and one or more pumps are in corrective maintenance, one more extra day is required to start the preventive maintenance until the conditions are fully satisfied.

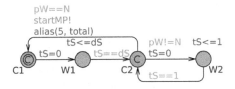

Fig. 2. Frequency of preventive maintenance.

When all requirements for preventive maintenance (the model in Fig. 2) are satisfied, it sends a message to the model in Fig. 3. This model will select the next pump for preventive maintenance. The model will choose the model with the longest time without maintenance. It is not necessary to check the status of the equipment before the request (MP[cL!]) since it only happens when the equipments are operational.

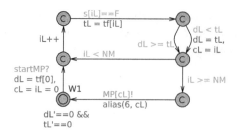

Fig. 3. Equipment select to preventive maintenance.

Models in Fig. 4 describe maintenance and product sales loss. The model in Fig. 4a is updated by any change from the pump status and estimates the amount of time that these pumps are stopped for corrective or preventive maintenance and computes the product sales loss when there is three or more pumps stopped. The automata in Fig. 4b measures the final costs after 7,300 days (20 years). It counts the costs from preventive maintenance, corrective maintenance, product sales loss, and total costs (sum of all sources).

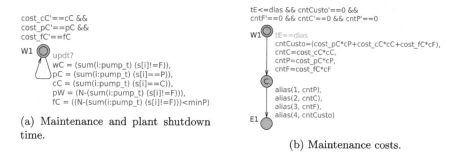

(a) Maintenance and plant shutdown time.

(b) Maintenance costs.

Fig. 4. Estimation of maintenance costs.

Where wC represents the quantity of equipment in operation, pC the amount of equipment under preventive maintenance, cC the amount of equipment under corrective maintenance, and fC is a binary variable that represents the state of product sales.

4 Results Analysis

The preventive maintenance interval (dS in Fig. 2) was modified to identify the best time among preventive maintenance actions (lower costs and higher availability). To this, the variable (dS) changed from 1 to 50 days. The model (XML file) was changed by R [26] to create different files verified by separate instances of UPPAAL. By the end of the analysis, were identified the minimum cost and maximum availability.

To extract information was made an DLL called from UPPAAL [27,28] to print the values as illustrated in Figs. 2, 3 and 4a. It gets the time when starting the preventive maintenance, the pump selected, and the costs. For data analysis the following expressions were verified in UPPAAL to obtain maintenance costs and production loss over 2^{15} simulations of 20 years:

1. E[<=days;runs]   (max: End.cntP): Preventive maintenance costs.
2. E[<=days;runs]   (max: End.cntC): Corrective maintenance costs.
3. E[<=days;runs]   (max: End.cntF): Production loss costs.
4. E[<=days;runs]   (max: End.cntCusto): Total costs.

Results are exposed in Fig. 5a and 5b. Figure 5a indicates the costs related to preventive maintenance. The higher the frequency of preventive maintenance, the higher the cost of failure prevention. Figure 5b shows the influence of the preventive maintenance interval on corrective maintenance costs. The longer the interval between preventive maintenance, the higher the cost involved in corrective maintenance.

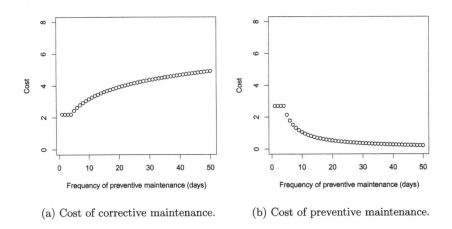

(a) Cost of corrective maintenance.   (b) Cost of preventive maintenance.

Fig. 5. Maintenance costs.

Production sales loss and total costs are displayed in Fig. 6a and 6b. Figure 6a revealed that costs related to the production shutdown increase when the frequency of preventive maintenance is lower. Manufacturing downtime costs have a higher dispersion compared to others. Figure 6b shown the total costs and the uncertain limits to different preventive maintenance intervals. Longer the preventive maintenance intervals, bigger the uncertain costs. The lowest costs happen when the preventive maintenance interval is approximately ten days.

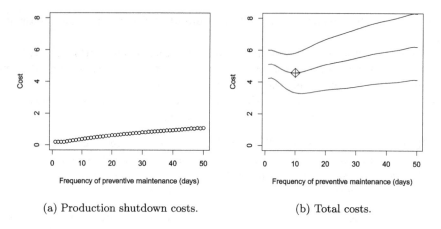

(a) Production shutdown costs.

(b) Total costs.

Fig. 6. Costs of maintenance totals and production shutdown.

Maximum availability is performed to a higher frequency of preventive maintenance, as seen in Fig. 7 but giving higher total costs when compared to preventive maintenance intervals near to 10 days. Table 1 shows a comparison between project criteria: availability or costs.

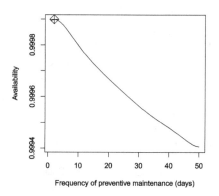

Fig. 7. Availability.

If availability is the priority, the costs will increase 12% over the 20-years. The minimum cost happened when performed 33 maintenance events per year (see Fig. 8a), or, one preventive maintenance performed every 11 days (66 days to each pump), including one day to start maintenance by the unavailability of the security criteria (see Fig. 8b).

Table 1. Comparison between strategies.

	Maximal availability	Minimal costs
Frequency of PM (days)	2	10
Costs (millions)	5.12	4.57
Availability	99.99	99.98
Production shutdown (20 years)	0.74	1.47

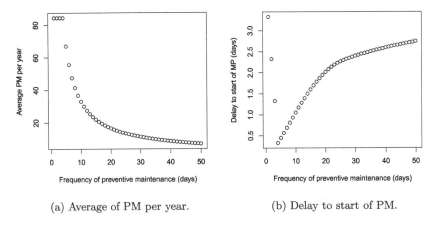

(a) Average of PM per year.   (b) Delay to start of PM.

Fig. 8. Average and delay of preventive maintenance.

All pumps are equals and, as predicted, had the same percentage of preventive maintenance (16.7%). The influence of each equipment on pump failure indicated that 45% of the failures were due to seal, 25% of bearing, 21% of shaft, 9% of motor and impeller registered some failures below 1%.

4.1 Influence of Corrective Maintenance Parameters

To estimate the influence of the T_{CM} on the optimal interval of preventive maintenance, different average values of T_{CM} (range from 80 to 600 h) were tested. No changes were made in the dispersion of the frequency distribution. All of the steps described before were repeated.

Figure 9 shows no longer optimal preventive maintenance frequency when performing T_{CM} below 380 h (in this case study). So, reducing the T_{CM} repair should grow up the preventive maintenance intervals.

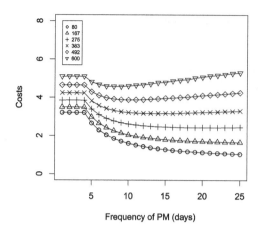

Fig. 9. Relationship between costs and T_{CM}.

5 Conclusion

The complexity of industrial equipment has been increasing with new technologies, and, for this reason, new methods could help to analyze the reliability of factories. For this purpose, it proposed the use of a stochastic model checker to analyze maintenance strategies to minimize the errors resulting from the analysis of these complex systems.

The goal of the paper is to identify the optimal frequency of preventive maintenance of systems through the modeling, simulation, and verification techniques applied to stochastic timed automata. This kind of solution permits a formal approach to the decision-making process, growing up the reliability. Behind that, suggest a sequence to a solid system modeling to reliability analysis.

The recommended solution examined through a study with serial and parallel devices produced consistent results. The case study selected represents a real industry situation. In the way, this contribution could increase the learning about the reliability industry, mainly when combined with the investigation of the mechanical, controllers, and communication protocols behaviors. It assumed that using continuous, discrete, and stochastic approaches can assist the preventive maintenance analysis strategies to complex systems.

The results are aligned with the reliability theory and get detailed behavior of the fails. It compared different industry goals: maximal availability and minimal costs. Both goals are important to the decision making maintenance program.

The next step to this research is built models to predictive approach study and, because of the growth of complexity, apply algorithms to accelerate the optimal point detection.

References

1. Albrice, D., Branch, M.: A deterioration model for establishing an optimal mix of time-based maintenance (TbM) and condition-based maintenance (CbM) for the enclosure system, Kansas (2015)
2. Ahmad, R., Kamaruddin, S.: An overview of time-based and condition-based maintenance in industrial application. Comput. Ind. Eng. **63**(1), 135–149 (2012)
3. Jasiulewicz-Kaczmarek, M., Gola, A.: Maintenance 4.0 technologies for sustainable manufacturing-an overview. IFAC-PapersOnLine **52**(10), 91–96 (2019)
4. Kunz, G., Perondi, E., Machado, J.: Modeling and simulating the controller behavior of an automated people mover using IEC 61850 communication requirements. In: IEEE International Conference on Industrial Informatics (INDIN), Art. no. 6034947, pp. 603-608 (2011). https://doi.org/10.1109/INDIN.2011.6034947
5. Leão, C.P., Soares, F.O., Machado, J.M., Seabra, E., Rodrigues, H.: Design and development of an industrial network laboratory. Int. J. Emerg. Technol. Learn. **6**(SPECIAL ISSUE.2), 21–26 (2011). https://doi.org/10.3991/ijet.v6iS1.1615
6. Silva, M., Pereira, F., Soares, F., Leão, C.P., Machado, J., Carvalho, V.: An overview of industrial communication networks. Mech. Mach. Sci. **24**, 933–940 (2015). https://doi.org/10.1007/978-3-319-09411-3-97
7. Groover, M.P.: Automation, Production Systems, and Computer-Integrated Manufacturing, 3rd edn. Prentice Hall Press, New York (2007)
8. Antosz, K.: Maintenance-identification and analysis of the competency gap. Eksploatacja i Niezawodność **20**, 484–494 (2018)
9. Sobaszek, Ł., Gola, A., Swic, A.: Time-based machine failure prediction in multi-machine manufacturing systems. Eksploatacja i Niezawodność - Maint. Reliabil. **22**(1), 52–62 (2020)
10. Loska, A., Paszkowski, W.: SmartMaintenance - the concept of supporting the exploitation decision-making process in the selected technical network system. In: International Conference on Intelligent Systems in Production Engineering and Maintenance, pp. 64–73. Springer (2017)
11. Kunz, G., Machado, J., Perondi, E., Vyatkin, V.: A formal methodology for accomplishing IEC 61850 real-time communication requirements. IEEE Trans. Ind. Electron. **64**, 6582–6590 (2017)
12. Kunz, G., Perondi, E., Machado, J.: A design strategy for obtaining reliable controllers for critical mechanical systems. Mechatronics **54**, 186–202 (2018)
13. David, A., Larsen, K.G., Legay, A., Mikucionis, M., Poulsen, D.B.: Uppaal SMC tutorial. Int. J. Softw. Tools Technol. Transf. **17**, 397–415 (2015)
14. Behrmann, G., David, A., Larsen, K.: A tutorial on Uppaal. In: Formal Methods for the Design of Real-Time Systems (2004)
15. Behrmann, G., David, A., Larsen, K.G., Pettersson, P., Yi, W.: Developing UPPAAL over 15 years. Softw.: Pract. Exper. **41**, 133–142 (2011)
16. Sen, K., Viswanathan, M., Agha, G.: Statistical model checking of black-box probabilistic systems. In: CAV. LNCS, vol. 3114, pp. 202–215. Springer (2004)
17. Younes, H.L.S.: Verification and planning for stochastic processes with asynchronous events. Ph.D. thesis, Carnegie Mellon (2005)
18. Basile, D., Beek, M.H., Ferrari, A., Legay, A.: Modelling and analysing ERTMS L3 moving block railway signalling with simulink and Uppaal SMC. In: Larsen, K., Willemse, T. (eds) Formal Methods for Industrial Critical Systems. FMICS 2019. Lecture Notes in Computer Science, vol. 11687 (2019)

19. David, A., Jensen, P.G., Larsen, K.G., Mikučionis, M., Taankvist, J.H.: Uppaal Stratego. In: 21st International Conference on Tools and Algorithms for the Construction and Analysis of Systems TACAS (2015)
20. Cassez, F., David, A., Fleury, E., Larsen, K.G., Lime, D.: Efficient on-the-fly algorithms for the analysis of timed games. In: 16th International Conference CONCUR (2005)
21. David, A., et al.: On time with minimal expected cost! In: 12th International Symposium on Automated Technology for Verification and Analysis ATVA (2014)
22. Abernethy, R.: The New Weibull Handbook: Reliability and Statistical Analysis for Predicting Life, Safety, Supportability, Risk, Cost and Warranty Claims, 5th edn. R.B. Abernethy, North Palm Beach (2006)
23. Kline, M.B.: Suitability of the lognormal distribution for corrective maintenance repair times. Reliab. Eng. **9**(2), 65–80 (1984). ISSN 0143-8174
24. Alexander, D.C.: Application Of Monte Carlo simulations to system reliability analysis. Texas A&M University. Turbomachinery Laboratories (2003). http://hdl.handle.net/1969.1/164018
25. Morris, S.F.: Failure rate estimates for mechanical components (2020). https://reliabilityanalyticstoolkit.appspot.com
26. R Core Team: R: a language and environment for statistical computing. R Foundation for Statistical Computing, Vienna, Austria (2019). https://www.R-project.org/
27. Jensen, P.G., Larsen, K.G., Legay, A., Nyman, U.: Integrating tools: co-simulation in UPPAAL Using FMI-FMU. In: 22nd International Conference on Engineering of Complex Computer Systems ICECCS, Fukuoka, pp. 11–19 (2017)
28. Cassez, F., Aledo, P.G., Jensen, P.G.: WUPPAAL: Computation of Worst-Case Execution-Time for Binary Programs with UPPAAL. In: Aceto, L., Bacci, G., Bacci, G., Ingólfsdóttir, A., Legay, A., Mardare, R. (eds.) Models. Algorithms, Logics and Tools (2017)

Experimental Investigation of the Effect of Mass Load on Flight Performance of an Octorotor and Dodecarotor UAV

Şahin Yildirim[1(✉)], Nihat Çabuk[2], and Veli Bakircioğlu[2]

[1] Department of Mechatronics Engineering, Engineering Faculty, Erciyes University,
38039 Melikgazi, Kayseri, Turkey
`sahiny@erciyes.edu.tr`

[2] Vocational School of Technical Sciences, Aksaray University, Bahçesaray Mh.,
68100 Aksaray, Turkey

Abstract. In this study, the flight performance of a universal Vertical take-off and landing (VTOL) unmanned aerial vehicle (UAV) with 12 and 8 rotors was examined under load and no-load conditions. Thanks to its universal structure, experimental studies with 8 and 12 rotor UAVs were performed on the same platform and under the same conditions. In addition, the controller parameters were kept the same for both vehicle types in order to observe the effect of the mass load only. Hierarchical PID controllers are used as the controller architecture to control the orientation and position of the both vehicles. The flight performance of both vehicles was examined in three stages as take-off, trajectory and landing. Settling time, rise time parameters and position errors were used to benchmark the flight performance of the both vehicles. The results show that the performance of the 12-rotor vehicle is superior, especially in terms of trajectory tracking performance although the take-off and landing performance is very close to each other for both vehicles. This was observed more clearly under load conditions.

Keywords: Dodecarotor · Flight Performance · Octorotor

1 Introduction

Vertical take-off and landing (VTOL) air vehicles are unmanned aerial vehicles (UAV) that perform their flights by changing the angular speeds of their rotors. Since they can land and take off vertically these vehicles are becoming increasingly common in areas such as military, recreational, agricultural and transport. In all areas of use, these vehicles must be loaded in different amounts depending on the purpose.

In literature, several studies have been conducted on VTOL air vehicles. These studies are mostly related to the control [1–4], while partly on the design of these vehicles [4–8]. W. Zhu et al. [2] proposed a finite-time controller for the quadrotor aircraft to achieve hovering control in a finite time. They stated that compared to the classical PD control, it has been shown that the finite time control can improve the closed-loop system's dynamical performances. In addition, studies on design, control and flight performance analysis were also conducted [9–11]. China E. Lin et al. [12], developed a hybrid vehicle model by installing four short-arm internal combustion engines and four long-arm dc motors. Longer flight times and greater load capacity are made to enable the goals of this study stated that they have achieved. Most of these studies were based on simulation. There are also studies on the examination of the behavior of these vehicles under load. Angelis et al. [13] proposed a novel control strategy for stabilizing the dynamics of a multirotor carrying a suspended load. They stated that according to results of numerical simulations show the effectiveness of the proposed methodology and its suitability for practical application in an operative scenario by considering model uncertainties, and control system implementation features. A. R. Godbole and K. Subbarao [14] studied about the mathematical modeling and control of an unmanned aerial system with a payload suspended using a cable.

In this study, autonomous flight tests of two types of VTOL air vehicle both under load and under no load were performed. The first of these vehicles is a conventional eight-rotor vehicle and the second is a non-conventional design with twelve rotors. The second vehicle has twelve rotors which can be controlled independently, with eight rotors in the upper plane and four rotors in the lower plane. With and without load for both vehicles, four autonomous flight experiments conducted with a rectangular trajectory and the flight data of the vehicle were compared. The results of this comparison are thought to contribute to the design of a VTOL UAV for any purpose.

The presentation of the study is as follows; after the introduction section, the experimental system under investigation in this study is introduced in Sect. 2. Subsequently, in Sect. 3, experimental studies and the graphical results are given. Finally, in Sect. 4, the obtained results are discussed and evaluated.

2 System Description and Flight Condition

The universal VTOL drone system is given in Fig. 1. As can be seen from the figure, four of the motors are placed in the lower plane and eight of the motors in the upper plane. Each of these motors is independently controlled by the controller. Thus, different configurations can be obtained.

On the same drone system, the different drone models are obtained by disabling some motors seen in Fig. 1 as software. The motors numbers disabled according to the models are given in Table 1.

A detailed mathematical model was obtained for the system in our previous study [15]. UAVs can be controlled in two different ways as manual and autonomous. In autonomous control, the vehicle may follow a desired trajectory. The controller calculates the required forces and torques for the motion of the vehicle according to the desired trajectory and the roll-pitch angle limits.

Control of the UAV is primarily about the control of four basic variables: the Euler angles which are roll-pitch-yaw angles, and the altitude of the vehicle. These four variables used to control the position and speed of the drone. In the experiments, the controller with the same parameters and the same trajectory was used for autonomous flights. The results of the experimental study are significantly affected by weather conditions. The experiments were conducted in an environment where the air temperature was 15 °C and

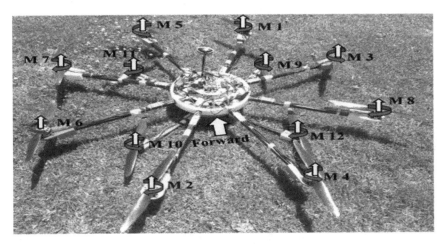

Fig. 1. Universal VTOL drone.

Table 1. The universal VTOL drone models according to motor numbers.

Vehicle type	Dodecarotor	Octorotor
Disabled motors	–	9, 10, 11, 12

the wind was 11 km/h in the southwest direction. The number of satellites connected to the GPS was 11 and horizontal dilution of precision value was 0.91 which are the important parameters that affect the position accuracy of the vehicle in autonomous flight [16, 17].

The mass of the load used in flight experiments is approximately 6.5 kg, and Fig. 2 shows the vehicle with load. With this load, the total mass of the universal drone system is become approximately 20 kg. Here, the mass used for the load was kept constant and fixed for both vehicles so that the effect of the rotor numbers could be observed standalone. In addition, the control parameters were kept the same for both loaded and unloaded vehicles in flight experiments. Figure 3 shows the controller structure. The gains used for the controllers were tuned empirically for the system without load.

Flight planning consists of three parts and the analyzes were performed accordingly. In the first stage of flight planning, the vehicle was first asked to rise to a height of 5 m relative to the ground. In the second part, it was requested to follow a rectangular trajectory while maintaining its height and to change the orientation during the first two turns and to return to the starting point without changing the orientation of the vehicle in the last turn. Finally, in the third stage, it was asked to land to the starting ground point. The results obtained from the flight experiments are presented in the next section.

Fig. 2. System with load.

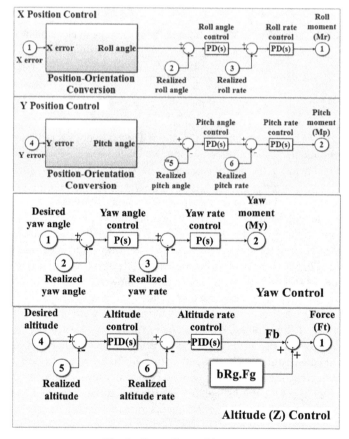

Fig. 3. Controller architecture

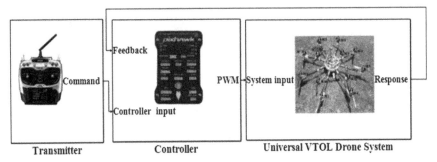

Fig. 4. Schematic representation of the info flow through the system.

3 Experimental Results of the Flight Performance

Several experimental studies were performed in this study, however only the most typical results are presented in this paper. Experiments are performed for an autonomous flight. In autonomous flight, by receiving the autonomous flight mode command from the radio transmitter, the vehicle follows the trajectory in the flight control card according to the controller algorithm. Depending on the determined speed and acceleration limits, pulse width modulation values for electronic speed controllers are generated by the controller during trajectory tracking of the vehicle. The response of the system is based on instantaneous data such as positions, Euler angles, rates and accelerations both from the internal inertial measurement unit on the control card and the externally available global positioning system module. This instantaneous data is entered as feedback to the controller and trajectory tracking is performed. Figure 4 shows the info flow through the system.

The results of autonomous flight performed for both vehicles both under load and without load are presented graphically as seen in Fig. 5, Fig. 6, Fig. 7 and Fig. 8. In the graphics, take off stage that is the first stage of the flight was marked with light red field, trajectory tracking stage that is the second stage of the flight was marked with light blue field and landing stage that is the final stage of the flight was marked with light green field.

The turnings in the rectangular trajectory are marked with red field. The purple field in the graphs indicate the instant of the motion beginning after the last turning to the starting point without changing orientation of the vehicle. Finally, settling margin for the take off stage was marked with light black field. This representation is also used in the following graphics in the paper.

In Fig. 5 and Fig. 6, results obtained from the flight experiments of the Octorotor, In Fig. 7 and Fig. 8 obtained from the flight experiments of the Dodecarotor with no load and with load conditions were graphically illustrated, respectively.

When the graphs are examined, for the no load condition, the rise and settling time in the Dodecarotor are observed as 4.25 and 7.88 s for the first flight stage, respectively, whereas they are observed as 5.42 and 9.91 s for Octorotor. When the second part of the flight is examined in terms of errors of roll and pitch angles, it is seen that the maximum absolute error of Octorotor is about 2 and 4° and it is about 4.7 and 5° for Dodecarotor,

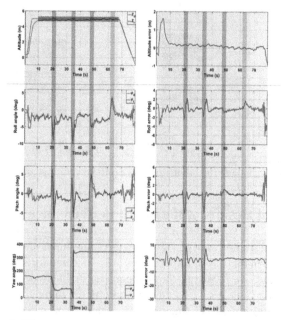

Fig. 5. Flight performance graphics of the Octorotor with no load.

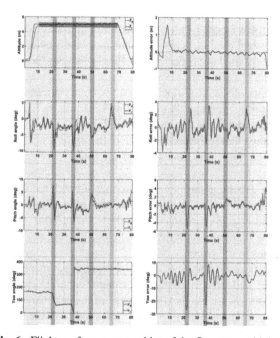

Fig. 6. Flight performance graphics of the Octorotor with load.

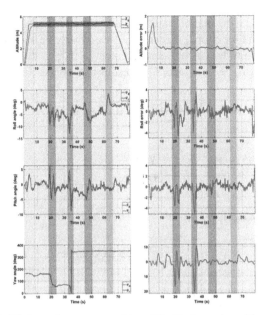

Fig. 7. Flight performance graphics of the Dodecarotor with no load.

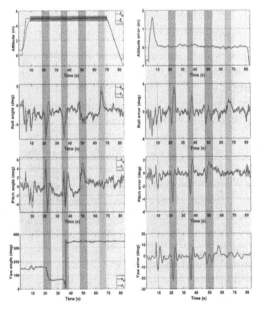

Fig. 8. Flight performance graphics of the Dodecarotor with load.

respectively. In the load conditions, the rise and settling time in the Dodecarotor is observed as 4.48 and 9.09 s for the first flight stage, respectively, whereas it is observed as 6.57 and 11.37 s for Octorotor. For second part of the flight time, in terms of errors of roll and pitch angles, the maximum absolute error of Octorotor is about 3.8 and 5.1° and it is about 4.8 and 4.6° for Dodecarotor, respectively. Furthermore, the maximum trajectory error for altitude which is approximately 1.5 m is seen to be close to each other for both systems. The summary of the results mentioned above is given in Table 2.

Table 2. Summary of the flight performance results from experiments performed for both vehicles.

Flight Stages	Parameters [Units]		Vehicle type			
			Dodecarotor		Octorotor	
			Loaded	No Load	Loaded	No Load
Take off	Settling time [s]		9.09	7.88	11.37	9.91
	Rise time [s]		4.48	4.25	6.57	5.42
Trajectory tracking	Max. abs. error	Altitude [m]	Approx. 1.5		Approx. 1.5	
		Roll [deg]	4.8	4.7	3.8	2
		Pitch [deg]	4.6	5	5.1	4

4 Conclusion

In this paper, the flight performance analysis of two different UAVs with 12 and 8 rotors was examined under load and no-load conditions. In terms of rising time and settling time, dodecarotor air vehicle showed better flight performance than octorotor. In the context of roll and pitch angle errors, although, octorotor better performance than dodecarotor in no load condition, the performances occurred close to each other in with load condition.

On the other hand, in terms of maximum altitude and yaw angle errors and also mean errors, which are the actual controlled references, it is seen that Dodecarotor performs better than Octorotor. When the landing performance, which is the most critical stage for UAVs, is examined in the terms of the roll-pitch angle, the dodecarotor is acting more aggressively to maintain its altitude than octorotor, but still compromised the yaw angle.

When the no-load flight condition and the loaded flight condition are evaluated among themselves in both vehicle types, in the context of the increase of maximum absolute roll and pitch errors when the vehicles are loaded, the error increase in the octorotor was greater than that of the dodecarotor. Thus, it can be said that the vehicle with dodecarotor has a better performance in the increase of load.

References

1. Ma, T., Wong, S.: Trajectory tracking control for quadrotor UAV. In: 2017 IEEE International Conference on Robotics and Biomimetics (ROBIO), pp. 1751–1756. IEEE (2017). https://doi.org/10.1109/ROBIO.2017.8324671

2. Zhu, W., Du, H., Cheng, Y., Chu, Z.: Hovering control for quadrotor aircraft based on finite-time control algorithm. Nonlinear Dyn. **88**(4), 2359–2369 (2017). https://doi.org/10.1007/s11071-017-3382-8
3. Martinez Alvarez, A., Lozano Espinosa, C.A.: Nonlinear control for collision-free navigation of UAV fleet. SN Appl. Sci. **1**(12), 1 (2019). https://doi.org/10.1007/s42452-019-1606-x
4. Burggräf, P., Pérez Martínez, A.R., Roth, H., Wagner, J.: Quadrotors in factory applications: design and implementation of the quadrotor's P-PID cascade control system. SN Appl. Sci. **1**(7), 1–17 (2019). https://doi.org/10.1007/s42452-019-0698-7
5. Anweiler, S., Piwowarski, D.: Multicopter platform prototype for environmental monitoring. J. Clean. Prod. **155**, 204–211 (2017). https://doi.org/10.1016/j.jclepro.2016.10.132
6. Brischetto, S., Ciano, A., Ferro, C.G.: A multipurpose modular drone with adjustable arms produced via the FDM additive manufacturing process. Curved Layer. Struct. **3**, 202–213 (2016). https://doi.org/10.1515/cls-2016-0016
7. Lin, C.E., Supsukbaworn, T.: Development of dual power multirotor system. Int. J. Aerosp. Eng. **2017**, 1–19 (2017). https://doi.org/10.1155/2017/9821401
8. Vu, N.A., Dang, D.K., Le. Dinh, T.: Electric propulsion system sizing methodology for an agriculture multicopter. Aerosp. Sci. Technol. **90**, 314–326 (2019). https://doi.org/10.1016/j.ast.2019.04.044
9. Kotarski, D., Piljek, P., Brezak, H., Kasać, J.: Design of a fully actuated passively tilted multirotor UAV with decoupling control system. In: 2017 8th International Conference Mechanical and Aerospace Engineering ICMAE 2017, pp. 385–390 (2017). https://doi.org/10.1109/ICMAE.2017.8038677
10. Bucki, N., Mueller, M.W.: A novel multicopter with improved torque disturbance rejection through added angular momentum. Int. J. Intell. Robot. Appl. **3**(2), 131–143 (2019). https://doi.org/10.1007/s41315-019-00093-4
11. Ömürlü, V.E., Kirli, A., Büyükşahin, U., Engin, Ş.N., Kurtoğlu, S.: A stationary, variable DOF flight control system for an unmanned quadrocopter. Turkish J. Electr. Eng. Comput. Sci. **19**, 891–899 (2011). https://doi.org/10.3906/elk-1007-579
12. Lin, C.E., et al.: Engine controller for hybrid powered dual quad-rotor system. In: IECON 2015 - 41st Annual Conference of the IEEE Industrial Electronics Society, pp. 001513–001517. IEEE (2015). https://doi.org/10.1109/IECON.2015.7392315
13. de Angelis, E.L., Giulietti, F., Pipeleers, G.: Two-time-scale control of a multirotor aircraft for suspended load transportation. Aerosp. Sci. Technol. **84**, 193–203 (2019). https://doi.org/10.1016/j.ast.2018.10.012
14. Godbole, A.R., Subbarao, K.: Nonlinear control of unmanned aerial vehicles with cable suspended payloads. Aerosp. Sci. Technol. **93**, 105299 (2019). https://doi.org/10.1016/j.ast.2019.07.032
15. Yıldırım, Ş, Çabuk, N., Bakırcıoğlu, V.: Design and trajectory control of universal drone system. Measurement **147**, 106834 (2019). https://doi.org/10.1016/j.measurement.2019.07.062
16. Langley, R.B.: Dilution of precision. GPS World **10**, 52–59 (1999)
17. Freimuth, H., König, M.: Planning and executing construction inspections with unmanned aerial vehicles. Autom. Constr. **96**, 540–553 (2018). https://doi.org/10.1016/j.autcon.2018.10.016

Indoor GPS System for Autonomous Mobile Robots Used in Surveillance Applications

Philip Coandă, Mihai Avram, Victor Constantin(✉), and Bogdan Grămescu

Mechatronics and Precision Mechanics Department, University Politehnica of Bucharest, Bucharest, Romania
victor.constantin@upb.ro

Abstract. Development of urban areas has created a need for stable, reliable and cost-efficient indoor GPS systems based on regularly available technologies. Such systems already exist and are in use. This paper proposes such a system with an emphasis on ease of use, long-term reliability or maintenance free through network redundancy techniques as well as an open-source architecture. The proposed architecture uses a Bluetooth/Wi-Fi's RSSI (Received Signal Strength Indicator) parameter as relative to multiple mobile and fixed nodes to determine the distance to an object. The nodes, both mobile and fixed, are purpose-built for the object of this study and feature multiple other peripherals, as described in the paper. The basic implemented model of these nodes is also described in the paper.

Keywords: Indoor positioning system · Received signal strength · Mobile robots

1 Introduction

The need for such a system comes with the continuous development of buildings and their spreading on even larger surfaces, which makes using the GPS system impossible. Whether it is office buildings, storage buildings or industrial workshops, such a location system can provide valuable data that can be further used for better resource management (e.g.: locating machines in real time, location of people and implicitly warning in emergencies).

Another important feature of such a system is security, the system being local, without its connection with equipment whose data flow cannot be controlled by the user. Structurally, the system is intended to be made of fixed and movable nodes. Fixed nodes are those with known position and the movable nodes are represented by entities or objects that must be tracked or located that have the movable node on them [1–4]. The mobile node, whose position is not known, will be located according to its position relative to the fixed nodes. Thus, a position relative to the position of the fixed nodes with an initially known position is desired. Figure 1 shows the basic proposed schematic of the system.

Today, localization has become much easier to achieve, given the latest technological advances. No long ago, GPS was the only standard which could provide such functionalities (Global Positioning System), things have evolved and today we can use

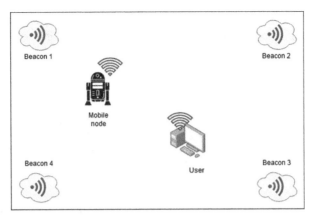

Fig. 1. Representation of beacons, mobile node and the user in the proposed system

alternative methods for location. The location is no longer dependent on large, bulky, power-consuming equipment and is accessible to anyone.

Although GPS is still today the most widely used standard for outdoor location, for indoor applications it does not work properly. Buildings and shopping centers have become larger and larger and are made of materials that attenuate radio waves, making it almost impossible to connect to an outdoor satellite.

In this situation, the indoor localization techniques became necessary with the increasing dimensions of the buildings of today and at the same time new solutions had to be developed, considering that the use of the GPS standard is either expensive for the application in question or it is impossible for constructive reasons.

2 State of the Art

RSSI is a term that comes from the field of telecommunications and represents a way of measuring the intensity of the received signal. Normally, this is a parameter that is not widely known to users, but which is very important in measuring and analyzing the received signal. RSSI is a relative index that is used to measure signal quality. The IEEE 802.11 standard specifies that RSSI can take values from 0 to 255, each equipment manufacturer being responsible for defining its maximum RSSI value. For example, Cisco uses values between 0 and 100, and Atheros between 0–60. This aspect is highly important in the interpretation and use of measured data [5]. A localization system using RSSI can be made with different wireless standards. This paper will present three standards, these being Bluetooth, Wi-Fi and LoRa/LoRaWAN.

Bluetooth is a wireless standard for transmitting data over short distances. It uses shortwave radio waves in the range 2.4–2.485 GHz. Bluetooth operates at frequencies between 2402–2480 MHz, or 2400–2483.5 MHz. It divides the information into packets which are then transmitted on one of the 79 channels. Each channel has a bandwidth of 1 MHz.

The Bluetooth standard was created to replace the data transmission through cables, while also wishing to have low power consumption. The maximum distance for the transmission of information is classified in classes (ex BLE4.0), and each class corresponds to a certain energy consumption. It should also be considered that radio waves are subjected to alteration depending on the environment in which they are propagated. For example, inside, there are large attenuations due to objects in the environment [6].

Wi-Fi is a radio technology that is based on the IEEE 802.11 standard. Devices that can use Wi-Fi technology can connect to each other or to the internet through a wireless access point. Such an access point has a signal area of about 20 m inside and larger outside, in the open field. There are different versions of Wi-Fi technology, each with different maximum distances, different speeds and different operating frequencies. The most used of these are the 2.4 GHz radio band and the 5.8 GHz radio band. Each of these bands is divided into several channels and each channel can be used by multiple networks at the same time [5].

LoRa/LoRaWAN is a long-range radio communication technology that uses unlicensed radio bands in Europe (169 MHz, 433 MHz, 868 MHz) and the US (915 MHz) and allows long-distance communication (more than 10 km in rural areas) with low power consumption. This makes it a technology to consider in long-distance IoT systems, maritime platforms, mining and logistics. LoRaWAN is the network in which LoRa operates. To make it easier to understand, a simple analogy can be made between LoRaWAN and Wi-Fi, the former being the basic standard for IoT applications. LoRaWAN is responsible for the management of the data, the channels, the transmission frequency and the required power [1].

Using one or more of the above technologies and different filtering techniques for the acquisitioned data, in paper [4], the following results are obtained for indoor positioning systems that can be seen in Table 1 where Ac. l. err/m is actual localization error per meter and Av. l. err/m is the average localization per meter:

Table 1. Positioning results of paper [4]

Method	5 × 5 m		10 × 10 m		100 × 100 m	
	Ac. l. err/m	Av. l err/m	Ac. l. err/m	Av. l err/m	Ac. l. err/m	Av. l err/m
Classic RSSI	1.4422	1.8543	2.1587	1.9229	27.7492	26.8520
Correction RSSI	0.3162	0.3375	0.8495	0.6325	7.3246	6.7493

3 System Overview

The proposal is a lightweight indoor positioning system with low power Bluetooth modules which uses RSSI parameter to determine the position.

In the present application, for a simpler determination of the parameters required for such a system, a limited number of fixed nodes (four) and a single mobile node were used. The latter is represented by a mobile robot that will be controlled by the user. The change of position and its updating is intended to be displayed on a real-time graph.

For the realization of software programs, it is desired to use the following programming languages:

- Python - implementation of programs that will run on a fixed workstation (server). The program creates a socket server that will exchange information with the ESP-12E module. This could be ordered in this way and send the serial readings to a Bluetooth module integrated in the mobile nodes. The whole assembly represents the mobile node.
- C++ - the program running on the ESP-12E module was made and aims to communicate between the user and the fixed node and at the same time reads the data transmitted through the serial communication port by the HM-10 Bluetooth module. The ESP-12E module can be programmed in several programming languages, some of which are: C, MicroPython, CircuitPython, C++. For versatility and considering the number of libraries available, the C++ language was chosen.

The proposed solution consists in the realization of a system containing a network of nodes, some of them having a known position, and others, mobile, having an unknown position to be determined (Fig. 2).

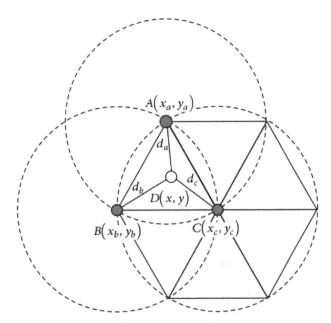

Fig. 2. Trilateration principle

The mobile nodes move in the emission zone of the fixed nodes and by measuring the received signal intensity between the two nodes (the fixed node and the mobile node), the distance between the fixed and the mobile node is determined at a certain time.

For a better understanding of the positioning systems and the technologies used, three solutions have been studied that will be presented later in the paper.

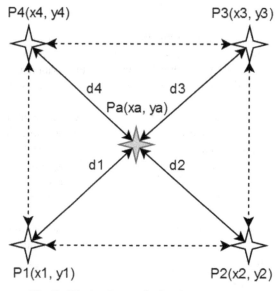

Fig. 3. Block scheme of a four-beacon system

According to [2–4] the estimation of the distance using the received signal power is based on the estimation of the distance of one node relative to another node that transmits a signal by processing the intensity of the received signal power. The node that transmits the signal is fixed and its position is known. These nodes are called anchor nodes or beacon nodes. If the intensity of the power received by a node that is at a certain distance from a beacon node is known, then the power at distance d is given by the formula (1):

$$Pr(d) = A - 10 * n_p * log(d)[dBm] \qquad (1)$$

where Pr is the power received at the distance d relative to the beacon node, A is the power received at a unit distance from the beacon node, and n_p is the transmission factor, the latter being different depending on the environment. Parameters A and np are constant in the external environment, but they can vary considerably over short distances in the internal environment. Therefore, inside, the area is divided into several smaller areas, with each area having the parameters A and np with a constant value. Parameter values can be determined experimentally for each area and, once known, the distance between a node and a beacon node can be expressed by the formula (2):

$$d = 10 - Pr + A * 10 * n_p[m] \qquad (2)$$

In order to be able to predict the location of a node in a two-dimensional area, the distance must be measured from at least three beacon nodes that are not on the same line. With the locations of the three known beacon nodes, the location of the node is determined by constructing three circles having the center of the beacon node and the radius equal to the distance from each beacon node to the node whose position it wants to be located. Thus, using the following set of Eqs. (3) whose representation can be found in Fig. 3, you can find the position of the node.

$$(x_a - x)^2 + (y_a - y)^2 = d_a^2$$
$$(x_b - x)^2 + (y_b - y)^2 = d_b^2 \quad (3)$$
$$(x_c - x)^2 + (y_c - y)^2 = d_c^2$$

where (x_a, y_a), (x_b, y_b), (x_c, y_c) are the coordinates of the known nodes, d_a, d_b, d_c, the distances from each beacon node to the node whose position is needed to be determined, and (x, y) represents the coordinates of the node whose position is desired to be located.

In order to determine the position in 3D space, it is necessary to use four beacon nodes that are not located in the same plane. In this case, the position of the node can be found by the intersection of the four spheres that are formed from the measurement of the four distances of the beacon nodes to the node whose position is desired to be located.

To have a better understanding of the localization principle, a simulation was done using MATLAB. There is represented the point Pa which can be found at the intersection of the four circles, their radius being the distance to the corresponding point, as per Fig. 4.

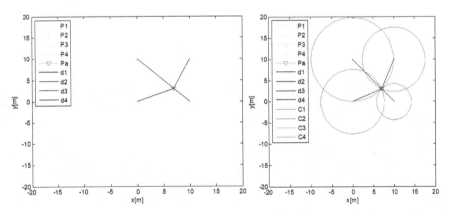

Fig. 4. Representation of circle intersection with corresponding distance-radius from mobile node to fixed beacons.

4 Mobile and Fixed Nodes

In order to be able to operate, the system needs at least three fixed nodes with known position and at the same time exploitable. These are necessary to be able to apply the algorithm for determining the distance (3) between the fixed and the mobile nodes, by which we can calculate the distances $d1, d2, d3$, respectively $d4$ specific to each mobile node.

Once the three required minimum distances have been determined, the above-mentioned trilateration algorithm (3) can be applied. This allows us to determine the position of the mobile node (Fig. 5) relative to the other fixed points whose position is known, using the distance between the mobile node and each point separately. As the number of fixed nodes grows, the position determination is more accurate. Because we are in the plane, the z axis representing the height is not considered, being equal to zero.

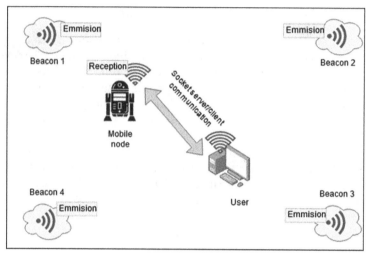

Fig. 5. Block scheme of communication pattern between fixed nodes, mobile nodes and user

The mobile nodes are specifically designed for this application, as presented in Fig. 6. The main components are the ESP12 module which I needed in order to be able to communicate with the node and a BLE 4.0 module which is used to create the fixed and mobile beacons network.

The programming of the ESP module it is desired to be done using an FTDI-style (USB to TTL converter) and the communication between the BLE4.0 module and the ESP module is done by serial communication. To be able to have a mobile node, a propulsion system is needed. In this case, two motors and a motor driver must be implemented on the board to be able perform the movement.

The fixed beacons are represented by the BLE4.0 modules. These modules can be programmed via AT (ATtention) commands in order to have different behavior, taking into consideration the application. A good characteristic of this modules is the low power consumption.

As it can be seen in Fig. 7, a demo version for the mobile node was developed in order to be able to start testing with good initial results regarding connectivity and ease of use.

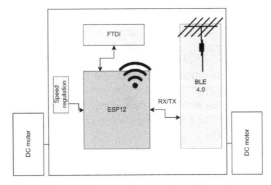

Fig. 6. Schematic design of mobile node

Fig. 7. Test version of mobile node

5 Conclusions and Future Development

In this paper, it was proposed the schematic of an indoor positioning system using bluetooth low energy modules and RSSI parameter to calculate the position. Further development involves the development of a dedicated software program to be able to collect, analyze and filter the data and to provide the necessary means to be able to present the data in an understandable form.

Considering the advantages of an indoor positioning system, it is desired to implement and test it to determine and improve the current limitations of the system. The modules from which the system is made are easy to reproduce on a large scale using both tried and tested and novel 3D printing technologies [7], so it makes it suitable for applications where it is necessary to implement it quickly and efficiently, without high material costs. The system can also be easily implemented for education purposes [8].

References

1. de Carvalho, S., Joel, J.P.C., Alberti, A.: LoRaWAN—A low power WAN protocol for internet of things: a review and opportunities. In: 2017 SpliTech. IEEE (2017)

2. Goutham, A., Thanikaiselvan, V.: Improving the performance of RSSI based indoor localization techniques using neural networks. In: 2018 Second International Conference on Electronics, Communication and Aerospace Technology (ICECA). IEEE (2018)
3. Xiaolong, S., Shenbqi, Y., Jian, H.: Improved localization algorithm based on RSSI în low power Bluetooth network. In: 2016 2nd International Conference on Cloud Computing and Internet of Things (CCIOT). IEEE (2016)
4. Jianqiao, X., Qin, Q., Zeng K.: A distance measurement wireless localization correction algorithm based on RSSI. In: 2014 Seventh International Symposium on Computational Intelligence and Design, vol. 2. IEEE (2014)
5. IEEE Computer Society LAN/MAN Standards Committee: IEEE standard for information technology-telecommunications and information exchange between systems-Local and metropolitan area networks-specific requirements part 11: Wireless LAN medium access control (MAC) and physical layer (PHY) specifications. IEEE Std 802.11 (2007)
6. Miller, A., Brent, A., Chatschik, B.: Bluetooth Revealed: The Insider's Guide to An Open Specification for Global Wireless Communication. Prentice Hall PTR, Upper Saddle River (2001)
7. Besnea, D., Dinu, E., Moraru, E., Spanu, A., Rizescu, C., Constantin, V.: Experimental researches regarding the manufacturing of new thermoplastic materials used in additive technologies. Materiale Plastice **56**(1), 167–170 (2019)
8. Mogoş, R., Bodea, C., Dascălu, M.: Technology enhanced learning for Industry 4.0 engineering education. Revue roumaine des sciences techniques **63**(4), 429–435 (2018)

Incorporating Inteco's 3D Crane into Control Engineering Curriculum

Frantisek Gazdos(✉) and Lenka Sarmanova

Faculty of Applied Informatics, Tomas Bata University in Zlin, Nam. T.G. Masaryka 5555, 760 01 Zlin, Czech Republic
gazdos@utb.cz

Abstract. The paper describes the Inteco 3D Crane as a suitable scale-model to support practical training of students in control engineering and related fields. It also contains an illustrative case study controlling the payload position in 3D space in a robust way using the systematic polynomial approach, including also experimental identification of the crane dynamics in all three axes. As such, it can serve as an inspiration for both academic staff and students dealing with related fields of study.

Keywords: Control engineering · Inteco 3D crane · Modelling · Robust control · Polynomial approach

1 Introduction

Control engineering represents a multidisciplinary field of study placing high demands on the students – prospective engineers in this field. They have to understand, in a broader context, mathematics, physics, informatics, electrical engineering, instrumentation and measurements, signal processing and others. Knowledge of these subjects together with appropriate practical training then enables designing safe, reliable and efficient control systems – "brain" of most engineering applications. Practical training in this field realized in a special laboratories equipped with real-time scale process models has proven to be beneficial for this purpose [1–3]. The labs usually contain some typical processes to prepare prospective control engineers for various types of systems and conditions. Here, students are forced to work with real hardware, identify it, implement suggested control algorithms and face real challenges, such as stochastic disturbances, signal saturations, nonlinearities, etc. Consequently, they are more confident under real conditions in practice. One such laboratory has been presented in the author's previous publication [3]. This contribution aims at introducing possibilities of one specific scale model acquired for this lab recently, namely 3D Crane by Inteco [4]. Crane control, including modelling, identification and various control strategies is a popular scientific subject, e.g. [5] can serve as a starting survey paper, various control strategies related particularly to the discussed Inteco 3D Crane can be found e.g. in the references [6–11]. Here, in this paper, the crane possibilities are discussed, followed by an illustrative case study controlling the payload position in 3D space in a robust way using the systematic polynomial approach [12–14], including also experimental identification of the crane dynamics in all 3 axes. As such, this paper can serve as an inspiration for both academic staff and students struggling with control engineering and related fields of studies.

The paper starts with general description of the crane and its possibilities, followed by a detailed state-space mathematical model of this system. Next section illustrates control system design for the problem of payload position control, including also experimental identification and derivation of controllers based on the algebraic approach using polynomials and some results from the robust control theory. The paper concludes giving results of real-time control of this system and some final remarks.

2 Inteco 3D Crane

2.1 General Description

The 3D Crane of the Polish company INTECO [4] represents a non-linear electromechanical system with a complex dynamics which can be used to create challenging control problems [15]. It is a laboratory-scaled model which can be controlled from a PC using a common A/D-D/A board and MATLAB software libraries. Basic set-up presented in Fig. 1 consists of a payload hanging on a pendulum-like lift-line wound by a motor mounted on a cart, thus the payload can be lifted and lowered in the z-direction. The cart is placed on a rail and it is capable of horizontal motion along the rail in the y-direction using next motor. Finally the rail with the cart can also move horizontally in the x-direction using another motor. As a result, the payload attached to the end of the lift-line can move freely in 3 dimensions and it is driven by 3 DC motors, which represent 3 possible control inputs (manipulated variables) of this system, controlled by PWM. The model enables to measure 5 state variables that represent possible controlled outputs – the cart coordinates on the horizontal plane together with the lift-line length and, in addition, also two deviation angles of the payload. All the variables are sensed with identical measuring encoders with resolution 4096 pulses/rotation, enabling to measure the payload deviation with accuracy 0.0015 rad [15].

A typical control task represents payload position control in the 3D space, more advanced control problems can take into account also oscillation of the payload during the motion with the aim to suppress it as much as possible. These tasks typically include identification of the system(s), validation of the model(s), controller(s) design, testing and optimization, either in the SISO or MIMO set-ups.

Fig. 1. Inteco 3D Crane – detail

2.2 Mathematical Model

Let us assume the scheme of the crane according to Fig. 2 below [15, 16].

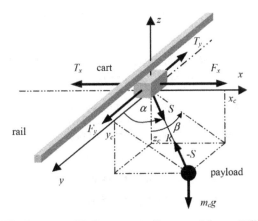

Fig. 2. Inteco 3D Crane – coordinates and forces [15]

Further, let us denote 5 measured variables as:

- x_w, y_w – distance of the rail with the cart from the center of the construction frame, and distance of the cart from the center of the rail, respectively,
- R – length of the lift-line,
- α, β – angle between the y-axis and the lift-line, and between the negative direction on the z-axis and the projection of the lift-line onto the x-z plane,

and other used physical quantities as:

- m_c, m_w, m_s – payload, cart and moving rail mass,
- x_c, y_c, z_c – coordinates of the payload,
- S – reaction force in the lift-line acting on the cart,
- F_x, F_y, F_R – forces driving the rail with cart, the cart along the rail and controlling the length of the lift-line, respectively,
- T_x, T_y, T_R – corresponding friction forces.

Then the payload position can be described by the equations:

$$x_c = x_w + R\sin(\alpha)\sin(\beta), \quad y_c = y_w + R\cos(\alpha), \quad z_c = -R\sin(\alpha)\cos(\beta). \tag{1}$$

Next, the following symbols are used in the sequel:

$$\mu_1 = \frac{m_c}{m_w}, \quad \mu_2 = \frac{m_c}{m_w + m_s},$$

$$u_1 = \frac{F_y}{m_w}, \quad u_2 = \frac{F_x}{m_w + m_c}, \quad u_3 = \frac{F_R}{m_c}, \tag{2}$$

$$T_1 = \frac{T_y}{m_w}, \quad T_2 = \frac{T_x}{m_w + m_c}, \quad T_3 = \frac{T_R}{m_c},$$

$$N_1 = u_1 - T_1, \quad N_2 = u_2 - T_2, \quad N_3 = u_3 - T_3.$$

If we denote the state variables of the 3D crane model as:

$$x_1 = y_w, \quad x_2 = x_1' = y_w', \quad x_3 = x_w, \quad x_4 = x_3' = x_w',$$
$$x_5 = \alpha, \quad x_6 = x_5' = \alpha', \quad x_7 = \beta, \quad x_8 = x_7' = \beta', \tag{3}$$
$$x_9 = R, \quad x_{10} = x_9' = R',$$

and introduce other auxiliary variables (where g denotes the gravity constant):

$$s_n = \sin(x_n), \quad c_n = \cos(x_n),$$
$$V_5 = c_5 s_5 x_8^2 x_9 - 2x_{10}x_6 + gc_5c_7,$$
$$V_6 = 2x_8(c_5x_6x_9 + s_5x_{10}) + gs_7,$$
$$V_7 = s_5^2 x_8^2 x_9 + gs_5c_7 + x_6^2 x_9, \tag{4}$$

then, it is possible to finally derive the following non-linear state-space model of the crane dynamics, consisting of 10 interdependent differential equations [15]:

$$x_1' = x_2, \quad x_2' = N_1 + \mu_1 c_5 N_3,$$
$$x_3' = x_4, \quad x_4' = N_2 + \mu_2 s_5 s_7 N_3,$$
$$x_5' = x_6, \quad x_6' = \left[s_5 N_1 - c_5 s_7 N_2 + \left(\mu_1 - \mu_2 s_7^2\right) c_5 s_5 N_3 + V_5\right]/x_9,$$
$$x_7' = x_8, \quad x_8' = -(c_7 N_2 + \mu_2 s_5 c_7 s_7 N_3 + V_6)/(s_5 x_9),$$
$$x_9' = x_{10}, \quad x_{10}' = -c_5 N_1 - s_5 s_7 N_2 - \left(1 + \mu_1 c_5^2 + \mu_2 s_5^2 s_7^2\right) N_3 + V_7, \tag{5}$$

which shows complexity and non-linearity of this model.

3 Control System Design

In this paper, due to the limited space, only the problem of payload position control is illustrated. It is based on the decentralized control approach, starting with experimental identification of the motion dynamics in all 3 axes first and then followed by controller design which uses the systematic algebraic approach with polynomials leading to the pole-placement problem solution. This is addressed in a robust way ensuring satisfying control performance not only on simplified identified models but on the real system as well, which is proved experimentally.

3.1 Experimental Identification

For identification of the crane dynamics in all 3 axes x, y, z, e.g. a simple technique of step-responses evaluation can be used. For this purpose, a step change of input signal in different operating points was injected into the system and the responses were recorded, processed and evaluated. This was done for each of the 3 DC motors separately, obtaining corresponding responses in all 3 axes. The z-axis was identified in both directions as recommended in the literature [15], i.e. up, denoted $z+$ and down, labeled as $z-$. Obtained responses (shifted to zero, normed to unit step input signal and averaged) for the axes x and y are presented in the graphs of Fig. 3 below (in blue), together with the resultant identified models (in red). Results for the z-axis provided similar match but are omitted due to the limited space of this conference paper.

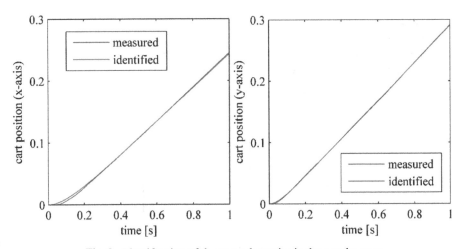

Fig. 3. Identification of the crane dynamics in the x and y axes

With respect to the obtained responses from the identification, the following simplified model was suggested to describe the crane dynamics in all the axes:

$$G(s) = \frac{b_0}{s(s+a_0)} = \frac{k}{s(Ts+1)}, \qquad (6)$$

which is represented by an integrative system with a given gain k and time-constant T. As shown in the presented graphs above, this model describes the system dynamics sufficiently. In order to find the unknown parameters of the model, the MATLAB function *fminsearch* was used, minimizing the sum of squared differences between the modelled and measured responses. Results of this optimization procedure, i.e. the identified parameters k and T for all 4 cases are given below:

$$G_x(s) = \frac{0.273}{s(0.108s + 1)}, \quad G_y(s) = \frac{0.308}{s(0.055s + 1)},$$
$$G_{z+}(s) = \frac{0.137}{s(0.032s + 1)}, \quad G_{z-}(s) = \frac{0.115}{s(0.036s + 1)}. \tag{7}$$

These models described by transfer functions are further used for control system design, i.e. to design suitable compensators in each axis to control the payload position in the 3D space of the crane.

3.2 Design of Controllers

Control system design here is based on the systematic algebraic approach using polynomials leading to the solution of polynomial equations [12–14]. It allows to find both suitable controller structure and its parameters and it naturally leads to the pole-placement problem which can be solved by many different approaches [17]. Control set-up suggested in this work is sketched in Fig. 4 below consisting of a system G to be controlled by means of a compensator with two parts – feedback controller Q and feedforward controller (filter) R. The variables y, w, u and v denote controlled variable, its reference (set-point) signal, control input (manipulated variable) and disturbance, respectively. This configuration will generally enable lower overshoots of the controlled variable and provide more realizable values of the control input.

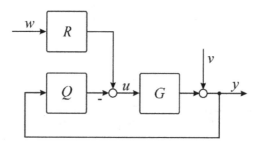

Fig. 4. Control set-up

Further suppose that both the plant and the controller(s) can be approximated by transfer functions $G(s)$, $Q(s)$ and $R(s)$ with coprime polynomials $\{b(s), a(s)\}$, $\{q(s), p(s)\}$ and $\{r(s), p(s)\}$ according to (8)–(9) while satisfying (10)-(11).

$$G(s) = \frac{b(s)}{a(s)} \tag{8}$$

$$Q(s) = \frac{q(s)}{p(s)}, \quad R(s) = \frac{r(s)}{p(s)}, \tag{9}$$

$$\deg a(s) \succ \deg b(s) \tag{10}$$

$$\deg p(s) \geq \deg q(s), \quad \deg p(s) \geq \deg r(s), \tag{11}$$

Basic requirements for the control systems above are formulated as:

- stability,
- asymptotic tracking of the reference signal,
- disturbance attenuation,
- inner properness.

From the scheme of Fig. 4 and assuming (8)–(9), it is easy to derive following relationships between the controlled variable $y(t)$ ($Y(s)$ in the complex domain) and the input signals $w(t)$ and $v(t)$ ($W(s)$ and $V(s)$ similarly):

$$\begin{aligned} Y &= \frac{GR}{1+GQ}W + \frac{1}{1+GQ}V = \frac{br}{ap+bq}W + \frac{ap}{ap+bq}V \\ &= \frac{br}{d}W + \frac{ap}{d}V = TW + SV \end{aligned} \tag{12}$$

Here, d denotes a characteristic polynomial of the closed loop defined as:

$$ap + bq = d \tag{13}$$

and symbols S and T represent important transfer functions of the loop known as the sensitivity function and the complementary sensitivity function respectively.

Similarly, it is straightforward to derive the formula (14) for the control error $e(t)$ ($E(s)$ in the complex domain):

$$E(s) = \frac{1}{d}\big[(d - br)W(s) - apV(s)\big]. \tag{14}$$

From (12) it is clear that the control system of Fig. 4 will be stable if the characteristic polynomial $d(s)$ given by (13) is stable. This polynomial equation, after a proper choice of the stable polynomial $d(s)$, is used to compute the unknown feedback controller polynomials $q(s)$ and $p(s)$. Roots of the characteristic polynomial $d(s)$ are known as poles of the closed loop. Their proper placement influences not only stability of the loop but also the achieved control quality, i.e. settling-time, overshoots, control input course etc. Therefore the so-called pole-placement problem is a natural part of the polynomial approach to control system design.

Further assume the identified model of the plant is in the form of (6), i.e. an integrative system with general transfer function (8), specified more as:

$$G(s) = \frac{b(s)}{a(s)} = \frac{b(s)}{s \cdot \tilde{a}(s)} \tag{15}$$

and let us also suppose, as it is a common case, that both the reference signal $w(t)$ and disturbance $v(t)$ can be approximated well by step functions, in the complex domain:

$$W(s) = \frac{w_0}{s}, \quad V(s) = \frac{v_0}{s} \tag{16}$$

for some reals w_0, v_0. Then, substituting (16) into the control error (14) yields

$$E = \frac{1}{d}\left[(d-br)\frac{w_0}{s} - ap\frac{v_0}{s}\right] = \frac{1}{d}\left[(d-br)\frac{w_0}{s} - \tilde{a}sp\frac{v_0}{s}\right] = \frac{1}{d}\left[(d-br)\frac{w_0}{s} - \tilde{a}pv_0\right] \tag{17}$$

which shows that step disturbances will be compensated automatically, thanks to the integrative nature of the plant (15). Moreover, in order to ensure also asymptotic tracking of the reference signal, the formula (17) shows that the term $(d-b.r)$ must be divisible by the "s" term, i.e. $(d-b.r) = t.s$ must hold for some unknown polynomial $t(s)$. This relation can be rewritten into the form of (18) which enables to find the unknown numerator polynomial $r(s)$ of the feedforward controller $R(s)$ when the characteristic polynomial $d(s)$ is properly prescribed.

$$ts + br = d \tag{18}$$

Then, the unknown polynomials of both controllers Q, R (9), namely $q(s)$, $p(s)$ and $r(s)$ can be calculated from the couple of polynomial Eqs. (13), (18), i.e.:

$$\begin{aligned} \tilde{a}sp + bq &= d \\ ts + br &= d \end{aligned} \tag{19}$$

for a given characteristic polynomial $d(s)$.

The inner properness of the control system is satisfied if all its parts (transfer functions) are proper. With regard to the strictly proper plant transfer function (8), (10), proper controllers (9), (11) and taking into account the solvability of polynomial Eqs. (19), it is possible to derive the following formulas for the degrees of the unknown polynomials:

$$\begin{aligned} &\deg q(s) = \deg \tilde{a}(s), \quad \deg p(s) = \deg \tilde{a}(s), \quad \deg r(s) = 0, \\ &\deg d(s) = 2\deg \tilde{a}(s) + 1, \quad \deg t(s) = 2\deg \tilde{a}(s), \end{aligned} \tag{20}$$

when seeking the simplest controllers structures. For the identified model of the crane dynamics in all the axes (6) then holds: $\deg q = 1$, $\deg p = 1$, $\deg d = 3$ and $\deg t = 2$; consequently the controllers take the form:

$$Q(s) = \frac{q(s)}{p(s)} = \frac{q_1 s + q_0}{p_1 s + p_0}, \quad R(s) = \frac{r(s)}{p(s)} = \frac{r_0}{p_1 s + p_0} \tag{21}$$

which shows that the feedback compensator $Q(s)$ is represented by a filtered PD controller while the feedforward regulator is just a filtered P-controller.

3.3 Pole-Placement Problem

For practical computation of the controller's polynomials q, p and r it is necessary to choose a suitable stable polynomial d appearing on the right side of the polynomial Eqs. (19). This is the so called pole-placement problem mentioned earlier, e.g. [17]. Therefore we are seeking suitable poles p_i of the designed loop to fulfill given requirements. Then d can be expressed as (22) for some poles, i.e. its roots p_i. As a result, the control design procedure transforms to the optimization problem of finding the right poles providing the required control quality.

$$d(s) = \prod_{i=1}^{\deg d} (s - p_i) \qquad (22)$$

Here, it is suggested to have this polynomial in the following simple form:

$$d(s) = (s + \alpha)^3 \qquad (23)$$

for some positive parameter $\alpha > 0$; although this choice is the simplest it will not only guarantee stable aperiodic control process (all closed-loop poles will be equal to $-\alpha$) but it also enables to tune the loop simply by one single parameter α. Solving the polynomial Eqs. (19) for such choice of the characteristic polynomial (23) provides the following formulas for the unknown coefficients of the controllers (21):

$$p_1 = 1, \quad p_0 = 3\alpha - a_0, \quad q_1 = \frac{3\alpha^2 - a_0 p_0}{b_0}, \quad q_0 = r_0 = \frac{\alpha^3}{b_0}, \qquad (24)$$

depending on the identified parameters b_0, a_0 of the controlled system (6) and, on the "tuning" parameter α introduced above.

As the Inteco 3D crane is non-linear and multivariable in nature, having quite complex dynamics, while the suggested identified models of dynamics in all the axes are quite simple, it is further suggested to optimize the tuning parameter α with respect to robustness of the designed loop, in order to cope with the nonlinearities and possible disturbances. Peak gain of the sensitivity function frequency response given by the infinity norm H_∞ is a classical measure of the loop robustness [18]. Therefore it is suggested here to use it as tool to optimize the one tuning parameter α. The sensitivity function is, according to (12), given as:

$$S = \frac{1}{1 + GQ} = \frac{ap}{ap + bq} = \frac{ap}{d} \qquad (25)$$

then, the criterion describing the robustness reads:

$$J = \|S\|_\infty = \sup_\omega |S(j\omega)| \qquad (26)$$

where ω indicates the frequency. Values of this criterion depending on the tuning parameter α for the identified model G_x (7), describing x-axis dynamics, are presented in Fig. 5 below – general overview first, followed by more detailed inspection.

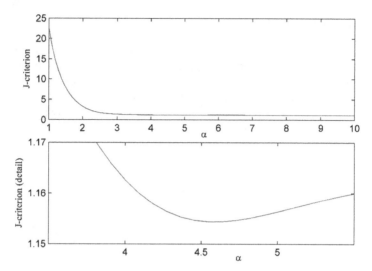

Fig. 5. Norm H_∞ of the sensitivity function S with α

As can be seen from the graph, sensitivity of the loop decreases rapidly with α first and then it does not change significantly. Moreover, closer inspection reveals minimum of the graph indicating least sensitive, i.e. most robust setting of the tuning parameter. This specific setting is further used to calculate the controllers' coefficients (24) and for both simulation testing and real-time control of the Inteco 3D Crane. Similar graphs can be obtained also for the other identified models from (7) providing these optimal values of the tuning parameter α for controllers in all the axes:

$$\alpha_x = 4.65, \quad \alpha_y = 9.06, \quad \alpha_{z+} = 15.39, \quad \alpha_{z-} = 13.86, \tag{27}$$

which finally yields the following resultant controllers:

$$Q_x = \frac{8.49s + 39.77}{s + 4.68}, \quad R_x = \frac{39.77}{s + 4.68}; \quad Q_y = \frac{14.89s + 131.96}{s + 8.87}, \quad R_y = \frac{131.96}{s + 8.87};$$
$$Q_{z+} = \frac{57.06s + 851.4}{s + 14.92}, \quad R_{z+} = \frac{851.4}{s + 14.92}; \quad Q_{z-} = \frac{60.86s + 825.8}{s + 13.57}, \quad R_{z-} = \frac{825.8}{s + 13.57}. \tag{28}$$

3.4 Real-Time Control Results

The resultant controllers (28) were tested by simulation means first, providing good results, and finally by real-time experiments on the Inteco 3D Crane. Some of the experiments, namely the position control in the x and y axes are presented in Fig. 6 and Fig. 7 below. More experiments including control in all 3 axes will be, due to the limited space of this paper, showed in the presentation and prospective extended version of this paper.

As can be seen from the graphs, the control response is stable with only minor overshoots and the set-point positions are tracked relatively quickly. Moreover, control

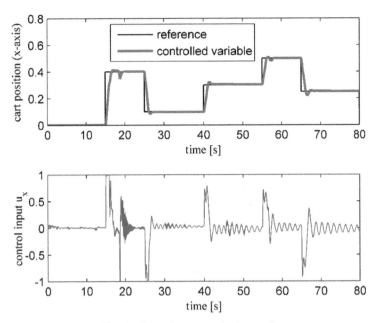

Fig. 6. Control response in the x-axis

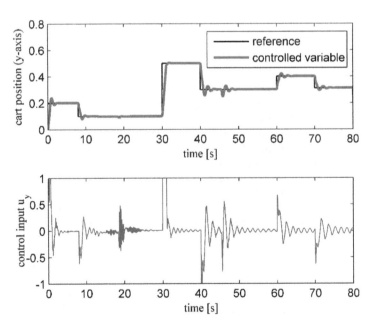

Fig. 7. Control response in the y-axis

input values are in a reasonable range and disturbances are attenuated also successfully (disturbance in time 17 s for x-axis and 45 s for y-axis were simulated by gently slamming into the payload by hand), which confirms robustness of the designed loop.

4 Conclusion

This paper describes the Inteco 3D Crane as a suitable scale-model to support practical training of students in control engineering and related fields of study. This eletromechanical model, nonlinear and multivariable in nature, can be used to create both simple and complex control challenges for prospective control engineers. This work presents a case study, illustrating a position control of the payload in the 3D space of the crane. As such, it can help and inspire both teachers (to incorporate similar hardware into their control courses) and students (to understand the problems involved) in their "struggle" for better practical skills related to designing safe, reliable and efficient control systems.

In this paper, due to its limited space, only the position control problem is illustrated, however, authors are in the stage of preparing an extended version of this contribution, where also the more challenging problem of suppressing payload oscillations will be addressed, as another example.

References

1. Horacek, P.: Laboratory experiments for control theory courses: a survey. Annu. Rev. Control **24**, 151–162 (2000)
2. Leva, A.: A simple and flexible experimental laboratory for automatic control courses. Control Eng. Pract. **14**, 167–176 (2006)
3. Gazdos, F.: Using real-time laboratory models in the process of control education. In: Machado, J., Soares, F., Veiga, G. (eds.) HELIX 2018. LNEE, vol. 505, pp. 1097–1103. Springer, Cham (2019). https://doi.org/10.1007/978-3-319-91334-6_151
4. INTECO. http://www.inteco.com.pl. Accessed 30 Jan 2020
5. Abdel-Rahman, E.M., Nayfeh, A.H., Masoud, Z.N.: Dynamics and control of cranes: a review. J. Vib. Control **9**(7), 863–908 (2003)
6. Nguyen, Q.T., Vesely, V.: Robust decentralized controller design for 3D crane. In: Fikar, M., Kvasnica, M. (eds.) Process Control 2011, pp. 485–489 (2011)
7. Trajkovic, D.M., Antic, D.S., Nikolic, S.S., Peric, S.L., Milovanovic, M.B.: Fuzzy logic-based control of three-dimensional crane system. Autom. Control Robot. **12**(1), 31–42 (2013)
8. Petrehus, P., Lendek, Z., Raica P.: Fuzzy modeling and design for a 3D Crane. In: IFAC ICONS 2013, pp. 479–484. IFAC (2013)
9. Vazquez, C., Fridman, L., Collado, J., Castillo, I.: Second-order sliding mode control of a perturbed-crane. J. Dyn. Syst. Measur. Control **137**(8) (2015)
10. Aksjonov, A., Vodovozov, V., Petlenkov, E.: Three-dimensional crane modelling and control using euler-lagrange state-space approach and anti-swing fuzzy logic. Electr. Control Commun. Eng. **9**(1), 5–13 (2016)
11. Pauluk, M.: Optimal and robust control of 3D crane. Przeglad Elektrotechniczny **92**(2), 206–212 (2016)
12. Hunt, K.J.: Polynomial Methods in Optimal Control and Filtering. Peter Peregrinus Ltd., London (1993)
13. Kucera, V.: Diophantine equations in control – a survey. Automatica **29**(6), 1361–1375 (1993)

14. Anderson, B.D.O.: From Youla-Kucera to identification, adaptive and nonlinear control. Automatica **34**(12), 1485–1506 (1998)
15. 3D Crane User's Manual. Inteco Ltd., Krakow (2017)
16. Sarmanova, L.: Modelling, identification and control of an Inteco 3D Crane. Master's thesis, Tomas Bata University in Zlin, Faculty of Applied Informatics, Zlin (2019)
17. Kucera, V.: The pole placement equation – a survey. Kybernetika **30**(6), 578–584 (1994)
18. Skogestad, S., Postlethwaite, I.: Multivariable Feedback Control: Analysis and Design. Wiley, Chichester (2005)

Design of Laser Scanners Data Processing and Their Use in Visual Inspection System

Ivan Kuric, Matej Kandera, Jaromír Klarák(✉), Miroslav Císar, and Ivan Zajačko

Faculty of Mechanical Engineering, Department of Automation and Production Systems, University of Žilina, Univerzitná 8215/1, 010 26 Žilina, Slovakia
{matej.kandera,jaromir.klarak}@fstroj.uniza.sk

Abstract. Visual product inspection is an important element, whether in-process or output product inspection in today's industry. The present paper describes an approach to visual inspection, based on the usage of line laser scanners. Their task is to obtain coordinates of individual geometric profile elements of the scanned object in the form of raw data in .csv output. Paper also describes the design of processing and sorting of these raw data and creating output from them in a format suitable for subsequent processing. In order to enable the subsequent evaluation of the processed data and the visual representations generated therefrom, a procedure for combining the individually obtained images of the surface of the rotating object into a single linear image is also proposed. The described procedure can accelerate and simplify the whole process of visual inspection of determined objects.

Keyword: Visual inspection · Laser scanners · Image processing

1 Introduction

Visual inspection, whether direct or indirect, is one of the valuable tools of non-destructive product inspection methods. Visual inspection with appropriate lighting is often one of the first methods for detecting and locating suspect areas on inspected objects. In the past, it was a highly subjective approach to quality control and there was no documentation to describe its progress. The success of the result depended mainly on the trained operators, the cleanliness of the environment, the properties of the test object, the quality of the optical devices and the correct illumination (when using an industrial camera as a recording device) of the investigation area. These parameters are of course very important even today, but the equipment used is much more sophisticated.

Another approach to obtaining data from real objects except for cameras is to use line laser scanners. The advantage of using scanners is a fact, that they work on the principle of triangulation of a red light beam, which they emit to the object to be examined, thereby partially eliminating the above-mentioned need for additional illumination. They are also less affected by variations in ambient light intensity. For this reason, in some cases, the use of the so-called "Black box" is not necessary as when using cameras.

Visual inspection is one of the basic methods of non-destructive testing. In the visual inspection process, the worker follows the prescribed procedures, which can range from simple inspection of objects with the naked eye to detect external defects to the performance of various measurements to ensure compliance with the required standards. Nowadays, optical systems include real-time imaging and various analysis using computers [1–3].

The first step to enable visual inspection with analysis using a computer is to obtain the input data correctly. Different types of files can be used as input data, in visual inspection, as the name implies, this data will be in the form of a visual representation of the scanned object or its selected parts [4].

2 Capture Method and Data Export

In the research described in this paper, a line laser scanner was chosen to capture the surface of the selected object. The object to be scanned is an automobile tire (Fig. 1), in the described case specifically its sidewall. A micro epsilon scan-CONTROL 2600-50 device, with resolution up to 640 dots per line was used to scan the product. Scan-CONTROL Configuration Tools and 3D View software were used along with this line scanner, which is an interactive 3D visualization software designed for displaying and exporting 3D captured data during the scanning, supplied directly by the manufacturer [5].

Table 1. Achieved scan resolution.

Resolution	Value [mm]
Axis x	≈0,100
Axis y	≈0,060
Axis z	≈0,004

Mentioned sensors operate on the principle of triangulation, in which they can provide a resolution capability in the z-axis up to 0.004 mm. This value represents the resolution of the sensitivity of the sensing point from the device sensor (Fig. 1).

The captured data was exported from Microepsilon 3D View in two data file types. As pictures in.png format (Fig. 2) and also in the form of coordinates of individual points saved in.csv format. The.csv file contains the position of the individual scanned points of the scanned object profile in the Cartesian coordinate system in the x, y and z axes (Fig. 1). The exported.png format serves a basic illustration of the surface that has been scanned and stored in a specific.csv file.

Fig. 1. The scanned object and detail of the selected area during the scanning process.

3   Processing of Data from Scanning

The obtained.png files were used to illustrate scans of the tire surface while the data source for further processing was the.csv file.

For generating the visual content, the z-coordinate of the points obtained from the scanning was used. One scan contained approximately 15.5 mil. points classified in 25000 lines aligned in the y-axis with the number of scanned points per line ranging from 580 to 630. During scanning process, 11 images of individual tire sidewall sections were taken, as shown in Fig. 2.

Fig. 2. Illustrations of the tire surface generated by the software from the scanner manufacturer.

These scans are generated by micro-epsilon software that offers a basic visual preview of the scanned surface. In this case, the output data does not contain geometric information, it is only a 2D representation of a surface scan with recognizable geometric shapes.

When processing data in.csv format, the image was first generated from an unclassified point cloud. Described algorithm is based on the generation of images based on raw z-axis data to produce an image with a high degree of interference as it is displayed in Fig. 3 [6, 7, 10].

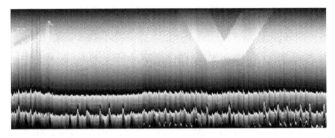

Fig. 3. An image generated from raw z-axis data with a high degree of interference.

In order to achieve the desired image quality and to allow further work with data, it is necessary to sort the image into a 640 × 25000 matrix. The value of 640 is characterized by a maximum number of points scanned in a line (x-axis) and the value of 25000 indicates the number of such lines which are consecutively scanned (y-axis). The points were classified using x coordinates of each point and its position definition in the x coordinate were continuous values from −25 to 27 mm. The classification was performed using algorithm based on the following Eq. (1):

$$X_p = \left\lfloor \left| \frac{X_i - f_{\min}(X)}{f_{\max}(X) - f_{\min}(X)} \right| \right\rfloor * 639 \in \mathbb{N} \tag{1}$$

which determined the position of individual points X_p based on the overall scan, which is based on finding of maximal $f_{\max}(X)$ and minimal $f_{\min}(X)$ values from data by the position value in the x axis. Subsequently, this image was generated and is shown in Fig. 4, in which the detectable geometry of the tire scan is suitable for further processing [13, 14].

By further processing is meant the possibility of utilization associated with the application of the comparative method to compare etalon data and a scan of the controlled product. The checked product can be searched for anomalies or differences to the etalon sample. In this case, this method of detecting anomalies was chosen because of its universality and verification of the theory, where the emphasis was also placed on finding errors or objects that are not commonly encountered in practice or that the system has not yet encountered.

The sorted image shown in Fig. 4 is further processed. The main feature of this state of processing is the recognizable tire sidewall surface geometry. Another assumption is to scan the entire product. The problem was the hardware that did not allow to capture

Fig. 4. The processed image generated from the sorted scanned data.

more than 25,000 lines. In this case, attention was directed to the number of scans per second, with approximately 250 scans per second, when this number actually fluctuated around the specified value. As a result, mentioned parameter was set based on a compromise between the scan time (approximately 170 s) and the result image resolution and sharpness. In this way, the amount of line scans obtained was not sufficient to scan the entire tire due to the mentioned hardware limitations. The creation of a complete scan had to consist of combining several scans into one. The resultant image was created by combining scans 1, 2 and 4 as shown in Fig. 2. Scan number 1 was set as a base to which scan number 2 and then this continuous image was supplemented with a portion of scan 4. The connection was made on the principle of pairing of border areas of 2000 lines. Defining a match was realized by searching for a global minimum (2).

$$E_{min} = \sum (a_i - b_i)^2 \qquad (2)$$

Individual scans were connected to each other on the basis of such a match. The scan boundaries are shown in Fig. 5, where scan 2 and scan 4 were combined and as a result a continuous scan of the entire perimeter of the tire sidewall was created with a smooth transition.pixel filter applied

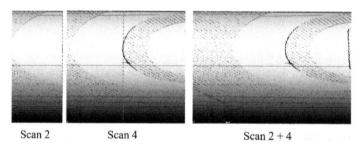

Scan 2   Scan 4   Scan 2 + 4

Fig. 5. Preview of continuous scan created from partial scans 2 and 4.

Scan 1 was added to the continuous scan, where only the missing area was added to the parts from scans 2 and 4. Because of the completion, it was necessary to determine the boundaries from both sides of the scan and add this part to the continuous scan 2 + 4. produced a continuous 1 + 2 + 4 scan. The final continuous scan is then suitable for detecting anomalies in the geometry of the inspected object, which may be deemed inappropriate under predefined conditions and thus affect the overall quality of the inspected product. Anomaly highlighting is generated by comparing etalon data that is guaranteed to be error-free and inspected data that has a certain probability of error.

Before comparing an etalon and an inspected sample, it is necessary to locate and pair the selected area on both scans. When a match is found, it overlaps and produces an image with highlighted differences. Due to the non-constant scanning of the points, microscopic displacements occur, especially at the border regions. Another non-constant value is the scanned edges in the run-in area in the product rotation direction. At these parts of the geometry, the beams reflected at the wrong angle, which meant that the geometry points forming some of edges were not sensed. These are depicted as black areas in the images. The areas highlighted in this way do not match when compared and thus create line differences in the compared image, which are filtered for better entity contrast. As a result, only the difference area with a larger pixel cluster is highlighted than the 4 × 4 pixel filter applied [8, 9, 12].

Fig. 6. Demonstration of anomalies occurring on the scanned object.

In Fig. 6. Two specific areas with geometric differences that indicate a possibility of the anomaly are highlighted in white. In this case, the scanned object would be evaluated as a failure, if it were an application directly in the industry. Subsequent to categorizing the product according to the type and severity of the defect, an evaluation would be made as to whether the product is suitable for repair and if there is a possibility to eliminate the defects or to be eliminated from the manufacturing process.

4 Conclusion

Visual inspection as a non-destructive method of product control is one of the basic methods of detecting external inaccuracies. The article presents a way of raw line laser scanner data processing, suitable for further finding and highlighting anomalies on a specific object with specific properties. In the case of this paper, it is an automobile tire and its sidewall in particular. The described method of obtaining data for further processing is based on the surface scanning with a line laser scanner. For further processing of acquired data, it is possible to create the system in two different ways. First is intelligent fault-finding using trained artificial neural networks, and the other is a comparison method between an etalon and an inspected object. In the comparative method, the system is resistant to the occurrence of unknown anomalies. These anomalies are highlighted without the system being adapted to look for specific types of geometric anomalies. Following the mentioned reasons, this comparative method was chosen for the described task. Differences in images can mean two possible states, either an anomaly or entities occurring on the object, but which do not have a constant position for each product. With the sensing parameters set, satisfactory resolution (Table 1) in the z and y axes was achieved, but these parameters were limiting the resolution in the x-axis. Increasing the resolution in the x-axis would be possible by scanning a smaller product width or by using different scanning devices with a higher number of scanned points. The limitation of this system is the compromise between scan quality, scan time and computational performance, which also depend on evaluation hardware used.

Acknowledgement. This article was made under the support of APVV project – APVV-16-0283 Research and development of multi-criteria diagnosis of production machinery and equipment based on the implementation of artificial intelligence methods.

References

1. Mix, P.E.: Introduction to Nondestructive Testing, 2nd edn. John Wiley, Hoboken (2005)
2. Tlach, V., Ságová, Z., Kuric, I.: Collaborative assembly task realization using selected type of a human-robot interaction. Proc. Eng. **40**, 541–547 (2019)
3. Kuric, I., Bulej, V., Saga, M., Pokorny, P.: Development of simulation software for mobile robot path planning within multilayer map system based on metric and topological maps. Int. J. Adv. Robot. Syst. **14**(6) (2017)
4. Dodok, T., Čuboňová, N., Kuric, I.: Workshop programming as a part of technological preparation of production. Adv. Sci. Technol. Res. J. 111–116 (2017)
5. Kuric, I.: New methods and trends in product development and planning. In: 1st International Conference on Quality and Innovation in Engineering and Management (QIEM), pp. 453–456, Cluj Napoca (2011)
6. Košinár, M.; Kuric, I.: Monitoring possibilities of CNC machine tools accuracy. In: 1st International Conference on Quality and Innovation in Engineering and Management (QIEM), pp. 115–118, Cluj Napoca (2011)
7. Sage, A., Melsa, J.: System Identification, p. 221. Academic press, New York (1971)
8. Filinov, M.V., Fursov, A.S., Klyuev, V.V.: Approaches to the assessment of the residual resource of technical objects. Control. Diagn. (8), 6–16 (2006)
9. Eykhoff, P.: System Identification: Parameter and State Estimation, p. 555. Wiley-Interscience, New York (1974)
10. Eykhoff, P., (ed.): Trends and Progress in System Identification, 402 p. Oxford, England, Pergamon (1981)
11. Graupe, D.: Identification of System, p. 302. R.E.Krieger Publ. Company, New York (1976)
12. Isermann, R.: Fault-Diagnosis Systems: An Introduction from Fault Detection to Fault Tolerance, p. 475. Springer, New York (2006)
13. Costa, B.S.J.: Fuzzy fault detection and diagnosis. In: Handbook on Computational Intelligence, 2 Volumes. Volume 1: Fuzzy Logic, Systems, Artifical Neural Networks, and Learning Systems, pp. 251–288. World Scientific Publishing Co. Pte. Ltd. (2016)
14. Sapietová, A., Sága, M., Kuric, I., et al.: Application of optimization algorithms for robot systems designing. Int. J. Adv. Robot. Syst. **15**(1) (2018)

Monitoring System of Taekwondo Athletes' Movements: First Insights

Tudor Claudiu Tîrnovan[1], Pedro Cunha[2(✉)], Vítor Carvalho[3], Filomena Soares[2], Camelia Avram[1], and Adina Aștilean[1]

[1] Department of Automation, Faculty of Automation and Computer Science, Technical University of Cluj-Napoca, Cluj-Napoca, Romania
`Claudiu.Tirnovan@student.utcluj.ro`, {`Camelia.Avram, Adina.Astilean`}`@aut.utcluj.ro`
[2] Department Industrial Electronics, Algoritmi Research Centre, Minho University, Braga, Portugal
`id5514@alunos.uminho.pt, fsoares@dei.uminho.pt`
[3] School of Technology & Algoritmi Research Centre, IPCA & University of Minho, Braga, Portugal
`vcarvalho@ipca.pt`

Abstract. Taekwondo practice has been increasing worldwide, however, the methods of analyzing the performance of the athletes have not evolved in the same proportion. The evaluation methods used are performed manually, making the process slow and ineffective. Regarding that, is presented in this paper a system able to get the 3-axis acceleration and gyroscope measurements of the Taekwondo athlete's movements from a MPU-6050 sensor via Wi-Fi communication. The developed WPF Application (Windows application) provides the interface to add the athlete and training session's data, and al-lows to view in real time the session's values through plots. The data obtained is saved into a SQL database. The application will be used by Taekwondo trainers to evaluate the performance of the athlete's movements in real time in training environment.

Keywords: Taekwondo · Martial arts · Sensor · Accelerometer · Gyroscope · Arduino

1 Introduction

In sports the performance of the athlete is a key to obtain better results, whether as a team or individually.

Taekwondo is an ancestral martial art with more than 1500 years. The movements that characterize this martial art are head-height kicks, spinning and jumping kicks performed using the hands and the legs [1]. In Portugal the practice of Taekwondo began with the Grand Master David Chung Sun Yong [2]. Since then the number of athletes federated in Taekwondo practice has been increasing, having more than 4500 athletes nowadays [3]. The popularity of Taekwondo practice has also been increasing worldwide, becoming an Olympic sport in the year 2000 [4].

To evaluate the performance of the athletes is a difficult task for coaches in any sport. Considering this, over time, with the development of technology, systems have been developed to assist coaches in evaluating athletes' performance. With emphasis on sports with high social impact and financial capacity. In Taekwondo despite the technological development there has been no development of technological tools to aid the evaluation of the performance of the athletes in training environment. The athlete's performance evaluation during the training is currently made manually by the coach. He/she usually uses on time visual evaluation or videos of the athletes' training sessions, time consuming tasks that hinders the quick feedback from the coach to change and adapt the training process.

The system presented in this paper is part of a project that aims to develop a real-time prototype to assist the performance of Taekwondo athletes during training sessions. A windows application allows to visualize and store the athletes training session's data by using a mini Wi-Fi board based on ESP-8266 [5] and a triple axis accelerometer and gyroscope (MPU-6050) [6] to collect and send data to the windows application and a SQL database to store the data.

This paper is organized in six sections. In the second section, the state of the art is presented. In the third section, hardware architecture, describes the components used. In the fourth section, software architecture, details the Windows application, database and Arduino code. The fifth section presents the scenarios developed. In the sixth section are enunciated the conclusions and future perspective of work.

2 State of the Art

The evaluation of the performance of athletes in sport has been raising the interest of the scientific community, through the development and adaptation of new technologies that facilitate this process. According to other works, there are already systems that analyze the performance of athletes in sports as swimming, tennis, cycling or martial arts. However, there are few available technological tools for assisting Taekwondo trainings.

Wearable devices are valuable instruments for the improvement of sports performance. However the existing systems are still limited [7].

In [8] the authors agreed that magneto inertial technology is a reliable tool to improve athlete's performance, the training specificity and to prevent injuries. These sensors measurements can be used to estimate temporal, dynamic and kinematic parameters.

Smart sensors and sensor fusion allows to study the impact suffered by the athlete. In [9], the authors demonstrated the use of smart sensors and sensor fusion in biomedical applications and sports areas, promoting a reflection about techniques and applications to process physical variables associated with the human body. The application can be used in areas related to rehabilitation, the athlete's performance development, among others.

In [10], the authors agreed that the impact signals combined with IMU may be a reliable way of scoring, whilst heart rate measurement enables monitoring of the athlete's physical state. The technique used consists in integrating a "non-invasive" sensor system into Taekwondo clothes. The impact is measured using pressure sensors, thin film piezo resistive force and accelerometers. The communication between the sensor and the computer is based on Bluetooth and it was discovered a limitation of bandwidth using this transmission protocol.

A system to evaluate the performance of the Taekwondo athletes in real-time, as an auxiliary of the training sessions, is presented in [11], aiming to provide a faster feedback process, and contributing to a most effective evolution of the athlete performance. The system is composed by a 3D camera Orbbec Astra, a computer and a software developed for the purpose. The 3D camera through the depth sensor allows to obtaining the Cartesian coordinates of several joints of the human body in the real world. The software allows recording and visualizing the athlete's personal data as well as the training performance in real time, through numerical values display and/or graphical profiles of movements' velocity and the coordinates of the athlete's hands and feet.

3 Hardware Architecture

Taekwondo is a sport of contact and agility in which athletes perform fast and wide movements. The required hardware should then have low weigh and small dimensions.

Fig. 1. Developed prototype: a); b).

Figure 1 presents the developed prototype it uses a Wi-Fi board with a battery shield along with an accelerometer and gyroscope, powered by a Lithium-ion Polymer Battery 3.7 V 320 mAh. The total weight of the prototype is 25 g, measuring 60 mm in length and 40 mm in width.

For the processing and transmission of data was chosen the Wemos D1 mini a Wi-Fi board based on ESP-8266 (Fig. 1a), with 11 digital input/output pins and 1 analog input [5]. This board is compatible with Arduino, one of the most used boards for research and development of prototypes. So the development environment (IDE) and code used to program it was the same used with Arduino.

The sensor chosen to obtain the acceleration and gyroscope data was the GY 521 MPU 6050 (Fig. 1b), which is a three axis gyroscope and acceleration module, with standard communication IIC. The acceleration range is between ± 2 and ± 16 g and the gyroscope range is between $+250$ to $+2000$ °/s [6].

The system includes a battery shield [12] which allows to choose between powering the system through a battery or through a USB charger, as well as allowing to charge the battery when the system is being powered by the USB charger (Fig. 1b). A rechargeable Lithium-ion Polymer battery was selected (3.7 V, 320 mAh, hose size is $3 \times 30 \times 40$ mm).

In Fig. 2 presents the Wemos D1 mini Wi-Fi board and the GY 521 MPU 6050 sensor connection diagram. The digitals D1 and D2 pins of Wemos D1 can receive SDA (the line to send and receive data) and SDL (the line that carries the clock signal) signals of the sensor. The SDA and SCL are two simple bidirectional wire bus, part of I2C protocol. In fact, that protocol combines features of UARTs and SPI protocols. Some of the advantages of transmitting data via I2C are the noise resistance, a long track record in the field, the compatibility with a number of processors with integrated I2C ports and the ease to emulate in software by any microcontroller.

Fig. 2. Wemos D1 mini Wi-Fi board and the GY 521 MPU 6050 connection diagram.

The data transfer consists in establishing a communication between one master and one slave, to exchange 8-bit bytes. The others components can use the bus only after the master release the SDA line and SCL value is high. When SCL value is low, no device can change its value, except at start and stop bits [13]. The operating voltage supply of the sensor is 3.3 V, for that, the VCC pin of the sensor was connected to 3.3 V pin of the board.

4 Software Architecture

The components of the software architecture are the Arduino code, the developed Windows application and the SQL database (Fig. 3).

The Windows application creates a user friendly interface which helps the user to insert new data, to get the sensor readings and to work with the measurements. All data are saved into a SQL database.

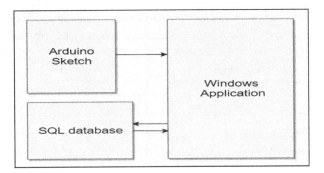

Fig. 3. Software architecture

4.1 Arduino Sketch

The developed code aimed at transmitting the data obtained by the GY 521 MPU 6050 sensor, using the "MPU6050.h" library by ElectronicCats [14] (Fig. 4).

```
sensor.initialize();

sensor.setXAccelOffset(-3733);
sensor.setYAccelOffset(166);
sensor.setZAccelOffset(778);
sensor.setXGyroOffset(82);
sensor.setYGyroOffset(83);
sensor.setZGyroOffset(-30);
```

Fig. 4. Setting offsets

In order to assure correct reading values from the GY 521 MPU 6050 the calibration offsets were defined. For that, it was used the Arduino code that returns calibration offsets of the GY 521 MPU 6050, developed by Luis Rodenas [15]. For the calibration to occur, it is needed to call in the code the `initialize()` function and setting the offsets resulted from calibration sketch (Fig. 4). Then the sensor is ready to read the acceleration and gyroscope values. The acceleration values were converted in g using Eq. 1.

$$xAccel = \frac{xSensorRead}{16384} * 10 \qquad (1)$$

The next step was the establishment of the connection to the Internet via Wi-Fi and to create the client side using the "ESP8266WiFi.h" library [16]. Once the Wi-Fi connection is accomplished, the board is trying to connect to server side (windows application) in order to send the measurements. In the loop function, the application is trying to verify if there is a connection between the client and the server side. If the result is false, the application calls `client.connect(host,port)` in order to connect to server, where host is the IP address where the windows application is running. Once the `client.connected()` returns true, the client will start to send byte by byte strings as JSON objects with the acceleration and gyroscope data (Fig. 5).

```
![{"Xaccel":-10},{"Yaccel":0},{"Zaccel":0},
{"Xgyro":-4228},{"Ygyro":262},{"Zgyro":4420}]=
```

Fig. 5. Data in JSON format

As the server reads the information line by line and the data sent can be received on more than one line, the JSON string will be between to marks: "!" and " =". The marks are used to notify the server where is the start and the end of the JSON object.

4.2 Windows Application

The Windows application was developed using Visual Studio 2019 IDE. It is a WPF application (Windows Presentation Foundation). The interface of the Windows application has three different scenarios: General, Track Data and Graphic (Fig. 6). The user can interact with the system mainly in three ways.

One of them is to add the athlete and training session movement data and save both information, scenario General. Another way is tracking data by selecting the athlete, identify the movement and begin de training session, reading the MPU 6050 sensor data and save it into the database, scenario Track Data. The third way is to plot the data from a chosen athlete training session previously saved in the database, scenario Graphic.

In scenario General, first tab in Fig. 7, the user can create a new athlete and/or movement completing the specific boxes and pressing the Save button. In this tab, the user has the possibility to add athletes and movements into database.

The scenario Track Data, the "Track Data" tab in Fig. 7, allows the user to choose an athlete and a movement, then press the "Start Session" button to the application start

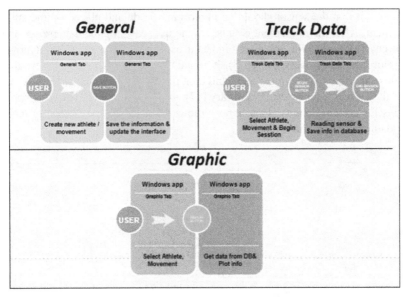

Fig. 6. WPF application usage scenarios

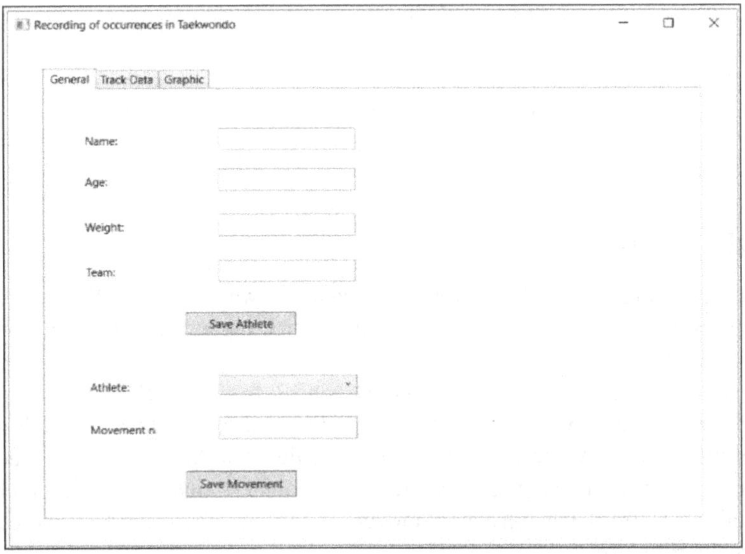

Fig. 7. WPF application interface (General Tab)

to process the data from the MPU 6050 sensor. The application will accept the request connection from the ESP-8266 and will start listen the port to receive data via socket. Once recreated the JSON object, the information followed by the movement id will be inserted into a SQL database using SQL commands. The session will stop when the user press Stop Session button. In that moment the server close the connection to ESP-8266.

It is important that the sensor should be in working mode and placed on the athlete in that moment. After the movement is finished, the user can stop the data processing.

In scenario Graphic, the 3rd tab of Fig. 7, the user can chose an athlete and movement and to plot the data processed (Fig. 8). It contains two dropdown lists where the user can choose an athlete and a movement and then plot the data stored in the database. To plot the data it was used LiveCharts library [17], an open source library, which provides functions for creating charts. This library allows to plot charts lively using incoming data updating at every new value.

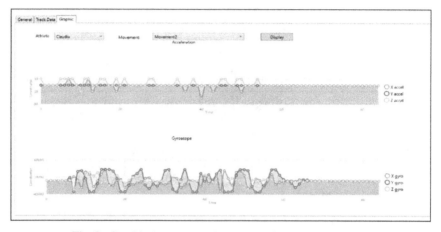

Fig. 8. Graphical movement data output of the training session

4.3 SQL Database

For a correct storage of the data, a relational database was created in Structured Query Language (SQL). The SQL database consists in three tables: Athlete, Movement, and Coordination. The structure can be viewed in Fig. 9.

Athlete table contains five fields which cannot be null: "ID", "Name", "Age", "Weight" and "Team". Coordinates table contains a datetime field, "TimeStamp", 6 fields for coordination values, as doubles and one field for "IdMovement", used to referee to a specific movement. The Movement table contains "IDmov", "MovementName" and "IDath". An athlete can have more movements, a movement can have more athletes and more coordination data.

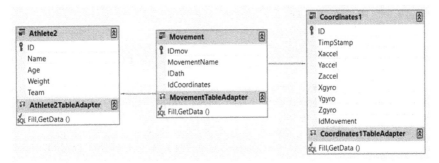

Fig. 9. Database architecture

5 Final Remarks

Taekwondo has been gaining more practitioners in recent years, however the monitoring of the training techniques have been maintained over time. They are mainly performed manually because there are few technological tools developed to assist during training. So, it is of at most importance to develop an economically affordable and non-invasive tool to assist the practice of Taekwondo in real-time [18].

Following this trend, the system presented was developed to assist the coach in the evaluation of the athlete performance, through the reading of the 3-axis accelerometer and gyroscope values of the athlete movement joints. Presenting the collected values in charts at the end of the training session and by consulting previously recorded sessions. The system gives a fast feedback of the performance of the athlete, allowing a rapid correction or adaptation of the training techniques in the sense of more efficient evolution of the athlete's performance.

The prototype module includes the Wemos D1 mini Wi-Fi board, the 3-axis accelerometer and gyroscope MPU 6050, the battery shield and battery, along with software and a computer with Wi-Fi. The presented system intends to be an affordable and easy-to-use technological solution.

The prototype was tested in laboratory environment. Next step is to test the system in training sessions with Taekwondo athletes. This will validate the system in terms of robustness and efficiency for the hardware and data transmission.

In the near future, a larger number of athletes' movement data will be acquired in order to create a dataset of standard movements used in Taekwondo to be used in training deep learning models so that the system can automatically identify and quantify the movements performed by the athlete during the training session.

Acknowledgment. This work has been supported by FCT – Fundação para a Ciência e Tecnologia within the R&D Units Project Scope: UIDB/00319/2020.

References

1. Kazemi, M., et al.: A profile of olympic taekwondo competitors. J. Sports Sci. Med. **5**(CSSI), 114–121 (2006)

2. Taekwondo History. http://www.dojangabilio-costa.com/home. Accessed 30 May 2019
3. Federated Practitioners - Statistics – Tables. http://www.idesporto.pt/conteudo.aspx?id=103. Accessed 30 May 2019
4. About the Federation. http://www.worldtaekwondo.org/about-wt/about-wt/. Accessed 30 May 2019
5. Wemos D1 mini. https://wiki.wemos.cc/products:d1:d1_mini. Accessed 10 Mar 2019
6. Triple Axis Accelerometer & Gyro - MPU-6050. https://components101.com/sensors/mpu 6050-module. Accessed 10 Mar 2019
7. Li, R.T., Kling, S.R., Salata, M.J., Cupp, S.A., Sheehan, J., Voos, J.E.: Wearable performance devices in sports medicine. Sports Health **8**(1), 74–78 (2016). https://doi.org/10.1177/1941738115616917
8. Camomilla, V., Bergamini, E., Fantozzi, S., Vannozzi, G.: Trends supporting the in-field use of wearable inertial sensors for sport performance evaluation: a systematic review. Sensors **18**(3), 873 (2018). https://doi.org/10.3390/s18030873
9. Mendes Jr., J.J.A., Vieira, M.E.M., Pires, M.B., Stevan Jr., S.L.: Sensor fusion and smart sensor in sports and biomedical applications. Sensors **16**(10), 1569 (2016). https://doi.org/10.3390/s16101569
10. Amaro, B., Antunes, J., Cunha, P., Soares, F., Carvalho, V., Carvalho, H.: Monitoring of bioelectrical and biomechanical signals in taekwondo training: first insights. In: Tavares J., Natal Jorge, R. (eds.) VipIMAGE 2017. ECCOMAS 2017. Lecture Notes in Computational Vision and Biomechanics, vol 27. Springer, Cham (2018). https://doi.org/10.1007/978-3-319-68195-5_46
11. Cunha, P., Carvalho, V., Soares, F.: Development of a real-time evaluation system for top taekwondo athletes SPERTA. In: SENSORDEVICES 2018, The Ninth International Conference on Sensor Device Technologies and Applications, pp. 140–145. IARIA, Venice, Italy (2018)
12. Battery Shield. https://wiki.wemos.cc/products:d1_mini_shields:battery_shield. Accessed 13 Jun 2019
13. I2C Manual. https://www.nxp.com/docs/en/application-note/AN10216.pdf. Accessed 13 Jun 2019
14. MPU6050 Arduino Library. https://github.com/ElectronicCats/mpu6050. Accessed 18 Apr 2019
15. Arduino script for MPU-6050 auto-calibration. https://42bots.com/tutorials/arduino-script-for-mpu-6050-auto-calibration/. Accessed 18 Apr 2019
16. ESP8266WiFi library. https://arduino-esp8266.readthedocs.io/en/latest/esp8266wifi/readme.html. Accessed 18 Apr 2019
17. LiveCharts. Link. https://lvcharts.net/App/examples/wpf/start. Accessed 30 Mar 2019
18. Cunha, P., Carvalho, V., Soares, F.: Real-time data movements acquisition of taekwondo athletes: First Insights. In: International Conference on Innovation, Engineering and Entrepreneurship, pp. 251–258. Springer, Cham, June 2018. https://doi.org/10.1007/978-3-319-91334-6_35

Performance Evaluation of the BioBall Device for Wrist Rehabilitation in Adults and Young Adults

Bárbara Silva[1], Ana Rita Amorim[1], Valdemar Leiras[1], Eurico Seabra[1](✉), Luís F. Silva[1], Ana Cristina Braga[2], and Rui Viana[3]

[1] Department of Mechanical Engineering, School of Engineering, University of Minho, Campus of Azurém, Guimarães, Portugal
eseabra@dem.uminho.pt
[2] Algoritmi Research Center, School of Engineering, University of Minho, Campus of Azurém, Guimarães, Portugal
[3] Faculty of Health Sciences, Fernando Pessoa University, Porto, Portugal

Abstract. The BioBall device, developed for wrist function rehabilitation, allows the execution of movements with a greater level of control. A component was added to make it possible to read the range of motion of the patient's wrist. Thus, this project came up intending to evaluate the performance and suitability of the BioBall device for wrist function rehabilitation. To this end, a technical evaluation was performed in order to verify the consistency and accuracy of BioBall. Technical tests consisted of amplitude readings taken by the device, either in automatic or manual mode, "Passive Exercise" and "Physical Exercise" respectively. From the results obtained in this analysis it was concluded that the BioBall device, besides functioning correctly, can collect the angular data with consistency and it is suitable to follow the evolution of a patient's rehabilitation. In order to evaluate whether the BioBall device performs reproducible and repeatable range of motion measurements, a test-retest was performed on healthy subjects (with the space of one week). The maximum amplitude of each wrist movement of the participants was measured. The results of the statistical analysis showed that, although there is some variation in the amplitudes obtained between the test and the retest, the device has good reliability in measuring the range of motion of the wrist.

Keywords: BioBall device · Movement · Rehabilitation · Wrist

1 Introduction

The wrist is a highly complex joint mechanism, associated with a wide range of movements. In addition to providing support and flexibility to manipulate objects, the wrist allows the execution of tasks whose delicacy and strength vary widely.

In a healthy wrist, three groups of movements can be considered: flexion and extension (FLEX-EXT), cubital deviation and radial deviation (CD-RD), pronation and supination (PRO SUP). The limit range for each type of movement can vary depending on people's anthropometric characteristics. The amplitudes that can be expected in a healthy human vary between 60°–85° for flexion, 50°–80° for extension, 30°–45° for cubital deviation, 15°–30° for radial deviation, 80°–90° for pronation and 80°–90° for supination [1, 2]. These amplitude values are usually determined using a goniometer.

Several injuries or pathologies can affect the wrist. These have a high incidence in adulthood and occur mostly in females, as osteoporosis increases the fragility of bones, with greater susceptibility to bone fractures. Regarding the younger class, these disorders are mainly due to sports injuries. The causes of wrist injuries can be diverse, possibly due to isolated traumatic events, such as fractures and ligament injuries, or to repetitive activities (overuse injuries). The treatment depends on the state of the disorder and the anatomical-physiological characteristics of the patients. In most cases, patients are referred to physiotherapy after treatment [3, 4].

The recovery of the wrist function is essentially based on traditional procedures with mechanical and rudimentary devices. Although there are some devices that allow the patient to perform rehabilitation exercises with controlled movements, these do not always cover all types of wrist movements or are more directed to muscle strengthening.

The BioBall device is an adapted and improved version of the device already on the market designated PowerBall®. In its development, the format and vibratory characteristics of the PowerBall® for relaxation and proprioception were used, and tools were added to allow the rehabilitation of the wrist movements. Some changes were also introduced on the device in order to improve its performance and comfort when using it.

Thus, this project came up with the objective of testing the BioBall performance. For this, two types of test were outlined: a technical analysis of the device, in order to verify the device's repeatability and accuracy; a 'test-retest' using a heterogeneous population, in order to assess the reliability of the device in measuring the wrist range of motion (ROM). The results of the two test groups are discussed throughout this article, in view of some functional aspects of the device that may be improved.

2 The BioBall Rehabilitation Device

The BioBall project had as motivation the fact that PowerBall®, despite being a useful device for relaxation and proprioception, is not suitable for the rehabilitation of the wrist, but for a more advanced phase where muscle strengthening exercises are performed. In this way, the new device maintains the relaxation and proprioception system, having been adapted to a controlled vibration system. Additionally, includes a mechanism that allows to reproduce the basic movements of the wrist, both in passive and active mode, coupled with a support that provides stability and portability to the system.

In this way, the developed device can be divided into two systems, the system for relaxation and proprioception and the system for wrist rehabilitation. The final prototype of this device, made with elements off the shelf, is shown in Fig. 1.

2.1 Rehabilitation System for Wrist Movements

The objectives of rehabilitation are to restore complete joint movement and functional capacity in order to provide the patient's return to their daily activities. The interventions used by physiotherapists to achieve these goals can be classified as active, when the patient is obliged to play an active role in his rehabilitation, or passive, when the movements are induced in the patient without having to make any effort [5–7].

The rehabilitation system of the device consists of a system that reproduces the main movements of the wrist, assists its execution (passive rehabilitation) and creates resistance to them (active rehabilitation). For this, it was necessary to create a support capable of accommodating all the components and providing stability and portability to the device. The support, shown in Fig. 1 a) and b), basically consists of three bars, with rotating joints and without restrictions, where a position adjustment system was placed, which restricts the rotation of the bars, and a clamp fixation system for the support. At the top of the support is a box (controller) that contains all the electronic components and their connections, the necessary entrees, buttons and an LCD. This system also includes a stepper motor that will impulse the movement of the ball which, due to the eccentric metal bar with "S" shape attached to the motor shaft, allows to recreate the trajectories of the basic movements of the wrist. This metal bar has a system that allows the coupling and decoupling of the ball when necessary.

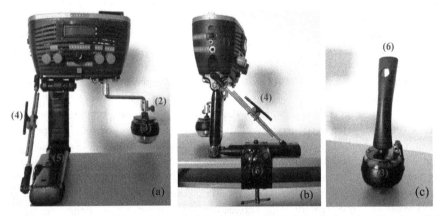

Fig. 1. Final prototype of the BioBall device: (a) front view of the device; (b) side view of the device; (c) ball attached to the handle for relaxation and proprioception exercises; (1) - Controller; (2) - Eccentric bar in "S" shape; (3) - Ball; (4) - Position adjustment bars system; (5) - Support system of the "clamping clamp" type; (6) - Handle.

Additionally, in order to allow the reading of the ROM of the patients' wrist and, consequently, monitor their evolution throughout the rehabilitation programs, a rotatory encoder was added to the device. The encoder is an electromechanical component that converts position to digital electrical signal, allowing the measurement of the rotational movement of a motor. It is usually composed of a disc with markings, a light-emitting component (LED) and a receiver on the opposite side of the disc, as shown in Fig. 2a).

A 1024 pulse per rotation rotary encoder (Yumo E6B2-CWZ3E) was attached to the stepper motor shaft, as shown in Fig. 2b). This encoder outputs gray code which can be interpreted using a microcontroller and find out which direction the shaft is turning and by how much.

Fig. 2. Inclusion of an encoder in the system: (a) schematic representation of the optical rotary encoder operation (Adapted from [8]); (b) connection of the encoder to the stepper motor.

2.2 Relaxation and Proprioception System

Proprioception, associated with sensorimotor control of the joints, is the term that describes the body's ability to feel and perceive itself. Trauma can lead to significant injury of the skin, ligament and muscle and can consequently interrupt the generation and transmission of appropriate proprioceptive stimuli [5]. A proprioceptive element is important in any rehabilitation program because it restores proprioceptive sensitivity and improves the functioning of the joint, reducing the risk of new injuries. Rehabilitation methods that use vibration and tactile stimulation of the skin and muscle-tendon receptors in the wrist are an important basis and have shown positive results in the sensory re-education process [9, 10].

In this way, a relaxation and proprioception system was integrated into the device. This system consists of a ball in which a vibration system is incorporated. It helps muscle relaxation and, at the same time, works as an instrument of controllable proprioception from which it is possible to extract data on the evolution of the patient's recovery. For this purpose, an outrunner brushless motor (Brushless outrunner motor A28L 920 kV) is integrated inside the ball, which allows to obtain controllable speeds. An eccentric was added to the axis of this motor, which creates an imbalance during its operation, causing the vibration of the ball. Due to the high characteristic speeds of the selected DC motor and the coupling of the eccentric weight on its axis, it is possible to reach vibrations in the order of 150 Hz. The ball also allows the assembly of a handle in order to be handled by the therapist (Fig. 1c).

2.3 Operation Modes

The support structure developed allows the execution of different positions by adjusting the bar system. In this way, besides allowing the simulation of all wrist movements, it is also adjustable to different patients, allowing a comfortable use of the device. As shown in Fig. 3a), it is possible to vary, for example, the height of the device up to about 70 mm. The upper bar of the support also has some degrees of freedom, with the possibility of placing the device vertically, horizontally or with some inclination, allowing a proper use by the patient, Fig. 3b).

With the ball coupled to the eccentric bar, which in turn is coupled to the motor shaft, it is possible to reproduce the movements of FLEX-EXT and CD-RD by grabbing

Fig. 3. Device adjustment mechanism that allows comfortable use by different patients: (a) variation in the height of the device; (b) variation in the inclination of the device.

the ball as shown in Fig. 4a) and b), respectively. Due to the nature of the PRO-SUP movements, the ball must be coupled directly to the motor shaft, without the need to use the eccentric bar. In this case, the device must be placed horizontally, and the ball must be grabbed as shown in Fig. 4c), where the angle between the arm and the forearm is 90°.

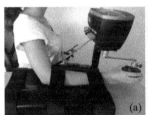  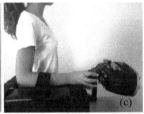

Fig. 4. Positioning of the device, forearm and hand of the patient in the different rehabilitation movements: a) FLEX-EXT movement; b) CD-RD movement; c) PRO-SUP movement.

2.4 Developed Programs

Four rehabilitation programs were developed in order to reproduce wrist movements as much as possible and to observe the evolution of patients. These programs were designated by: "Passive Exercise", "Active Exercise", "Physical Exercise" and "Mixed Exercise".

In the "Passive Exercise" two limit angles are defined for the amplitude of the desired movement (A1 in the anticlockwise rotation and A2 in the clockwise rotation) and a relative rotation speed for this movement, on a scale from 1 to 100. In this exercise, the device moves the ball, while the patient grabs it, helping him to perform the exercises without having to make any effort. In the "Active Exercise" a relative rotation speed is selected on a scale of 1 to 10 and the patient, grabbing the ball, has to create some resistance to the movement generated by the motor, thus exercising the muscles of the hand, wrist and forearm. The "Physical Exercise" program consists only of grabbing the ball and performing the movements freely to observe the maximum amplitude reached by each patient. To reconcile the advantages of rehabilitation exercises with proprioception exercises in the same program, a program designed "Mixed Exercise" was developed. In this program, it is possible to select the limit angles A1 and A2, the relative rotation speed and the vibration frequency of the ball. The operating mode is equal to the "Passive Exercise", but with the addition of vibration.

A program for relaxation and proprioception was also developed, involving vibration. This program allows the selection of the vibration frequency between 0 and 150 Hz. While the patient grabs the ball and moves it as far as possible or to the required point, the physiotherapist can increase or decrease the frequency of the ball vibration. This program can also be used to perform exercises freely, instructed by the physiotherapist, since it is possible to decouple the ball, obtaining greater freedom of handling. It also serves to relieve pain areas and, with the assembly of a handle, the effect for massage is also achieved.

3 Technical Analysis of the Device: Methodology and Tests

In order to calibrate and technically evaluate the suitability of the device, a series of tests were carried out to understand whether it works correctly, assessing its consistency and accuracy. Only two BioBall programs were used: "Physical Exercise" and "Passive Exercise" and geometric angle markings were made with a protractor.

3.1 "Passive Exercise" Test

Using the "Passive Exercise" program, an angular value was attributed to the device (A1 + A2) and a rotation speed, and the ball was expected to reach that angle. The angular values introduced were A1 = 0° and A2 = 90°, which corresponds to an angular range A1 + A2 = 90°. Since for high speeds the device presents more instability, the tests were performed only for a low rotation speed (relative speed of 3/100). For the angular interval inserted, the test was performed 5 times, for 5 min, in order to obtain an average value.

The values obtained (by the reading of the encoder) over this time were recorded in a program specifically developed for this purpose. Through a geometric analysis of the trajectory, it was intended to understand if the device was measuring the angular value correctly, that is, if the angular value of the trajectory obtained geometrically coincided with the angular interval initially introduced in the device. This geometric analysis consisted of using a pencil attached to the eccentric bar, which, due to the movement of the bar, marked the angular value on a millimeter paper.

To evaluate the performance of the encoder with the motor running, it was calculated the average values read by the encoder and it was recorded the maximum and minimum values obtained for the inserted angular interval. The standard deviation (SD) of the read values was determined to verify the dispersion of the measurements around the mean.

3.2 "Physical Exercise" Test

This test consisted of defining angular values that were marked geometrically on a millimeter paper, rotating the eccentric bar from a reference point to the desired angle.

The same angular values were marked in different zones (zones 1, 2, 3 and 4, represented in Fig. 5) of the area where it is possible to move the bar without the ball colliding with the support structure of the device.

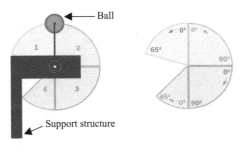

Fig. 5. Schematic representation of the range of motion of the bar. Zones 1, 2, 3 and 4 defined for the tests are represented in yellow, green, orange and blue, respectively. The angles were recorded in the direction shown by the colored arrows.

In this way, it was intended to determine whether the device measures angles with the same precision over its entire angular amplitude. For each angle, the procedure was repeated 5 times. In order to evaluate the encoder reading accuracy and its sensitivity when the motor is not running, the average values and the SD of the readings obtained in the different zones were also calculated.

3.3 Results

Regarding the "Passive Exercise" test, the statistical results are shown in Table 1. It was observed that the geometric measure of the trajectory coincided with the angular interval (A1 + A2) that was introduced in the device. The mean of the angular amplitude readings (86.89°) after 5 repetitions was a bit different from the inserted angular interval, with a maximum of 93.03° and a minimum of 79.81°. The SD was calculated from the read values for the 5 repetitions and it was obtained an average value of 2.99°.

Table 1. Experimental results of the angular trajectory in the "Passive Exercise" program for the relative speed of 3/100

A1 (°)	A2 (°)	Geometric (°)	Mean (°)	Maximum (°)	Minimum (°)	SD (°)
0	90	90	86.89	93.03	79.81	2.99

The results of encoder readings obtained in the test using the "Physical Exercise" program are shown in Table 2. It is possible to observe the average values obtained for each angle in different zones of range of motion of the bar, the global mean of the four zones, as well as the SD of those values.

Table 2. Experimental results obtained in the "Physical Exercise" program for the different zones of movement of the bar (the cells filled with '-' correspond to areas where this amplitude was not performed due to the support structure of the device)

Angle (°)	Zone 1 (°)	Zone 2 (°)	Zone 3 (°)	Zone 4 (°)	Global mean (°)	SD (°)
0	0.39	0.07	0.07	0.28	0.20	0.16
5	5.41	5.27	5.27	5.13	5.27	0.11
10	10.48	10.44	10.12	9.49	10.13	0.46
15	15.05	15.96	16.10	14.34	15.36	0.83
30	30.06	30.20	30.34	29.14	29.94	0.54
45	45.07	45.74	44.86	-	45.22	0.46
60	-	60.22	59.48	-	59.85	0.52
75	-	76.39	75.02	-	75.71	0.97
90	-	91.44	89.51	-	90.47	1.37

3.4 Discussion

The "Passive Exercise" test had as main objective to verify if the angular value of the trajectory obtained geometrically was coinciding with the angular interval initially inserted in the device. The results obtained showed that the geometric value of the amplitude obtained was practically equal to the value of the amplitude inserted in the device, which means that it is possible to perform rehabilitation exercises without any angular deviations that can be harmful to the patient.

On the other hand, the average of the readings obtained by the encoder during this test deviates a little from the value inserted in the device. These differences may be due to an insufficient acquisition rate in the reading of the encoder signal, or, possibly, to mechanical vibrations induced by the movement of the bar. However, these differences are not relevant for the performance of exercises with the "Passive Exercise" program, as reading ROM is not a fundamental aspect in these exercises.

Through the "Physical Exercise" test, the consistency of the reading was verified and maintained throughout the angular range of the device. The average values obtained for each angle showed a certain variation in the different zones of movement of the bar, however, this variation is not significant for the type of exercises to be performed. The results showed that repeatable measurements of an angle can be obtained for the total angular range of the device.

The variation of the readings recorded during the tests can be related to several factors, from the types of components used, as well as to its operational form. In general, this variation in readings is within previously reported ranges based on goniometer measurements. Goniometry results in quite high variations between observers, with SD varying between 3° and 10°, and intra-observer variations between 1° and 3° [11].

The BioBall device has proven to be a useful in controlling a defined angular value and it can obtain reliable measurements based on angular configurations introduced.

4 Reliability Analysis: Test-Retest

A measuring device must be reproducible and there must be repeatability in its measurements. In the case of the BioBall device, these two properties must be tested to see if the measurement component, achieved by incorporating an encoder, presents reliable measurements of the ROM of the patients.

A test-retest in healthy subjects was defined, in order to assess the reliability of the device. This chapter discusses the methodology adopted during the test-retest, all data treatment and statistical analysis carried out, the results obtained and their discussion.

4.1 Test-Retest Methodology

To perform this test, a sample was selected through a questionnaire and a test procedure was outlined for that sample. The results were statistically analyzed using the Microsoft Excel and the Statistical Package For Social Sciences (SPSS) programs.

Sample

The participants were selected through a questionnaire where information such as age, weight, height, dominant hand, occupation, among other aspects were collected. The sample consists of 61 healthy volunteers (27 males and 34 females), adults and young adults, aged between 18 and 35 years. Subjects who recently had injuries in the wrist, hands and fingers, who underwent surgery or suffered neuromuscular dysfunction were excluded from the study. Also, individuals who experienced pain while using the device or who were undergoing physical therapy during the time of the study were left out.

Procedure

The volunteers were initially informed about the procedures, postures and movements that would be required during the data acquisition process. Each selected participant was present in two sessions. The retest time was carried out 1 week later. In each session the individuals performed the FLEX-EXT, CD-RD and PRO-SUP movements of the wrist for the "Physical Exercise" program and the maximum amplitudes of each movement were recorded. All movements were repeated 3 times.

Statistical Analysis

A descriptive analysis was performed on the data with arithmetic mean and standard error of the mean (SEM) of the test and retest. In order to assess the stability of the means between the two sessions, a Student's T-test was performed for paired samples for each of the movements. To complement the results, reliability was determined by analyzing the two-way variance (two-way ANOVA) from which the intraclass correlation coefficient (ICC) was obtained and respective 95% confidence intervals (CI). Values of ICC below 0.5 indicate low reliability, values between 0.5 and 0.75 indicate moderate reliability, values between 0.75 and 0.9 indicate good reliability and values greater than 0.90 indicate excellent reliability [12].

Table 3. Results obtained in the test-retest analysis: mean of the maximum amplitudes obtained in the test and in the retest and SEM; ICC and IC obtained by the analysis between test and retest

Movement		Mean (°)	SEM (°)	ICC (95% CI)
Flexion	Test	104.46	1.89	0.799 (0.663–0.880)
	Retest	102.41	2.30	
Extension	Test	46.76	1.21	0.803 (0.671–0.883)
	Retest	48.68	1.35	
Cubital Deviation	Test	40.68	1.64	0.836 (0.726–0.902)
	Retest	44.06	1.64	
Radial Deviation	Test	31.37	1.15	0.854 (0.755–0.913)
	Retest	35.18	1.32	
Pronation	Test	97.46	2.29	0.756 (0.592–0.854)
	Retest	95.85	2.17	
Supination	Test	99.78	2.06	0.806 (0.675–0.884)
	Retest	103.13	2.48	

4.2 Results

The mean of the maximum amplitude obtained for each type of wrist movement in the test and retest and respective SEM are shown in Table 3.

The T-test showed that there are significant differences between the means of the test and the retest for cubital and radial deviation. There were no statistically significant differences between the means of the test and the retest of other wrist movements.

Still in relation to Table 3, it is also possible to observe the ICC and respective CI for each movement. The analysis of ICC values showed that the correlation between the test and the retest and, consequently, the repeatability of the device, is in average good. The analysis of the CI demonstrated that the correlation can vary from moderate to good, such as the flexion movement 0.799 (0.663–0.880), and from good to excellent, as in the case of the radial deviation 0.854 (0.755–0.913).

4.3 Discussion

Analyzing the average amplitude values obtained for each movement, it was found that there is a significant difference in the average amplitude of the FLEX-EXT movements when compared to the expected values presented in Sect. 1, normally determined by the goniometer. This difference was expected because the amplitude reading of the FLEX-EXT with the BioBall device presents a reference position different from the position using the goniometer. Trying to quantify this difference, calculations were made using AutoCAD software in which the average dimensions of an adult's hand, the dimensions of the ball and the eccentric bar were used. With this approximation, it was found that the difference between the two measurement methods is about 24°, which means that, if approximately 24° are subtracted to the average flexion value, it is obtained an average flexion within the expected values in a healthy individual (between 60° and 85°). On the other hand, if the average extension value is added approximately 24°, an average extension is obtained within the expected values (between 50° and 80°). Thus, it was verified that the device is suitable for measuring the amplitude of FLEX-EXT movements. Differences were also found in the mean values obtained for the movements of radial deviation, pronation and supination in relation to the expected values (15° to 30° and 80° to 90°, respectively). These differences can be justified by the difficulty in isolating the movements and, in the case of PRO-SUP, there may have been a compensation for rotation of the fingers when grabbing the ball.

Regarding the SEM obtained by the device for the different movements, it was found that the highest SEM value obtained was around 2.48°. This error, compared to maximum values of SEM obtained with a universal goniometer in other studies (about 4.65°) [13], is lower, which shows that the device has greater precision in measuring amplitudes than a universal goniometer.

The results of the t-test showed that there were no significant differences between the teste and retest means, except for the cubital and radial deviations. These differences can be justified by the fact that it is difficult to guarantee that the subjects perform movement compensation. Thus, in the case of cubital deviation the subjects could also have performed flexion of the wrist and extension, in the radial deviation.

The ICC values obtained for the reliability of the active ROM measurement were high, indicating a good reliability for all movements. The lower reliability values obtained in the present study can be attributed to the difficulty of handling the device, which made it difficult to identify the anatomical reference points, making it susceptible to the existence of differences in the subject's forearm position in each test. Also, the strength applied and the characteristics of the participants, such as skin flexibility and bone structure, fat and muscles, may have contributed to the variations in measurements.

Despite the difficulties and limitations, the obtained reliability values showed that there is a good reliability of the device when performing the movements. The errors obtained were not significant. Thus, it was possible to verify the repeatability of the device in the measurement of the patients' ROM, which is a great tool to assess the evolution of patients during the rehabilitation sessions.

5 Conclusions

This device can join two major aspects of rehabilitation: restoring joint movement and proprioception, with the aim of restoring the injured limb's functional capacity. It combines several functionalities observed in different equipment and rehabilitation techniques in just one device and has a support that provides portability, stability and suitability for use.

Tests carried out to check the accuracy, repeatability and consistency of the BioBall device showed positive results, allowing to conclude that this device is working according to planned and that it can be a useful device for controlling defined angular amplitude, as well as to measure random amplitudes.

The ICC values obtained showed that BioBall is a reliable device for measuring wrist movements, making it a good tool for assessing the evolution of patients' rehabilitation.

It was possible to validate and improve the idea of a low-cost wrist rehabilitation device that adds several functionalities, standing out from all the existing equipment on the market. However, in order to validate the device for the intended purpose and to conclude about its effectiveness in rehabilitation, a pilot test should be performed using patients with some type of wrist injury to observe if the BioBall contributes to the improvement of the patient's condition.

Acknowledgements. This work has been supported by FCT – Fundação para a Ciência e Tecnologia within the R&D Units Project Scope: UIDP/04077/2020.

References

1. Pina, J.A.E.: Anatomia Humana da Locomoção. 5th edn. Lidel- Edições Técnicas, Lda., Lisboa (2014)
2. Akdoğan, E., Aktana, M.E., Korua, A.T., Arslana, M.S., Atlıhana, M., Kuran, B.: Hybrid impedance control of a robot manipulator for wrist and forearm rehabilitation: performance analysis and clinical results. Mechatronics **49**, 77–91 (2018)
3. Llopis, E., Restrepo, R., Kassarjian, A., Cerezal, L.: Overuse injuries of the wrist. Radiol. Clin. North Am. **57**(5), 957–976 (2019)
4. Cooney, W.P.: The Wrist: Diagnosis and Operative Treatment, 2nd edn. Lippincott Williams & Wilkins (2010)
5. Hagert, E.: Proprioception of the wrist joint: a review of current concepts and possible implications on the rehabilitation of the wrist. J. hand **23**(1), 2–17 (2010)
6. Bruder, A.M., Taylor, N.F., Dodd, K.J., Shields, N.: Physiotherapy intervention practice patterns used in rehabilitation after distal radial fracture. Physiotherapy **99**(3), 233–240 (2013)
7. Karagiannopoulos, C., Michlovitz, S.: Rehabilitation strategies for wrist sensori-motor control impairment: from theory to practice. J. Hand Ther. **22**(2), 154–165 (2016)
8. Logicbus: Encoders. https://www.logicbus.com.mx/info_encoders.php. Accessed 23 Aug 2019
9. Cuppone, A.V., Squeri, V., Semprini, M., Masia, L., Konczak, J.: Robot-assisted proprioceptive training with added vibro-tactile feedback enhances somatosensory and motor performance. PLoS ONE **11**(10), e0164511 (2016)
10. Frank Rauch, M.D.: Vibration therapy. Dev. Med. Child Neurol. **51**(4), 116–168 (2009)
11. Hogeweg, J., Langereis, M., Bernards, A., Faber, J., Helders, P.: Goniometry-variability in the clinical practice of a conventional goniometer in healthy subjects. Eur. J. Phys. Med. Rehabil. **4**(1), 2–7 (1994)
12. Koo, T.K., Li, M.Y.: A guideline of selecting and reporting intraclass correlation coefficients for reliability research. J. Chiropr. Med. **15**(2), 155–163 (2016)
13. Santos, C.M., Ferreira, G., Malacco, P.L., Sabino, G.S., Moraes, G.F.S., Felício, D.C.: Intra and inter examiner reliability and measurement error of goniometer and digital inclinometer use. Locomotor Apparatus Exerc. Sports **18**(1), 38–41 (2012)

Smart Packages Tracking System

Camelia Avram[✉], Mihai Modoranu, Dan Radu, and Adina Aștilean

Technical University of Cluj Napoca, Cluj Napoca, Romania
camelia.avram@aut.utcluj.ro

Abstract. The paper presents the design and implementation of a system for tracking and identifying the packages sent by couriers, using various technologies such as GPS, GSM or RFID. The system structure is composed of a hardware and a software part. The two main objectives of this application are the tracking of the packages sent by courier, along with the generation of cost-efficient routes from the source to the destination. The application server, which handles the provision and management of information, has a layered architecture to increase the scalability of the application. Another feature of the system is the possibility of the customers to choose the most convenient transport variant from those exhibited by the Martins algorithm, based on the information about the storage centers present in the application and the connections between them. This feature makes the services personalized according to the client's needs, giving them value, which will return back through more orders and thanks.

Keywords: Cyber-physical Systems · Internet of Things · Mechatronic devices · Smart tracking

1 Introduction

The field of logistical services records continuous growth every year, having turnover of around billions of euros. Among the causes of this increase are the economic development and progress of electronic commerce. The courier services have become part of the life of each of us, and their importance is given by the necessity of the transport of goods. Whether we are talking about business or personal, there will always be a time when an object will have to be transported between two locations. Despite the importance and growth of the domain, the number of dissatisfied customers also increased.

In case a package is delayed or lost, the only information the companies know is the details of the parcel and the history of the storage centers through which it passed. Considering the number of machines that carry parcels, this granulation of the locations through which a parcel has passed is a real problem when a parcel must be located. The system presented in this paper is intended to be an efficient solution that reduces the number of lost packages and complaints, while increasing the confidence of customers in the companies that offer courier services.

2 State of the Art

The current concept of IoT refers to the use of Internet to interconnect in a system smart devices, automated systems, user applications and services, integrating different technologies [1]. The development of the telecommunications domain has allowed end users to utilize the last new technologies to interconnect IoT devices. Common IoT wireless communication technologies include Bluetooth, WiFi, ZigBee, NFC, LoRaWAN and RFID [2, 3]. By interconnecting smart objects of different kinds, intelligent solutions can be created for a large part of the annoyances encountered in today's society [4].

Some of the applications for which IoT has come up with robust solutions are in a strong connection with the wireless communications, with localization and object identification. Therefore, the Global Positioning System (GPS), the GSM system and services or the radio-frequency identification (RFID) technology, are more and more used in a variety of initiatives aiming to offer solutions to today's limitations [5]. With the advent of 5G technology, more devices will be able to connect to a single network [6], and the security of the applications will become a major challenge in this area. The authors of the research presented in [7], propose a series of technologies that should be used to secure the IoT systems and classifies them according to the field of use.

In articles [8–11] the authors propose several systems in which the mobile phone is seen as a tool for managing logistics information. As basic module, a GPS device, connected by Bluetooth technology to a mobile terminal, is used.

Other important category of applications refers to the items identification. For this purpose, one of the first product identification methods was that based on the bar codes. The main disadvantage of their utilization is the necessity to position the reader next to the code for scanning purposes. This is uncomfortable for large objects. A solution to eliminate the disadvantage could be the usage of radio-frequency identification method. The work presented in [12] describes such a system that would replace the bar codes used for packages sent by courier, a similar solution being proposed in [13]. Also, the authors describe a method of self-registering packages both when they pass through a storage center and when they arrive at their destination. In the case of giant cargo containers, [14], the use of RFID tags is by far superior to the use of bar codes. In [15], a machine tracking system based on RFID tags is proposed, another interesting way of locating vehicles being the using of RFID technology instead of GPS for vehicle tracking, [16]. The purpose of the system designed in [15] is to reduce the times when a traffic accident is detected by using RFID technology, video cameras and a network with an SDN-like architecture.

GPS, GSM and RFID, used separately, can solve many real and interesting problems, but if used together, they will be able to solve more complex problems by offering robust and comprehensive solutions. Complete tracking and identification solutions for supply chains are described in [17], for the determination of the best route to follow in [18, 19] or for monitoring [20–24] and transport security, in [25–33].

All the above proposed and implemented solutions demonstrate the utility of the new integrated concepts and technologies in a variety of applications. Moreover, with the increasing awareness of the Internet of Things concept, the number of ideas implemented in the logistics systems has significantly increased.

3 Package tracking system

The two main objectives of this application are the tracking of the packages sent by courier, along with the generation of cost-efficient routes from the source to the destination. Some of the most important functionalities of the application together with a brief description of the implementation are given below.

The system allows three types of users: the courier (who runs the car), the administrator, and the client. Couriers and customers can create their own accounts by accessing the registration page. One or more cars can be assigned to the courier account (operation done from the administrator account). Once accessing the site, the customer can select one of his packages and view either all the locations or only the last recorded location. The user has a page with all the storage centers of the courier company and another page where he can see a list of all his packages together with the related details, such as description, recipient, RFID code, destination address, and so on. The package record is controlled by the courier, who has three available buttons, with which he can change the system states. The three possible actions are: adding parcel, extracting parcel and confirming action. Each button has a LED attached indicating the intermediate point in which the system is located. If the add button is pressed, the LED corresponding to this button will be lit, and each packet passed through the RFID reader will be marked as added to the machine. After scanning all the packages that are added to the machine, the confirm button have to be pressed again. This action will launch the process of updating the information in the data-base. The same steps are similar in the case of extracting the packages from the car. Concurrently with the actions presented above, at a preset time, one of the scripts running on the development board transmits the location provided by the GPS module to the server.

To meet these requirements the system structure, composed of a hardware and software part, was designed as follows. The software part consists of a server, a database and a web application. The hardware component affixed to the courier machine has the roles of monitoring the location of the machine and keeping track of packages entering or leaving the car. This part of the system consists of several components: Raspberry Pi development board, GPS mode, GSM mode, RFID reader, three but-tons, three LEDs and a buzzer. The scripts running on the Raspberry Pi board are written in Python programming language.

The software part of the system deals with the storage of data provided by the hardware module and the interaction with the users. The components of the software module are the server, the database and the web application.

The server is built using Java programming language along with the Spring framework and deals with the flow and correctness of the data in the application. It exposes a number of REST endpoints, which the other modules can call to receive, send, delete and update information. The server is secured, meeting the specific needs.

Data storage is done in a non-relational database. Unlike a relational database, in which data is represented as tables with rows and columns, in the non-relational database, data is symbolized as JavaScript Object Notation (JSON). Choosing this solution the number of tables required for the application was reduced and some queries were streamlined.

The web application is built using the Angular framework and provides a series of graphical interfaces to interact with the system. Each user's action is reflected as a series of calls made by the web application on the server. In order to access some of the pages of the web application, the user must be authenticated with an account that has one of the administrator or client roles. The pages that can be accessed with-out creating an account in advance are those that allow to verify the location of a package using the RFID identification code, the page for viewing the storage centers and their locations, the page for registration and the one for authentication. If the user try to access another page, other than those listed above, without being authenticated with an account containing the necessary rights, the user will be redirected to the authentication page.

The courier account is stored on the Raspberry Pi card, in the source code. Therefore, every call to the server initiated by the hardware module located on the car must contain the driver authentication data in order to be accepted. All communications are encrypted.

Of all those who can benefit from the web application, the user with administrator rights has the most privileges. The only account with administrator role is created at the first running of the server. Administrator account assignments include viewing the locations of any package, regardless of the sender, adding new packages in the system and choosing a route from those offered by the routing algorithm, viewing company cars along with the packages contained, viewing, adding and shipping centers or the establishment and modification of costs between storage centers.

A graph based representation was used to generate possible routes between source and destination. The algorithm chosen to generate the most efficient routes between the source storage center and the destination is Martins algorithm, implemented on a non-oriented graph. In this graph, the nodes represent the storage centers of the courier company and the arcs correspond to the physical roads between centers. Three costs were considered for each arc, to be minimized: the distance, the travel time and the cost price.

4 System Architecture

The main objective of the package tracking system is to provide a robust solution for tracking packages sent by courier. The system provides a hardware module, which can be mounted on the courier machine, which will continuously transmit the location to the server. Some of the advantages offered by this type of application is the information and the confidence offered to the customers, along with the possibility of quickly identifying and correcting the errors that may appear during the transport. Identifying the packages handled by the driver of the courier vehicle, offers yet another dimension of the data used by the application, which makes the system more robust.

In Fig. 1 the system architecture is presented.

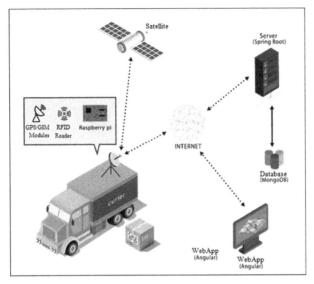

Fig. 1. The high level architecture of the tracking system

4.1 Hardware Architecture

The hardware component of the system, which is located in the courier machine, has the role of monitoring the location of the machine and keeping track of packages entering or leaving the car. The architecture of this component is shown in Fig. 2 and is composed of the Raspberry Pi development board, GPS module, GSM module, RFID reader, buzzer, buttons, LEDs and their connections. As can be seen in the architectural diagram, the development board is the command center that communicates with each component in order to respect the rules after which this system works.

Some of the pins provided by the Raspberry Pi board are general purpose input/output pins (GPIO pins). Some of the remaining pins are reserved for more complex communication interfaces, such as SPI, UART, I2C, and the rest of the pins are either power supply and output 3.3 V or 5 V, or are ground pins.

The pins to which the buttons are connected are set as input pins, and those for led and buzzers are set as output pins. This indicates the low complexity of the control mode, which is achieved either by changing the value of a digital output, or by setting an interrupt on an input pin. The other utilized modules communicate by synchro-nous serial interface (SPI) and asynchronous serial interface (UART). The GPS and the GSM modules are integrated on the same board, the communication being realized through UART. This component also contains two individual antennas for the two modules. Compared to the GPS/GSM module, the RFID card reader co-communicates via SPI, being connected to the Raspberry Pi development board.

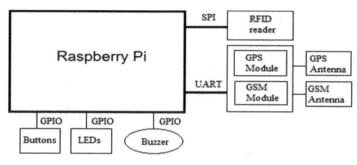

Fig. 2. The hardware architecture

4.2 Software Architecture

The application server, which handles the provision and management of information, has a layered architecture. This type of architecture offers high decoupling between modules, which increases the scalability of the application. In Fig. 3 is represented the corresponding structure of the server, having a presentation level, a logic level and a data level. Each level has access only to its lower level, to which it can request or provide data. The data exchange is made by the references of the lower level elements, contained by the upper level elements.

The presentation level situated at the top of this hierarchy exposes a number of terminal nodes where requests can be sent to provide or write data. The two clients accessing this level are the web client, represented by the web application and the client represented by the hardware module.

Each of the four controllers that make up the first level of the server has separate tasks depending on its field of activity. The UserController handles authentication and account creation requests. It also provides information about the users of the application.

The logic level is it divided into: manager level and services level. For a strong decoupling, the services and controllers should not access more than one element from the next level. As some desired actions of controllers may require information that would involve more services, it is necessary to add the sub-level managers. This sub-level does not affect the proposed architecture at all, but on the contrary, it offers scalability and better visualization of the data flows in the application.

A special service is the one called MartinsPlanner, which deals with generating the most efficient routes between two storage centers. When called, it collects all the data about the storage centers in the database, creating a non-oriented graph from them. Routes are found by applying the Martins algorithm to the graph created previously.

The data level consists of the database, where the information and the Repositories layer are stored. This layer contains four interfaces that are useful for creating Java objects, mapping information from the database to the previously provided data model. Through this mapping method, provided by Spring and MongoDB, queries for the database no longer need to be created, but only to define the methods corresponding to the queries from the four interfaces.

Requests to the server are made using the HTTP protocol, and the responses will be sent as JSON and will be mapped to the classes of models specified in the application.

The data model used to build the system is an important part of the implementation because it describes how information flows through the application. At the same time, this model offers an abstraction of the elements used in the system reality, making possible the use of specific algorithms with the role of solving problems digitally. Without a well-established model, the data flow would either be simplistic, which would be limited to primitive types, or would be an uncontrolled amalgam of data types, which would damage the application if it were to be scaled.

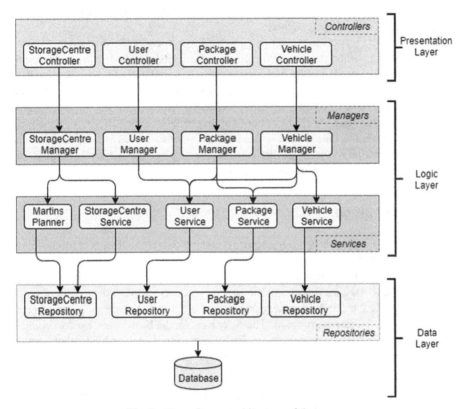

Fig. 3. The software architecture of the server

Each of the four controllers that make up the first level of the server has separate tasks depending on its field of activity. The UserController handles authentication and account creation requests. It also provides information about the users of the application.

The logic level is it divided into: manager level and services level. For a strong decoupling, the services and controllers should not access more than one element from the next level. As some desired actions of controllers may require information that would involve more services, it is necessary to add the sub-level managers. This sub-level does not affect the proposed architecture at all, but on the contrary, it offers scalability and better visualization of the data flows in the application.

A special service is the one called MartinsPlanner, which deals with generating the most efficient routes between two storage centers. When called, it collects all the data about the storage centers in the database, creating a non-oriented graph from them. Routes are found by applying the Martins algorithm to the graph created previously.

The data level consists of the database, where the information and the Repositories layer are stored. This layer contains four interfaces that are useful for creating Java objects, mapping information from the database to the previously provided data model. Through this mapping method, provided by Spring and MongoDB, queries for the database no longer need to be created, but only to define the methods corresponding to the queries from the four interfaces.

Requests to the server are made using the HTTP protocol, and the responses will be sent as JSON and will be mapped to the classes of models specified in the application.

The data model used to build the system is an important part of the implementation because it describes how information flows through the application. At the same time, this model offers an abstraction of the elements used in the system reality, making possible the use of specific algorithms with the role of solving problems digitally. Without a well-established model, the data flow would either be simplistic, which would be limited to primitive types, or would be an uncontrolled amalgam of data types, which would damage the application if it were to be scaled.

5 Implementation

Two scripts written in Python programming language run on the Raspberry Pi development board. The first script interacts with the SIM808 GPS/GSM module, from which it obtains the location that is transmitted to the server, and the other deals with the operating logic of the RFID card reader and the buttons made available to the courier. For the two scripts to work properly it is necessary that the hardware elements to be properly connected to the development board. The forty pins on the board were used to connect the hardware to the Raspberry Pi. Also, both simple input pins and pins that support more complex communication protocols have been used.

GPS/GSM Module

The GPS/GSM module uses the UART protocol and connects to the UART0 interface of the development board. In Table 1 are presented the pins used to interface the SIM808 module.

Table 1. Connecting the GPS/GSM module to Raspberry Pi

Pin GPS/GSM module	Pin Raspberry Pi
GND (ground)	Pin 6 – GND (ground)
RxD	Pin 8 – GPIO14 (UART0_RxD)
TxD	Pin 10 – GPIO15 (UART0_TxD)
VCC (5 V)	Pin 4 – 5 V (Power)

The supported communication protocol is asynnchronous and the two data lines, Tx and Rx, are cross connected to the serial port of the development board. The SIM808 module also supports a number of commands that helps finding and configuring the module states.

The algorithm used to control the SIM808 module:

> **Open** serial port no.#0; portul serial numarul 0;
> **Repeat:** each time interval
>   **While** GPS not connected:
>   /*Module intialization*/
>   //module verification
>   **Send** on serial port the „AT" comand;
>   **Wait** 1 second;
>   // GPS start
>   **Send** on serial port the „AT+CGNSPWR=1" comand;
>   **Wait** 1 second;
>   //Setting the receiving data mode
>   **Send** on serial port the „AT+CGNSSEQ=RMC"comand;
>   **Wait** 1 second;
>   //Get position
>   **Send** on serial port the „AT+CGNSINF"comand;
>   **Read** position;
>   **If** the position has the right form:
>   **Send** location to the server;
> **Until** the script is off;

RFID Reader

The RFID card reader, the buttons, the LEDs and the buzzer are controlled with a Python script. The SPI synchronous serial protocol is used to communicate with the RC522 integrator. The connection to the development board is made at the SPI0 port by

Table 2. PINs used to connect the RFID reader to the Raspberry Pi

Pin RFID reader	Pin Raspberry Pi
SDA	Pin 24 – GPIO8 (SPI0_CE0_N)
SCK	Pin 23 – GPIO11 (SPI0_SCLK)
MOSI	Pin 19 – GPIO10 (SPI0_MOSI)
MISO	Pin 21 – GPIO9 (SPI0_MISO)
IRQ	–
GND	Pin 9 – GND (ground)
RST	Pin 22 – GPIO25
VCC	Pin 1 – 3.3 V (Power)

seven wires out of the 8 provided by RC522. In Table 2 are presented the pins used to interface with the reader.

Also, three buttons were used with which the courier can change the system states. In Table 3 are presented the pins used to control the buttons.

Table 3. PINs used to connect the buttons to the Raspberry Pi

Button	Pins buttons	Pins Raspberry Pi
Button "Add"	Pin 1	Pin 12 – GPIO17
	Pin 2	Pin 30 – GND (ground)
Button "Delete"	Pin 1	Pin 40 – GPIO27
	Pin 2	Pin 30 – GND (ground)
Button "OK"	Pin 1	Pin 36 – GPIO22
	Pin 2	Pin 30 – GND (ground)

6 Simulation and Testing

The software part of the system resulting from the implementation consists of a Spring Boot server, a relational database and a web application made in Angular.

The display of all storage centers (from the list on the left) together with all their con-nections are available by clicking on the Storage Centers button, Fig. 4. These are located by latitude and longitude.

In Fig. 5 is presented the client page where he can visualize all his packages. These packages apear in a list and contains more details like: description, RFID code attached to the package, name of the receiver, adress for delivery and the time when the package were added. On this page the user can track all his deliverys.

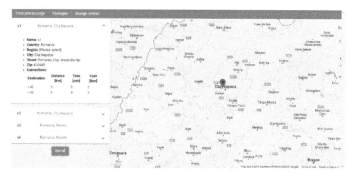

Fig. 4. Visualizing the storage center

Fig. 5. The user page for packages tracking

7 Conclusions

The main objective of this work was the design and implementation of a system for tracking and identifying the packages sent by couriers, using various technologies such as GPS, GSM or RFID. These interdependent technologies have been integrated in a single system that allows the continuous location of the courier machines and the identification of the packages that are introduced or removed from them, a method different from the classic one, used by the courier companies, where the packages are located only when I reach a checkpoint. The system is divided into the hardware area, where the packets are located and identified through GPS, GSM and RFID modules, and the software area consisting of a Spring Boot server, a non-relational database and a web application. The software part of the system deals with the storage, visualization and editing of the data in the system.

Also, another feature of the system is the possibility of the customers to choose the most convenient transport variant from those exhibited by the Mar-tins algorithm, based on the information about the storage centers present in the application and the connections between them. This feature makes the services offered to be personalized according to the client's needs, giving them value, which will return back through more orders and thanks. An experimental version of the system is available and all the imposed functionalities were successfully tested.

References

1. Gkhale, P., Bhat, O., Bhat, S.: Introduction to IOT. Int. Adv. Res. J. Sci. Eng. Technol. **5**(1), 41–44 (2018)
2. Greer, C., Burns, M., Wollman, D., Griffor, E.: Cyber-physical systems and internet of things. NIST Spec. Publ. **1900**(202), 10–13 (2019)
3. Salman, T., Jain, R.: A survey of protocols and standards for internet of things. Adv. Comput. Commun. **1**(1) (2017)
4. El Khaddar, M.A., Boulmalf, M.: Smartphone: the ultimate IoT and IoE device. In: Smartphones from an Applied Research Perspective, pp. 137–162. InTech Open (2017)
5. Rghioui, A., Oumnad, A.: Internet of things: visions, technologies, and areas of application. Autom. Control Intell. Syst. **5**(6), 83–91 (2017)

6. French, A.M., Shim, J.P.: The digital revolution: internet of things, 5G, and beyond. Commun. Assoc. Inf. Syst. **38**, 840–850 (2016)
7. Harbi, Y., Aliouat, Z., Harous, S., Bentaleb, A., Refoufi, A.: A review of security in internet of things. Wirel. Pers. Commun. **108**(1), 325–344 (2019). https://doi.org/10.1007/s11277-019-06405-y
8. Khraisat, Y.S.H., Al-Khateeb, M.A.Z., Abu-Alreesh, Y.K., Ayyash, A.A., Lahlouh, O.S.: GPS navigation and tracking device. Int. J. Interact. Mob. Technol. **5**(4), 39–41 (2011)
9. Zhou, Z., Ma, H., Lin, P.: Research on the key technologies of truck-tracking and navigation system based on GPS/3G. Trans Tech Publications (2011)
10. Verma, P., Bhatia, J.: Design and development of GPS-GSM based tracking system with Google Map based monitoring. Int. J. Comput. Sci. Eng. Appl. (IJCSEA) **3**(3), 33–40 (2013)
11. Vanmore, S.V., Jadhav, N., Nichal, S., Patil, M., Patil, A.: Smart vehicle tracking using GPS. Int. Res. J. Eng. Technol. **4**(3), 66–69 (2017)
12. Kis, P., Alexandru, M.: Real time monitoring and tracking system for an item using the RFID technology. Rev. Air Force Acad. **3**, 71–76 (2015)
13. Awadalla, M.H.A.: Real time shipment tracking system using RFID. Int. J. Comput. Technol. **17**(1), 7163–7180 (2018)
14. Al-Ani, M.S.: Packages tracking using RFID technology. Int. J. Bus. ICT **1**(3–4), 12–20 (2015)
15. Deng, J.: Architecture design of the vehicle tracking system based on RFID. Telkomnika **11**(6), 2997–3004 (2013)
16. Gopi, A., Rajan, D.P.R.L., Rajan, S., Renjith, S.: Accident tracking and visual sharing using RFID and SDN. Int. J. Comput. Eng. Res. Trends **3**(10), 544–549 (2016)
17. He, W., Tan, E.L., Lee, E.W.: A solution for integrated track and trace in supply chain based on RFID & GPS. In: IEEE Conference on Emerging Technologies & Factory Automation (2009)
18. Janssens, G.K., Pangilinan, J.M.: An empirical evaluation of martins' algorithm for the multi-objective shortest path problem (2011)
19. Zinari, S., Canalda, P., Spies, F.: WiFi GPS based combined positioning algorithm. In: IEEE International Conference on Wireless Communications, Networking and Information Security, pp. 684–688 (2010)
20. Prinsloo, J., Malekian, R.: Accurate Vehicle Location System Using RFID, an Internet of Things Approach. MDPI Sens. **16**(825) (2016)
21. Zaenurrohman, A., Marwanto, A., Alifah, S.: Temperature and humidity monitoring on IoT based shipment tracking. J. Telemat. Inf. **6**(1), 27–36 (2018)
22. Bhoyar, A., Varma, R.: GPS based real time vehicle tracking system for kid's safety using RFID and GSM. Int. J. Adv. Res. Ideas Innov. Technol. **4**(1), 290–292 (2018)
23. Kalid, K.S., Rosli, N.: The design of a schoolchildren identification and transportation tracking system. In: International Conference on Research and Innovation in Information Systems (ICRIIS) (2017)
24. Kumar, N.M., Dash, A.: The internet of things: an opportunity for transportation and logistics. In: IEEE International Conference on Inventive Computing and Informatics (ICICI), Coimbatore, India (2017)
25. Barros, C., Leão, C.P., Soares, F., Minas, G., Machado, J.: RePhyS: a multidisciplinary experience in remote physiological systems laboratory. Int. J. Online Eng. **9**(SPL.ISSUE5), 21–24 (2013)
26. Leão, C.P., et al.: Web-assisted laboratory for control education: Remote and virtual environments. In: Uckelmann, D., Scholz-Reiter, B., Rügge, I., Hong, B., Rizzi, A. (eds.) ImViReLL 2012. CCIS, vol. 282, pp. 62–72. Springer, Heidelberg (2012). https://doi.org/10.1007/978-3-642-28816-6_7

27. Gandibleux, X., Beugnies, F., Randriamasy, S.: Martins' algorithm revisited for multi-objective shortest path problems with a MaxMin cost function. French Italian Oper. Res. Soc. **4**, 47–59 (2006)
28. Kunz, G., Machado, J., Perondi, E., Vyatkin, V.: A formal methodology for accomplishing IEC 61850 real-time communication requirements. IEEE Trans. Ind. Electron. **64**(8), 6582–6590 (2017). Art. no. 7878522. https://doi.org/10.1109/TIE.2017.2682042
29. Silva, M., Pereira, F., Soares, F., Leão, C.P., Machado, J., Carvalho, V.: An overview of industrial communication networks. In: Flores, P., Viadero, F. (eds.) New Trends in Mechanism and Machine Science. MMS, vol. 24, pp. 933–940. Springer, Cham (2015). https://doi.org/10.1007/978-3-319-09411-3_97
30. Kunz, G., Perondi, E., Machado, J.: Modeling and simulating the controller behavior of an automated people mover using IEC 61850 communication requirements. In: IEEE International Conference on Industrial Informatics (INDIN), art. no. 6034947, pp. 603–608 (2011). https://doi.org/10.1109/INDIN.2011.6034947
31. Leão, C.P., Soares, F.O., Machado, J.M., Seabra, E., Rodrigues, H.: Design and development of an industrial network laboratory. Int. J. Emerg. Technol. Learn. **6**(SPECIAL ISSUE.2), 21–26 (2011). https://doi.org/10.3991/ijet.v6iS1.1615
32. Campos, J.C., Machado, J., Seabra, E.: Property patterns for the formal verification of automated production systems. IFAC Proc. Vol. (IFAC-PapersOnline) **17**(1 PART 1) (2008). https://doi.org/10.3182/20080706-5-KR-1001.4192
33. Campos, J.C., Machado, J.: Pattern-based analysis of automated production systems. IFAC Proc. Vol. (IFAC-PapersOnline) **13**(PART 1), 972–977 (2009). https://doi.org/10.3182/20090603-3-RU-2001.0425

Automatic Warehouse for Workshop Tools

Marco Ferreira[1], Miguel Rodrigues[1], and Caetano Monteiro[2](✉)

[1] Universidade do Minho, Guimarães, Portugal
[2] Algoritmi, Universidade do Minho, Guimarães, Portugal
cmonteiro@dem.uminho.pt

Abstract. Industry is becoming more and more automated and the deployment of robots for work improvement and production time reduction has been crucial in the evolution of the industry itself. At the storage level, the implementation of robots is practically mandatory. These robots have improved the speed, safety and organization of industrial warehouses. However, at the workshops, frequently, tools are still stored and organized manually on conventional shelves. In this work project it was studied the development of an automated warehouse based on a storage machine for automatic rental of VHS and/or DVD cassettes. It is intended to study the implementation of an automatic storage cabinet adapted for storing small tools and consumables, such as cutting inserts, screws, nuts, washers, etc., in a much faster and more organized way. The main functions of the storage machine were studied in order to solve the challenges placed by the new usage: the dimension limitations imposed by the cabinet shelves and the differences in weight of the new material to be stored and transported. In its original application, the handling robot is expected to carry about 550 g while a box full with screws can easily exceed 2 kg. This paper presents the solution proposed to solve this problem, and presents the envisaged future works.

Keywords: Automatic tool storage · Cartesian robot and automation · Mechatronics

1 Automatic Warehouses

Automatic warehouses are systems developed with the purpose of carrying out the automatic storing process of products, arranging them in proper conditions respecting the product specificities. These systems are based on robotic capability for transporting, placing and picking products, under the control of adequate software for operational control and data management. The main objectives to be achieved by warehouses are to deliver products to the customer quickly and efficiently, to organize the storage under computerized management, to obtain the responsiveness of a service provider, the quality, the efficiency and increase control of industry storage needs [1].

The warehouse studied and presented in this paper has, as main purpose, to serve as an experimental workbench for being used in a Laboratory facility for students and workers from domain of industrial mechatronics, in a similar purpose of [2, 3].

Warehouses present the following advantages and disadvantages:

Advantages:

- Minimum operator intervention;
- Fast operation execution;
- Possible reduction of storage area;
- Good organization and fast ability for products delivering;
- Easy stock control.

Disadvantages:

- High initial investment;
- Need for storage and infrastructures access adaptation.

Figure 1 presents a typical industrial warehouse arrangement.

Fig. 1. Industrial automated warehouse

Taking into account the technological evolution, the growing difficulties in handling and keeping track of an increasing number and of varied products, automated warehouses are becoming unavoidable in most industries for storing raw materials, finished products, consumables and tools. Figure 2 depicts the VIDEOMATIC 1432 rental cassette machine used to study its adaptation to an illustration physical model of an industrial warehouse for storing small tools and consumables, such as cutting inserts, screws, nuts, washers, and others.

Fig. 2. VIDEOMATIC 1432 rental cassette machine.

2 Project Framework

There are workshops whose tool storage consists in large rooms with many shelves and countless different tools [4–6]. The manual organization of a large number of tools and/or consumables easily becomes tedious and ineffective. Misplacement of tools in the shelfs is unfortunately frequent in workshops, resulting in extra time spent in searching activities or in their loss for practical purposes with the consequent need of buying replacement units. It is also difficult to keep under control the level of consumable stocks, leading to overstock or stock exhaustion. Having a film rental machine available in the University laboratory, these considerations motivated the realization of this project to transform the machine into an automatic tool magazine, developing a small warehouse for small tools and consumables like the ones used in a workshop.

2.1 VIDEOMATIC 1432 Film Rental Machine

This model of film rental machine has a warehouse cabinet with three service stations and two central service columns. The articles selected on the screens are delivered directly to the service column next to the user (see Fig. 2). Three users may be simultaneously selecting or renting articles. The two columns can also be used to quickly return the leased items. Figure 3 presents the overall dimensions of the machine.

There are also two service receptacles in the rear of the cabinet for the owner to withdraw or to incorporate new articles.

The storing cabinet shelves were originally designed to store DVD and VHS cassettes that have different dimensions. Table 1 indicates the dimensions of common VHS and DVD boxes. To provide room for these three types of boxes there are two different types of shelfs: type 1 (Fig. 4, left) accepting VHS 2 boxes, type 2 (Fig. 4, right) accepting either 1 VHS 1 box or two DVD boxes. The dimensions of these spaces are indicated in Table 2.

Table 1. Typical VHS and DVD box dimensions.

Type	Width [mm]	Length [mm]	Height [mm]
DVD	135	190	10
VHS 1	122	200	30
VHS 2	115	190	27

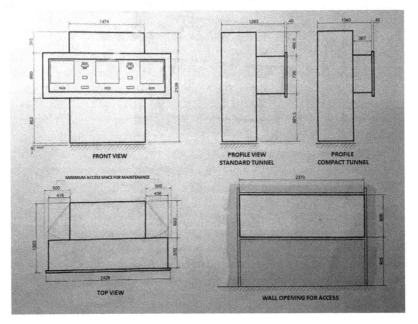

Fig. 3. VIDEOMATIC 1432 warehouse dimensions.

It must be noted that the length of the shelf spaces is shorter than the cassettes' length, and so the difference length will be in cantilever, allowing the pick-up action, explained below.

Table 2. Dimensions of the three types of shelf spaces.

Type	Width [mm]	Length [mm]	Height [mm]
1	145	150	37
2	110	150	37
3	145	150	15

Fig. 4. Shelf types in the VIDEOMATIC cabinet.

The cabinet shelf storage arrangement is shown in Fig. 5, were the transporting robot can be seen. It is a two axis Cartesian robot, driven by stepper motors (not seen in the figure). The handling device consists of a platform attached to the vertical axis, provided with 4 pneumatic actuated claws, working in two pairs that can be closed independently. The handling device can be moved in the Y direction to fetch (or to store) the individual boxes from the shelfs. This action is provided also by pneumatic actuator, which can assume three positions: left, central and right. When the platform is moving loaded, the claws must be in central position, the two pairs closed to tighten the box.

The box is pulled from the shelf by moving the open claw system left, to pick a box from the left wall (or right in the case of a box from the right wall); the left pair of claws closes over the box, the claw system moves to central position, opens again, repeats the movement left, then both claw pairs close and pull the box to the central position. The robot can now deliver the box to the chosen service receptacle, where an inverse sequence of actions permits delivering the box. The process of picking a box from the right wall or from the service receptacle is similar, and then the box will be stored in an empty shelf, also in the same manner.

Fig. 5. Shelf types in the VIDEOMATIC cabinet.

3 Adapting Problems

A VHS or a DVD cassette does not exceed 200 g while a box with the same volume filled with screws for example may exceed 2 kg. Some tests were done to verify the maximum pulling capacity of the handling device. In the original conditions for manipulating the cassettes, the maximum weight the handling device had to pull was about 550 g, much below the 2 kg envisaged. It was then decided to study the introduction of modifications to the handling device in order to be able to perform the new task.

3.1 Mechanical Analysis of the Problem

The handling device consists of 4 pneumatic actuated claws, two actuating at each side of the box. In the first picking action only two opposite claws can reach the box. Figure 6 shows a simple diagram of the forces actuating on the box during the pulling action. The box weight is equilibrated by a normal reaction force ($\vec{N}$ in Fig. 6), and, if a pulling or a pushing action is made, a friction force ($\vec{F}_\mu$) develops. The pulling system must be able to produce a puling force ($\vec{F}_{Robot}$ in Fig. 6) to equilibrate the friction between the box and the shelf base.

The equation system (1) permits the calculation of the force needed to pull the box out of the shelf.

$$\text{xx}: \begin{cases} \vec{F}_{Robot} + \vec{F}_\mu = 0 \\ \vec{N} + \vec{P} = 0 \end{cases} \iff \begin{cases} F_{Robot} = N \cdot \mu \\ N = m \cdot g \end{cases} \Rightarrow F_{Robot} = m \cdot g \cdot \mu \quad (1)$$

The present claw system also uses friction force to pull the boxes. At the beginning of the movement two claws actuating at opposite faces of the box must develop a clamping

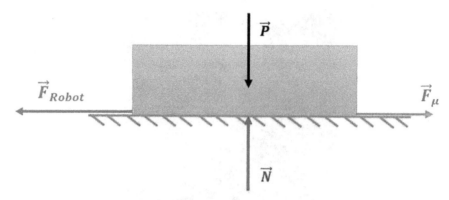

Fig. 6. Free body diagram.

force big enough to produce a friction force equal to $\vec{F}_{Robot}$ and hence to $\vec{F}_\mu$. The claws were designed to pull just a small mass of 200 g, and, considering a friction coefficient between the clamps and the box equal to the friction coefficient between the box and the shelf base, the clamping force must be about 10 N if the friction coefficient is 0.2, an acceptable value for the materials involved. For a 2 kg mass, under the same friction conditions, the clamping force required would be 100 N, an excessive force for the present system.

3.2 Alternative Mechanical Solutions

When analyzing the problem, some mechanical solutions arise. The first approach is to increase the friction coefficient between the clamps and the box, while keeping the friction coefficient between the box and the shelf at the same level, or lowering it.

To increase the friction between the clamps and the box, rubber inserts or stripes can be placed on the sides of the box and on the claps, and then a friction coefficient of about 1 will be attained; to lower the coefficient between the box and shelf base, a low friction material like Teflon can be considered: friction coefficients can be as low as 0.05. In this case the pulling force needed would be around 1 N for the same 2 kg box, and, with rubber inserts in the sides of the boxes, that would be the pulling force needed.

Another approach to solve the problem would be to modify the clamping claws in a manner that the pulling force would no longer depend on the friction coefficient between the claws and the box. That may be obtained by redesigning the claws and the box in order that during pulling (or pushing) the dragging force be transmitted by mechanical interference, with the claws closed. The needed force for pulling the box would then depend on the weight of the box and the friction coefficient between the box and the shelf base.

A steel bar with the largest cross section acceptable for the available shelfs (Fig. 4, left, box type 1), with the maximum cassette (VHS 1, Table 1) length of 200 mm, was considered to establish a target maximum pulling force needed for the handling device to exert. A 7.86 density steel bar and a 0.2 friction coefficient between the bar and the shelf base were considered. Table 3 summarizes the prevailing conditions in the system. A 16.5 N pulling force will be needed for picking up the bar.

Table 3. Maximum weight box: dimensions and pulling force.

Width [mm]	Length [mm]	Height [mm]	Volume [cm^3]	Weight [N]	Pulling Force [N]
145	200	37	1 073	82.7	16.5

4 Adapting Problems

The DVD and VHS boxes can be used, but only for low weight materials. Also the products inside the boxes should not be free to move or the equilibrium during handling will be affected.

Alternative boxes should use light weight, resistant non-expensive materials, like acrylic, PLA or ABS, for the outer case, and foam to accommodate components inside when needed. Figure 7 presents an example of a possible materialization of such a box, designed to store a ratchet wrench set. The volume of this particular box was found to be 912 mm^3. This particular box was designed with grooves on the sides for rubber insertion, enabling larger friction coefficient.

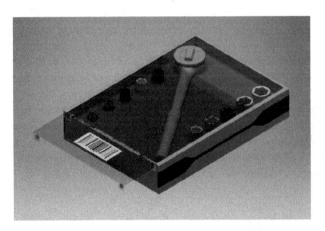

Fig. 7. CAD drawing of a tool box storage.

The box cover slides in a groove to open and close, and is maintained closed by neodymium magnets to prevent possible accidental opening inside the cabinet. Figure 8 shows an exploded view of the box arrangement. The cover also provides a space for the placement of the label with the box code bar, identifying it and its content.

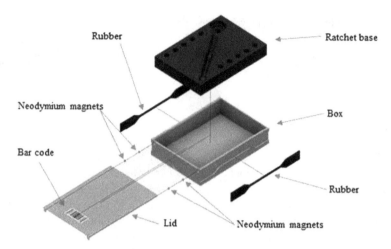

Fig. 8. Exploded view of box arrangement.

5 Conclusions and Future Directions

The work carried out so far allows the conclusion that the machine has enough power to be adapted into an automatic tool and components magazine. It is also possible to conclude that with minor modifications the machine will be able to support bigger weight boxes, allowing the storage of most small components and tools existing in the shop floor. The machine is a whole complex mechatronic system, thus proving support for design study students and industry workers from industrial mechatronics domain. In regard to future work, it must be said that the machine was running using the original devices, electronics and software. The database will be modified, in order to replace managing film rental information for tool storage and supply. A customized man machine interface is also into consideration, to allow new pick and place algorithms, to introduce new features or operations.

References

1. Gangala, C., Modi, M., Manupati, V.K., Varela, M.L.R., Machado, J., Trojanowska, J.: Cycle time reduction in deck roller assembly production unit with value stream mapping analysis. In: Rocha, Á., Correia, A.M., Adeli, H., Reis, L.P., Costanzo, S. (eds.) WorldCIST 2017. AISC, vol. 571, pp. 509–518. Springer, Cham (2017). https://doi.org/10.1007/978-3-319-56541-5_52
2. Leão, C.P., et al.: Web-assisted laboratory for control education: remote and virtual environments. In: Uckelmann, D., Scholz-Reiter, B., Rügge, I., Hong, B., Rizzi, A. (eds.) ImViReLL 2012. CCIS, vol. 282, pp. 62–72. Springer, Heidelberg (2012). https://doi.org/10.1007/978-3-642-28816-6_7
3. Barros, C., Leão, C.P., Soares, F., Minas, G., Machado, J.: RePhyS: a multidisciplinary experience in remote physiological systems laboratory. Int. J. Online Eng. **9**(SPL.ISSUE5), 21–24 (2013). https://doi.org/10.3991/ijoe.v9iS5.2756

4. MECALUX. https://www.logismarket.pt/armazens-automaticos/armazem-automatico-para-caixas-miniload/11226452-11707260-p.html. Accessed 19 June 2019
5. GelsonLuz. https://www.materiais.gelsonluz.com/2018/09/massa-especifica-do-aco.html. Accessed 19 June 2019
6. FESTO. https://www.festo.com/cat/pt-br_br/products_ESG. Accessed 19 June 2019

New Refinement of an Intelligent System Design for Naval Operations

M. Filomena Teodoro[1,2](✉), Mário J. Simões Marques[1], Isabel Nunes[3,4], Gabriel Calhamonas[4], and Marina A. P. Andrade[5]

[1] CINAV, Center of Naval Research, Naval Academy, Portuguese Navy, 2810-001 Almada, Portugal
mteodoro64@gmail.pt
[2] CEMAT, Center for Computational and Stochastic Mathematics, Instituto Superior Técnico, Lisbon University, 1048-001 Lisbon, Portugal
[3] UNIDEMI, Department of Mechanical and Industrial Engineering, Faculty of Sciences and Technology, New Lisbon University, 2829-516 Caparica, Portugal
[4] FCT, Faculty of Sciences and Technology, New Lisbon University, 2829-516 Caparica, Portugal
[5] ISTAR - ISCTE, Instituto Universitário de Lisboa, Lisboa Portugal, 1649-026 Lisbon, Portugal
maria.alves.teodoro@marinha.pt

Abstract. To better optimize the decision chain in a maritime disaster context, allowing a reduction both in time of support and costs, recently, the authors considered the facilities and high qualified staff of Portuguese Navy proposing a variant of the Delphi method, a method where the opinion of each element of a team is considered an important information source to contribute to a decision, implementing a system that prioritize certain teams for perform specific in incidents, taking into account the importance of each team that acts in case of a disaster occurrence.

Continuing the work developed recently and extending the technique used before, in the present manuscript, we propose an adjustment of each expert opinion weight when the Delphi method is applied. The weight of each expert does not not depend on years of expert experience exclusively but also from the type of such experience. Firstly, we have used the hierarchical classification, allowing to identify different patterns for experts with the "'same experience"'. Also discriminant analysis and multidimensional scaling revealed to be adequate techniques for this issue. The experience of each expert was evaluated using the proximity/distance between the individuals in the group of proposed experts and compared with the number of consensus presented.

We hope to have given an improvement to the optimization o the decision support system used in a maritime disaster context of the Portuguese Navy.

Keywords: Decision support system · Expert · Delphi method · Questionnaire · Catastrophe · Multidimensional scaling

1 Introduction

The Decision Support System (DSS) appears to be the result of new technologies, and optimal decisions are urgent in the short term. The DSS with the decision chain can be observed, so it can be optimized to improve the performance of tasks such as medicine, education, engineering, transportation [1–3]. They can also be found in marine environments. In the event of a disaster, DSS for naval operations can improve all stages of intervention: receiving information about the incident, what to do, direction of navigation for the incident, and advice on how to perform the action.

The THEMIS project, promoted by the Portuguese Navy, aims to design and implement a DSS capable of handling disaster relief operations in maritime situations. Some contributions using some new technology like augmented reality or user experience design to were proposed by the authors in [4–7].

We are particularly interested to build and implement a DSS with the ability to prioritize certain teams for specific incidents taking into account the importance of each team that acts in case of emergency, the sequence of tasks that should perform all possible orders to be given. As the author [1] claims, the Delphi method is an exceptionally useful method in which people's opinions are needed to reduce the lack agreement or incomplete knowledge. This method is especially important due the ability to build and organize group communication.

Being the Delphi method a structured group communication process where experts are anonymous and express their opinions on problems with uncertain and incomplete knowledge, accordingly with [8,9], it was used on the facilities and staff of the Portuguese Navy. The goal was to reach expert consensus on specific issues through an iterative process of response and feedback and analysis using simple statistics.

When the Delphi method is applied, the obtained results depend on the weighting of experts experience. In the [8,9], the chosen weights depend on the time of the expert service. In the present work, we evidence that these weights shall take into account not only the time of experience but the similarity of the experts opinion.

In Sects. 2 and 3 we describe details about the proposed methodology and some background. The evidence of the improvement issue need is provided in Sect. 4. Also are done some conclusions and suggestions.

2 Multidimensional Scaling

The Multidimensional Scaling(MDS) had its origin in the models of Psychology used at the end of the 19*th* century. In the scope of these first psychometric works, the authors of [10–12] stood out when they tried to understand the magnitude of stimuli in individuals, without the knowledge of its physical intensity. One of the pioneering works of MDS, according to [13] was due to [14] but it remained little used until it was revived and modernized in fifties by the author of [15,16], stimulated, in large part, by the development of modern digital computers -

which made the complex methodology computationally viable, especially in the multidimensional case, as well as in non-metric cases.

It is common to observe, among researchers from different areas of knowledge, some difficulties in measuring the structure of their objects of study. This occurs mainly when the structure of an object is latent or "hidden", when more than one underlying factor are appropriate for the interpretation of the data, or even when the answer given by the participants may not be determined by a single trace [17]. These characteristics are usually present in an object with a multidimensional structure, that is, when more than one underlying dimension is appropriate to collect the data. In this way, there is a risk that the interpretation will not be reliable because it does not reach all the attributes embedded in the data, since the experimenter's preconceptions may be reflected in the designation of verbal descriptors in the various dimensions, that is, the dimensions may being in the conceptual universe of the experimenter not in the perceptual universe of the observer [18]. In order to achieve an accuracy in the interpretation of the data of objects that present multidimensionality of parameters, the multidimensional scale technique (MDS) comes from a family of data proximity analysis techniques and has proved to be an important mathematical instrument for measurement. Usually, this technique has a quantitative and computational character and can be developed by different algorithms. According with [17], the technique for raising data from multidimensional scales consists of asking the participant to compare several objects (stimuli) in various features (parameters), concomitantly, and not just evaluate an object in a single stroke at a time. The proximity found between the objects will reflect the degree of similarity between them and will be obtained from the distance between the objects of study (stimuli), which will be represented graphically by means of points in a Euclidean space. Through the distances between the points (pairs of stimuli), the stimuli that are judged by the participants as similar will be represented by means of close points, as in a map, while the stimuli judged as not similar will be located with a greater spatial distance. In this way, from the formation of groups of stimuli judged as similar, it will be possible to identify the dimensions, that is, the coordinate axes used to locate a point in Euclidean space. One dimension allows you to view common characteristics shared by a significant amount of data. According to [19], the data produced by MDS are generally represented by two-dimensional diagrams, allowing for a better visualization, where any point can be defined in terms of a set of coordinates and the most distant distance. The next between two points is a straight line. However, many data require three, four or even five dimensions, so they are adjusted in two dimensions. For this type of visualization, there is a risk that the distance between the points (pairs of stimuli) is slightly incorrect and, so that the chosen solution has a good correspondence with the similarities between the objects originally evaluated, [20] proposed a measure of the degree of deviation between the distances and the dissimilarities observed called the stress function.

3 Delphi Method

This section begins with the characterization of the method and follows an approach to the process sound Delphi method was built with reference to the most basic principles, relevance and choice in Delphi research. Finally, some examples from distinct areas of knowledge where the Delphi method was applied.

3.1 Background

The ancient Greeks used the Delphos oracle to make the most important decisions in the country, such as planning a war or finding a new colony. Today, the oracle is not consulted to take a decision, but a panel of previously selected experts. The Delphi method, inspired by the Delphos oracle is generally recommended when mathematical models are not easy to use and personal judgment is appropriate. It was developed and was launched in the early 1960s by researchers Olaf Helmer and Norman Dalkey, under the aim of operational research. [21] point to four important characteristics of the Delphi method.

Anonymity: The anonymity of the answers and the fact that there are no physical meetings, gives the members of the expert group the freedom to express their opinions, while avoiding the influence of psychological factors such as the influence of their abilities. Persuasion, reluctance to leave the assumed position, the majority's dominance over minority views. Think, for example, of a group that includes well-known experts. It is possible for unknown experts to change opinions in accordance with statements from prominent experts to remove the statistical representation of the answers in the study. The authors of [22] say that anonymity ensures more objective answers and better results.

Repeatability: If you hold several rounds, experts can review responses in light of information received from the rest of the expert group.

Controlled feedback: At each iteration, experts receive information about the location of the other experts. This feedback is Irrelevant information is removed, and each member of the expert panel has access to their own answer and the group's answer. Feedback is presented and summarized through simple statistics (average, median, etc.) that summarize the answers of a group or person. The authors of [22,23] claim that anonymously controlled interactions between elements in a group is the determinant fact that distinguish the Delphi method from other personal interaction methods (eg meetings, committees or seminars).

Statistical treatment: The results can be presented in the form of a report or conclusion through the statistical representation of the answers from experts in each round. Simple measurements are usually used as the mean or median. At the end of each round, the statistical representation is elaborated so that the participants can better visualize their position before the group for the next round. In addition, the statistical representation provides the opportunity for the coordination team to follow the process of creating consensus among experts, which is the most important goal of the method.

In [24–26] we can find many examples of Delphi method applications. Some examples that show the importance of the Delphi method: natural sciences [1], pulp production [27], education [25], transportation [3], Social Policy [28], index construction of the car supply chain [29,30]. The author of [31] presents a literature review and several examples of the application of the Delphi method in the social sciences. In [32] is presented a compilation of published studies using the Delphi method. The authors of [21] present a compilation of dissertations and thesis where the Delphi method was used and synthesize and clearly present the diversity and flexibility of the Delphi method.

3.2 Delphi Method - Round I

This section briefly presents the first round of the Delphi method done in [8,9].

In order to continue using the Delphi method, in the previous stage of questionnaire application, a group of prospective experts was asked about age, gender, job grade, training class, type of experience in response to disasters (real versus training) and cumulative boarding time (less than 1 year, 1–3 years, 3–5 years, more than 5 years). Such data collection summarized the 52 tasks presented in Fig. 1.

Identify incidents	Repair electrical power system
Screen survivors / homeless and injured	Repair communications system
Provide 1st aid	Repair of lighting system [point of interest]
Census individuals	Repair mechanical energy production system
Identify location for [point of interest]	Repair power distribution system
Transport equipment to install [point of interest]	Recover basic sanitation
Mount [point of interest]	Create safety perimeter
Carrying severe injuries	Ensuring perimeter safety
Stabilize serious injury	Impose order and safety
Carry minor injured	Carry out order and safeguard of goodies and property
Rescue imprisoned victim	Carry out flight operations to transport material
Rescue victim from altitude	Perform flight operations for medical evacuations
Rescue victim in collapsed structure	Transport material [type] from [local] to [local]
Rescue isolated victim by land	Shift escort
Rescue single victim by air	Convey distribute food for the wounded
Rescue an isolated victim via water	Convey and Distribute food
Stabilize structures	Dead transport to morgue
Clear paths	Support the funeral ceremony
Build support structures for rescue	Status report
Fire fighting	Evacuate equipment to [point of interest]
Fighting floods	Evacuating population to [point of interest]
Carrying out shoring	Diving for minor repairs
Diving drinking water	Diving for rescue
Restoration of water supply	Evacuation of animals
Control of water leaks	Distribution of animal feed
Repair electric pumping system	Burying dead animal

Fig. 1. List of tasks. Source [8].

To understand the relative order (priorities) that each team should perform each particular task, the same group of experts filled a questionnaire ordering the importance rated from 1 (not important) to 6 (very important) about relief work for each existing team that can serve in a humantary disaster operation (see all 52 tasks in Fig. 1 and all possible teams in Fig. 2).

Reconnaissance	Water and Sanitation Technique Brigade
Search and Rescue - SAR	Mechanical Engineering Brigade
Search and Rescue Urban - SAR URB	Technical Brigade of Electricity
Search and Rescue Structures - SAR EST	Supply
Brigade Firefighting Technique	Food
Medical	

Fig. 2. Available teams. Source [8].

A total of 572 questions were grouped on a 6-point Likert scale (from 1 to 6). In the first round of the questionnaire, twelve experts with positions of lieutenant, naval commander and frigate commander (most common) were considered, all men aged 35 to 54 who had been on board for at least five years. Only 25% of all professionals have real/disaster preparedness experience. The remaining 75% of professionals have training experience.

To identify tasks that have reached consensus, you need to determine the Inter-Quartile Range (IQR), the difference between Q1 and Q3 records. IQR represents 50% of the observations closest to the median. An IQR of less than 1 means that at least 50% of the answers obtained fall within 1 point on the Likert scale [25].

To process the questionnaires data and classify each one as consensual/no consensual, was necessary to compute a weighted mean (WM), were the weights are based on the experience of each expert responded to with each team. These weights evaluate the individual time of service. Notice that the opinion of a more experienced expert has a more significant weight in the weighted mean of the importance of a certain team and vice-verse. See Table 1.

Table 1. Weights coefficient per each participant's level of experience.

Level of experience	Weight
1	0.1
2	0.5
3	0.9
4	1.1
5	1.5
6	1.9

For tasks with an IQR greater than 1, it was assumed that experts disagreed on the importance of a given team in performing a specific task. This will be how it will be assessed in the next questionnaire.

If the task IQR is less than or equal to 1, it was analyzed the frequency distribution. You can see that the distribution of the answers to the task is not consistent, but the time is approaching. In this case, a new criterion was proposed.

Not all questions are consistent when there is a non-consecutive significance level of 2 or more, which is 20% or more of the sum of weighted experience levels.

As per the proposed rule, the initial poll between the 572 question disagreed with the 290(50.7%) question, and the IQR of 282 was greater than 1. Only 8 between tasks with an IQR less than or equal to 1 did not meet the requirements of the proposed method. These questions totaling 290 will be considered in the next round.

4 Proposal

In the present section is provide the motivation to improve the weight chart in Table 1 applied in the first round of Delphi method to achieve WM. The starting weight depends on the individual service time (see Table 1). The statements of experienced experts place greater emphasis on the weighted average of the team's significance and vice-versa. However, experts specify in detail certain areas where the weight should be greater and those that were taken into account in the initial phase with a smaller coefficient. As preliminary approach, we have made a hierarchical classification of experts by task/team. The obtained dendograms have shown different profiles depending on the task. Discriminatory analysis confirmed this. Expert profiles are separated for each task/team. We propose to analyze the proximity of experts in a particular team. Some teams were tested and studied proximity measurements between experts using the PROXSCAL algorithm, of the MDS technique that minimizes normalized raw stress.

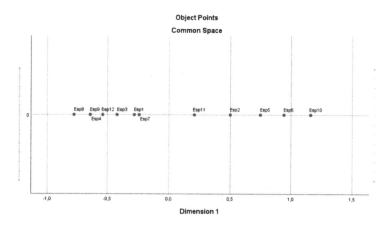

Fig. 3. Global position, dimension 1. The different experts appears in a different order for dimension 1. SAR-REC team.

We can give as example, the proximity of the twelve experts has a distinct pattern when we consider the SAR-URB, SAR-REC or SAR-EST. In a first and naive approach, we consider one dimension for simplicity. We can find in Figs. 3, 4 and 5 that different experts appears in a distinct order for different teams.

When we consider a 2D space (see in Figs. 6, 7 and 8), the results are in accordance with distinct similarities the experts, depending on the selected team. Clearly, the distribution and groups in the plan for the 3 teams is distinct.

For dimension 2, we can find in Table 2 the validation measures of Stress (a badness-of-fit measure for the entire MDS representation.) and others fit measures are displayed . The raw stress is small enough, the Tucker's coefficient of congruence is excellent in all 3 teams.

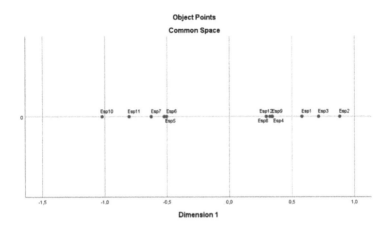

Fig. 4. Global position, dimension 1. The different experts appears in a different order for dimension 1. SAR-URB team.

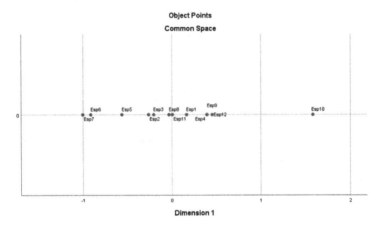

Fig. 5. Global position, dimension 1. The different experts appears in a different order for dimension 1. SAR-ST team.

Also, the measure of distance between experts per distinct teams has distinct patterns. The graphs of distance versus estimated proximity residuals can be found in Figs. 9, 10 and 11. These facts promote the idea that the Table 1 can be reformulated taking into account the similarity of the experts per each team.

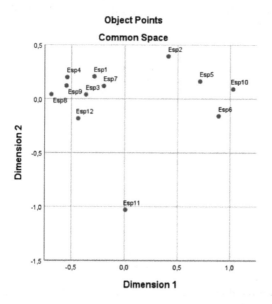

Fig. 6. Global position, dimension 2. The different experts appears in a different order for dimension 2. SAR-REC.team.

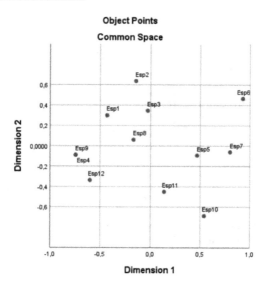

Fig. 7. Global position, dimension 2. The different experts appears in a different order for dimension 2. SAR-URB team.

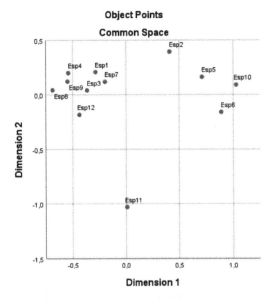

Fig. 8. Global position, dimension 2. The different experts appears in a different order for dimension 2. SAR-ST team.

Table 2. MDS applied to data from SAR-REC team, SAR-URB team and SAR-ST team. Stress and fit measures.

	SAR-REC	SAR-URB	SAR-ST
Normalized raw stress	0.07830	0.01832	0.00783
Stress I	0.08848	0.13535	0.08848
Stress II	0.17822	0.33304	0.17822
Stress III	0.01472	0.3747	0.01472
Dispersion Accounted For (DAR)	0.99217	0.98166	0.99217
Tucker's coeff. of congruence	099608	0.99080	0.99608

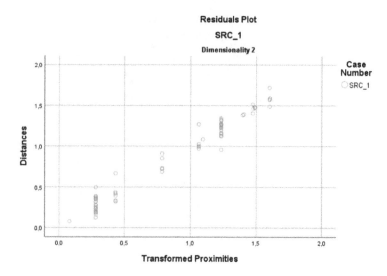

Fig. 9. Residuals of estimates proximities between experts (2D). The different experts appears distinct distances with each other when we consider different teams. SAR-REC team.

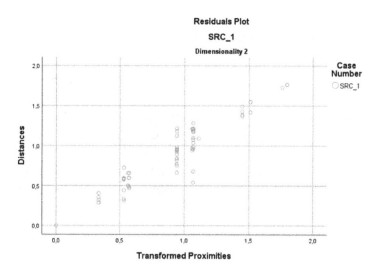

Fig. 10. Residuals of estimates proximities between experts (2D). The different experts appears distinct distances with each other when we consider different teams. SAR-URB team.

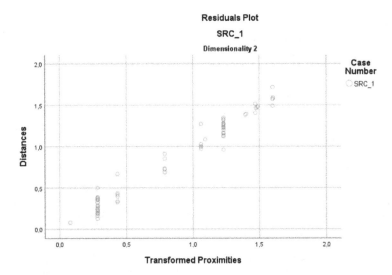

Fig. 11. Residuals of estimates proximities between experts (2D). The different experts appears distinct distances with each other when we consider different teams. SAR-ST team.

Acknowledgements. This work was supported by Portuguese funds through the *Center of Naval Research* (CINAV), Portuguese Naval Academy, Portugal and *The Portuguese Foundation for Science and Technology* (FCT), through the *Center for Computational and Stochastic Mathematics* (CEMAT), University of Lisbon, Portugal, project UID/Multi/04621/2019.

References

1. Powell, C.: The delphi technique: myths and realities. Methodol. Issues Nurs. Res. **41**(4), 376–382 (2003)
2. Reuven, L., Dongchui, H.: Choosing a technological forecasting method. Ind. Manage. **37**(1), 14–22 (1995)
3. Duuvarci, Y., Selvi, O., Gunaydin, O., Gür, G.: Impacts of transportation projects on urban trends in zmir. Teknik Dergi **19**(1), 4293–4318 (2008)
4. Simões-Marques, M., Correia, A., Teodoro, M., Nunes, I.: Building a decision support system to handle teams in disaster situations - a preliminary approach. In: Nunes, I. (ed.) Proceedings of Advances in Human Factors and System Interactions, AHFE 2017 Conference on Human Factors and System Interactions, Advances in Intelligent Systems and Computing, pp. 97–108. Springer, Cham (2018). https://doi.org/10.1007/978-3-319-60366-7_10
5. Nunes, I.L., Lucas, R., Simões-Marques, M., Correia, N.: Augmented reality in support of disaster response. In: Nunes, I. (ed.) Proceedings of Advances in Human Factors and System Interactions, AHFE 2017 Conference on Human Factors and System Interactions, Advances in Intelligent Systems and Computing, pp. 155–167. Springer, Cham (2018). https://doi.org/10.1007/978-3-319-60366-7_15

6. Marques, M., Elvas, F., Nunes, I., Lobo, V., Correia, A.: Augmented reality in the context of naval operations. In: Ahram, T., Karwowski, W., Taiar, R. (eds.) IHSED 2018, Advances in Intelligent Systems and Computing, pp. 307–313. Springer, Cham (2018). https://doi.org/10.1007/978-3-030-02053-8_47
7. Nunes, I., Lucas, R., Simões-Marques, M., Correia, N.: An augmented reality application to support deployed emergency teams. In: Bagnara, S., Tartaglia, R., Albolino, S., Alexander, T., Fujita, Y. (eds.) Proceedings of the 20th Congress of the International Ergonomics Association. IEA 2018. Advances in Intelligent Systems and Computing, pp. 195–204. Springer, Cham (2019). https://doi.org/10.1007/978-3-319-96077-7_21
8. Nunes, I., Calhamonas, G., Simões-Marques, M., Teodoro, M.: Building a decision support system to handle teams in disaster situations - a preliminary approach. In: Madureira, A., Abraham, A., Gandhi, N., Varela, M. (eds.) Hybrid Intelligent Systems. HIS 2018. Advances in Intelligent Systems and Computing, pp. 551–559. Springer, Cham (2019). https://doi.org/10.1007/978-3-030-14347-3_54
9. Simões Marques, J., Teodoro, M.F., Calhamonas, G., Nunes, I.: Applying a variation of delphi method for knowledge elicitation in the context of an intelligent system design. In: Nunes, I. (ed.) Proceedings of Advances in Human Factors and System Interactions, AHFE 2019 Conference on Human Factors and System Interactions, Advances in Intelligent Systems and Computing, pp. 386–398. Springer, Cham (2020). https://doi.org/10.1007/978-3-030-20040-4_35
10. Thurstone, L.: Multiple fator analysis. Psychol. Rev. **38**, 406–427 (1931)
11. Thurstone, L.: Primary Mental Abilities. University of Chicago Press, Chicago (1938)
12. Thurstone, L., Thurstone, T.: Fator Studies of Intelligence. University of Chicago Press, Chicago (1941)
13. Carroll, J., Green, P.: Psychometric methods in marketing research: part ii, multidimensional scaling. J. Mark. Res. **34**(2), 193–204 (1997)
14. Young, G., Householder, A.: Discussion of a set of point in terms of their mutual distances. Psychometrika **3**, 19–22 (1938). https://doi.org/10.1007/BF02287916
15. Torgerson, W.: Multidimensional scaling: i. theory and method. Psychometrika **17**, 401–419 (1952). https://doi.org/10.1007/BF02288916
16. Torgerson, W.: Theory and Methods of Scaling. Wiley, Chicago (1958)
17. Pasquali, L.: Instrumentos Psicológicos: Manual Prático De Elaboração. LabPAM, Brasília (1999)
18. Silva, J., Ribeiro-Filho, N.: Avaliação e Mensuração da Dor: Pesquisa. Teoria e Prática. FUNPEC, Ribeirão Preto (2006)
19. Sturrock, K., Rocha, J.: A multidimensional scaling stress evaluation table. Field Methods **12**, 49–60 (2000)
20. Kruskal, J., Wish, M.: Multidimensional Scaling. SAGE, Thousand Oaks (1978). https://doi.org/10.4135/9781412985130
21. Skulmoski, G., Hartman, F., Krahn, J.: The delphi method for graduate research. J. Inf. Technol. Educ. **6**, 1–24 (2007)
22. Lilja, K., Laakso, K., Palomki, J.: Using the delphi method. In: Kocaoglu, D. (ed.) Technology Management in the Energy Smart World, PICMET 2011, pp. 1–10. Piscataway (2011). https://doi.org/10.1063/1.5138006
23. Linstone, H., Turoff, M.: The Delphi Method-Techniques and Applications. New Jersey Institute of Technology, Newark, NJ (2002)
24. Gunaydin, H.: Impact of Information Technologies on Project Management Functions. Ph.D. dissertation, Chicago University (1999)

25. De Vet, E., Brug, J., De Nooijer, J., Dijkstra, A., De Vries, N.: Determinants of forward stage transitions: a delphi study. Health Educ. Res. **20**(2), 195–205 (2005)
26. Wissema, J.: Trends in technology forecasting. Res. Dev. Manage. **12**(1), 27–36 (1982)
27. Fraga, M.: A economia circular na indústria portuguesa de pasta, papel e cartão. Master thesis, FCT, Universidade Nova de Lisboa (2017)
28. Adler, M., Ziglio, E.: Gazing into the Oracle: The Delphi Method and its Application to Social Policy and Public Health. Kingsley Publishers, London (1996)
29. Azevedo, S., Carvalho, H., Machado, V.: Agile index: automotive supply chain. Eng. Technol. **79**, 784–790 (2011)
30. Azevedo, S., Govindan, K., Carvalho, H., Machado, V.: Ecosilient index to assess the grenness and resilience of the up stream automotive supply chain. J. Cleaner Prod. **56**, 131–146 (2013). https://doi.org/10.1016/j.jclepro.2012.04.011
31. Landeta, J.: Current validity of the delphi method in social sciences. Technol. Forecast. Soc. Chang. **73**(5), 467–482 (2006)
32. Rowe, G., Wright, G.: The delphi technique as a forecasting tool: issues and analysis. Int. J. Forecast. **14**(4), 353–375 (1999)

ICT4Silver: Design Guidelines for the Development of Digital Interfaces for Elderly Users

Nuno Martins[1(✉)], Sónia Ralha[2], and Ricardo Simoes[2]

[1] Polytechnic Institute of Cávado and Ave/ID+, Barcelos, Portugal
nmartins@ipca.pt
[2] Polytechnic Institute of Cávado and Ave, Barcelos, Portugal

Abstract. The Southwest European region (targeted by the SUDOE regional policy), will have by the year 2050 a total of 40% population over 60 years old, which is expected to be the oldest population in Europe. Therefore, the need exists to develop products and services which are well adapted to the needs of the elderly population. This investment, and the associated business, focused on the elderly, is usually termed the Silver Economy.

The ICT4SILVER project, supported by the Intereg SUDOE programme, is a consortium of 9 enterprises within the Southwest Europe territory (namely Portugal, Spain and France) which interfaced with circa 25 SMEs to support the final stage development of technological products aimed at the senior population.

Within the context of this project, a specific work has been conducted on studying user interaction and redesigning the graphical interfaces of three products. This paper describes that analysis and redesign effort as well as a broader proposal of design guiding principles (DGP) for digital interfaces aimed at the elderly. These DGP are divided in three types of guidelines: senior user characteristics, learning processes, and UX (user experience)/UI (user interface) design.

The main goal of this work was the application of the proposed DGP on the digital interfaces of three products in the scope the ICT4Silver project, namely: 'ArkeaOnLife', 'PhysioSensing', and 'Kwido'.

Keywords: ICT4Silver · Senior user · Digital interface design · User experience · User interface

1 Introduction

The work presented in this article is part of project ICT4Silver[1] and its main objective is to contribute to the development of digital interfaces, with a Design vision directed at the senior population.

---

[1] More information about the project at https://www.ict4silver.eu.

In 2050, SUDOE[2] will have the oldest population in Europe, with 40% of the population aged over 60 [1]. Faced with such evidence, there is a growing need to develop products and services adapted to the needs of the elderly, and consequently, the opportunity to invest in Silver Economy[3].

Based on this problem, the ICT4Silver project was created, which proposes responses to market opportunities and social challenges, and enhancing *ICT products (Information and Communication Technologies)* in the SUDOE space.

This project presents a research focused on products with digital interfaces aimed at the senior population, proposing solutions that include factors such as the understanding and acceptance of technology by this type of users. The importance of the designer, in this case, lies in the elimination of barriers between senior users and ICT products.

The study also focuses on the development of a relevant knowledge base, for designers and researchers, on the guiding principles of design for senior users.

The practical phase of the project was divided into two stages of development: the first involving the development of *design guiding principles (DGP)*; and the second, on the application of DGP in three different products.

In the first phase, the DGP are developed following three guidelines: the characteristics of senior users; learning; and the design of digital interfaces.

In the second phase, the redesign of the ICT4Silver project products was implemented, based on the developed DGP. The problems of the three products under study are structured, based on surveys and usability tests, as well as on the use of the DGP carried out. Based on the problems, possible redesign solutions were developed, also based on the established DGP, and on the analysis of competing products in the market.

The methodology adopted in product redesign is based on *Agile Design and Design Thinking* processes, with the objective of combining agile processes, empathy and creativity.

2 Technology, Senior User and Design

The state of the art, presented here, focuses on the themes of design, technology, and senior users, namely, ICT products with digital interfaces; senior users, in relation to their characteristics and limitations; and, finally, the importance of design in the relationship between the senior user and technology.

2.1 Technology's Relationship with the Senior User

Today, ICT is part of everyday life in developed societies and can also be of significant benefit to older people who are typically not comfortable adhering to this type of product, mainly because they experience great difficulties in using it.

These problems should make designers aware of the importance of knowing the characteristics of users, especially the elderly, in order to design inclusive solutions. Designers

---

[2] South-west European Territorial Cooperation Programme.
[3] Economy corresponding to the specific needs of the population above 60 years of age.

should understand that the process of designing products for the elderly requires knowledge of the user's attitudes and motivations, as well as their barriers and advantages of use [2].

There are several factors that influence the individual to adopt an ICT product. The socio-economic level and the geographical location are vital factors that should be considered.

Seniors from higher socioeconomic levels have more experience with technology and seniors in urban geographical areas have easier contact with technology products.

According to Hawthorn [2], the adoption of technology by older people depends on two important factors: the utility of the technology, and the ease with which they use it. The advantage of using the product is equally important for prompting adoption. If elderly users understand the advantages of using a given product, they will realize its usefulness. If the product is easy to use, the motivation to adopt it will also be positive [3].

Learning is another important factor, as is practice for acquiring experience. Technology is constantly evolving, and learning should always be considered and adapted.

Zajicek (2001, cited by [4]) adds that the cause of the poor relationship between the elderly and technology lies in human factors and the necessary faculties for the individual to be able to interact. For example, in a digital interface, the lack of visual acuity hinders interactivity and the understanding of information. The degradation of memory reduces the ability to build conceptual models, and the decrease in motor dexterity hinders the ability to perform tasks. These disabilities and difficulties reduce the elderly's confidence in dealing with technology and facing new challenges.

2.2 The Contribution of Design to the Relationship Between Senior User and Technology

In recent years, themes such as user, interface design, User Interface (UI) and User Experience (UX) have been the subject of in-depth research. Usability has become a very important parameter, leading its creators to the development of user-centered design.

In the development of the design process it is necessary to know the needs, preferences, skills, motivations and limitations related to user-pain. Czaja et al. [5] add to these factors the relevance of including and involving the senior user in the design process early and frequently. In the same line of thought, Newell et al. (cited by [2]) state that obtaining knowledge about the aging process is not sufficient for the work of a designer. It is not enough to know the common limitations and characteristics of a user group; it is necessary to interact with the user.

The aim of UI Design is to make interaction efficient and enjoyable. The user experience (UX Design) is influenced by the interface design choices. Interactions and design are not static, being influenced by the variation of technology systems, but also by the variations of their users. The variability of skills in the age group of the elderly hinders the designer's goal, becoming a challenging job [5]. For this reason, the designer plays a vital role.

2.3 The Senior User

The definition of the concept of a senior user is complex due to the variety of characteristics of this user group [3]. In order to understand the senior user, it is important to know the characteristics common to the ageing process, but also the individual characteristics.

According to Cancela: "The decline of different physiological structures takes place at different rates, with the ageing process and life expectancy differing between individuals" (cited in [6] p. 109). The aging process includes physiological, as well as psychological, social and economic characteristics. In aging, the body gradually changes, becoming more fragile in its ability to adapt to external conditions [6].

3 Design Guiding Principles and Their Application

Possible barriers in the adaptation of technological products by the senior user are created when there is no design research on this type of users - resulting in exclusive and unusable products.

The products under study, *ArkeaOnlife*, *PhysioSensing* and *Kwido*, are examples of products that present barriers to the senior user and that, through this study, new design solutions are presented in order to make them inclusive.

3.1 Guiding Principles for Digital Interface Design

The design guiding principles (DGP) presented in this study, are directed to the senior user and applied to products with digital interfaces. They are based on three fundamental points: characteristics and needs of the senior user; guiding principles of design learning; and guiding principles of design (UX and UI).

The guidelines presented refer to authors such as Czaja et al. [5] and Hawthorn [2], and also to design references for digital interface applications, such as —"Accessibility" Material Design [7]; "Make apps more accessible" [8]; "Introduction - Accessibility - Human Interface Guidelines" [9]; and, for the Web, the W3C Web Accessibility Initiative [10].

Characteristics of the Senior User. The senior user is a complex user who has, within the group of elderly, common but also individual attributes. The common characteristics of the aging process are preceptive, cognitive and psychomotor. The characteristics responsible for the individuality of the elderly are demographic and psychographic.

Demographic characteristics are related to race, ethnicity, education, health status and family structures. Different levels of education, increased age-schooling in the elderly, better health conditions, fewer young people providing care, different types of families and increased cultural and ethnic diversity are some of the factors related to demography that influence the individual characteristics of the elderly.

The psychographic characteristics are personality, motivation, attitude and interests. Individual characteristics are important in the exercise of technology adoption by the senior population. On the other hand, the low motivation of the elderly in adhering to technology, the openness to experience, fear and inhibition, as well as the stability of interests and prejudice associated with the elderly, hinder the adoption of technological products.

The common characteristics of the aging process provide comparative data, which help in the creation of parameters. These characteristics are preceptive, cognitive, and psychomotor, and are related to the usability of the product. The ease of use is affected by the way the individual understands and interacts with the interface.

Perceptive characteristics are related to vision and hearing and influence the ability to perceive and understand elements such as text, images, icons and buttons. Interaction is also affected by changes in visual abilities. Elderly people are slower to identify, understand and read. In hearing, their loss makes it difficult to identify and understand sounds.

In psychomotor characteristics, speed, direction and strength are affected by age, making it difficult to interact with a digital interface. The alteration of the sensitivity to touch and the understanding of the position of the body itself also hinder its usability.

Cognitive characteristics are altered in memory, attention and cognitive performance:

- The damage caused to the memory makes it difficult, in the short term and in the long term, to understand the mental maps of an interface.
- Changes in attention create barriers in the performance of tasks, even of a simple nature.
- Cognitive ability also changes, hindering responsiveness and the creation of new methodologies.

These are some of the characteristics common to the aging process and that should be taken into account by designers when building a product with inclusive digital interfaces.

Guiding Principles of Learning. We understand learning as any and all actions that involve training related to the use of a product, with digital interface. In this study, the following learnings are considered:

- trainings and classes, with teachers specialized in digital interfaces and ICT;
- tutorials (documents, videos, websites, etc.);
- instructions for use of the product;
- instructions and aids inserted in the digital product, as an integral part of the interface, such as: pop-ups, help menus and artificial intelligence.

In the guiding principles of learning, the user barriers in its adoption, the training process, and the characteristics of the support are highlighted.

In the training processes it is important to consider the environment, the trainer and the information support. The trainer should be knowledgeable about the programmatic contents but should also be aware of the characteristics and barriers of the elderly individual. In the tutorials, the contents should include design concerns, such as knowledge and visual heritage, content organization and sizes. Product instructions should also contain usability concerns and user barriers.

Guiding learning principles are important in the adoption of a product. If the senior user can understand the advantages, it can correspond to the motivation to use the product, but if the introduction is not adapted, its adoption will not be met.

Design Guiding Principles (UX and UI) of Digital Interfaces. Issues related to the adjustment of the interface to the elderly user and the design resources that can minimize the obstacles created by the aging process are addressed. Design guidelines are provided with input and output interface element approaches.

The guidelines presented focus on the following design and usability points:

- guiding principles regarding the graphic component (iconography, color, typography, text and messages, composition, graphics and images);
- guiding principles regarding interaction (usability, path and behavior, mental model, selection and activation of items, structure and architecture, interactive elements, transitions of interface pages, interaction aids and customization features);
- other guiding principles (audio, language, feature design and text input).

3.2 Redesign of Products Belonging to the ICT4Silver Project: ArkeaOnLife, PhysioSensing and Kwido

Given the aging of the population and the scarcity of products and services that meet the needs of the elderly, the ICT4Silver project has two main objectives: the first is to find answers to help in matters of health and to maintain the independence of the elderly, with special focus on rural areas under SUDOE; the second is to assist products and companies, in the SUDOE space, for entry into the Silver Economy market [1].

The ICT4Silver products proposed for analysis and development were: ArkeaOnLife, PhysioSensing and Kwido (see Fig. 1). These products were selected by combining their degree of maturity and availability of information for analysis.

ArkeaOnLife is a brand of Crédit Mutuel Arkéa, dedicated to connected living and home, for the independent user, or the user residing in specialized homes. This product is a smartwatch that allows emergency alerts, emergency calls and location of the elderly user.

PhysioSensing is a portable platform for balance and pressure, with visual bio feedback technology, through software.

Kwido is a mobile application that allows the sharing of information between the elderly user, the companies providing care, and associated health teams.

Throughout the ICT4Siver project, products were studied and tested with their users. The tests were carried out by institutions within the ICT4Silver consortium, in order to accelerate the entry of the products into Silver Economy.

In the development phase of this project, different methods and strategies were considered, in alignment with the work processes of *Agile Design* and *Design Thinking*, in order to allow a product to be developed with greater empathy and agility. The subscribed methodology was intended to allow optimum degrees of contact with the senior user.

Three phases of *Leam*[4] development were adopted: "thinking", "make" and "test". In the first phase of the methodology, "thinking", the possible chances of development

---
[4] Work processes based on *Agile Design* methodologies.

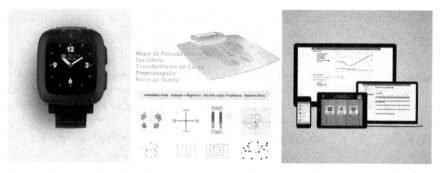

Fig. 1. ICT4Silver products (ArkeaOnLife, PhysioSensing e Kwido).

are structured. In the second stage, "make", a prototype is drawn up on the basis of the hypotheses and, finally, in the final stage, the proposal is made to the users.

In the first phase, the methodologies of *Design Thinking* are introduced: empathy, definition and idealization, in order to introduce a user-centered methodology.

In the "thinking" phase, the users and their characteristics are defined, through studies carried out during the course of the project. Usability tests and surveys are also analysed in order to understand user satisfaction [11, 12]; design; learning; and the technical limitations of the three products under study. The problems presented by users in usability tests are identified, as well as the difficulties examined, based on the guidelines presented during the development of the project [13, 14]. Studies were conducted on competing products (OscarSenior, Senior Phone, Ownfone, Beezie and Omate) in order to collect their attributes and solutions of UX and UI Design.

At the definition stage, brainstorming techniques and diagrams are used in order to gather all the ideas and indicate points of relevance for the project.

Finally, in the conception stage, *Workflow* is performed, with the architecture of the three different supports: clock, platform, software and mobile application.

The "make" phase is composed of the elaboration of a hypothesis and the creation of a *minimally viable product* (MVP). This product has the potential to be tested with its user, gathering the necessary feedback. The *Wireframes* of the three products were accordingly created, enabling the product's interactivity and usability to be understood; the first prototypes were then created, with a focus on determining aspects such as the positioning of the elements, navigation flow, priority of functionalities and contrast; and, finally, the identity of each brand was analyzed and the visual design of each product was prepared [15–17].

At this stage only a few screens are designed, as the aim is to test the MVP with the user and then, in a circular way, deepen and finalize the digital product. At this point of development, prototypes and mockups are produced.

In the "test" stage, the MVP is introduced, validating (or not) the product. If the product is not validated, it returns to the "make" stage. Usability tests and A/B tests can be performed.

Some of the elaborated considerations are presented in detail for each of the products under development. Such considerations are a selection of problems concerning the usability and design of the products, and subsequently, possible solutions.

ArkeaOnLife. In the ArkeaOnLife product there were problems such as the difficulty in reading the numbers, as well as the lack of information on the calendar and battery status. For the elderly, the information should be present in a very visible way to help them in reading and memory. In the elaboration of the new screens for the product, solutions were presented such as the addition of all the clock numbers and the creation of a second screen, with digital numbering and with calendar information and battery status; see Fig. 2. A third battery alert screen was also created.

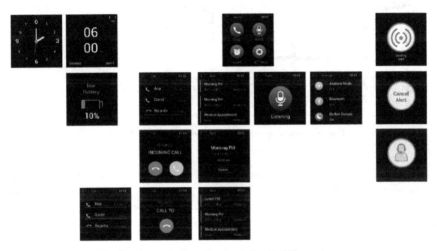

Fig. 2. New solutions for the ArkeaOnLife product.

PhysioSensing. In this product the main problems were the long menus without hierarchy and difficult to use interaction targets. The elderly has greater difficulty in concentration and visual selection, as well as motor difficulties in control, speed and direction. At the interface, the creation of menus with hierarchies and no levels of interaction was suggested. *Touch targets* and *pointer targets* were also changed and adapted to the visual and motor characteristics of the user. To help the working memory a site map and related elements were proposed to navigate back easily. The absence of contrast and the use of short-wave colors such as green and blue, also present levels of difficulty, 0thus a color palette based on the accentuated contrast was developed.

The creation of a fluid reading, using a constant and repetitive architecture (see Fig. 3) is proposed in order to avoid the constant use of cognitive abilities, such as concentration and memory.

Kwido. One of the problems in the Kwido interface was the extensive menu, with undirected and confusing contents. As a result, the different users (seniors, caregivers and health teams) found it difficult to locate the content aimed at them. In this product, an initial menu was thus created in order to organise the contents by type of user. Kwido also presents problems such as the excess of content, making it difficult to concentrate

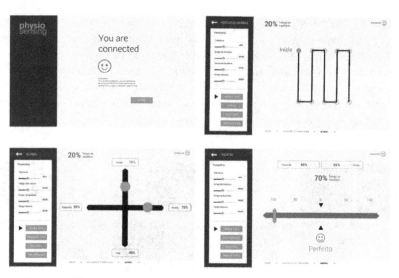

Fig. 3. New solutions for the product PhysioSensing.

and navigate fluidly; and targets of interaction with elements are too small and lacking in contrast and white space, which is reflected in a disorderly and hierarchical application. As a solution, the simplification of information was proposed, incorporating targeted content, buttons with appropriate dimensions and white space with hierarchies and contrast (see Fig. 4).

Fig. 4. Study of the Kwido product (on the right, the proposed new solution).

4 Conclusions

The goal of this project was to define and establish a design basis for supporting designers of digital products that meets the requirements as well as the needs of the elderly population.

There are currently in the market many interesting and promising technological products for the senior population which are actually not suitable for their intended market. This is due to physical and psychological characteristics, as well as some limitations, of the elderly. Considering this population segment is only expected to grow substantially in coming years, clearly there is both a societal need and a business opportunity in the creation and adaptation of technology-based products for the elderly. Through this project, we hope to not only help establish guidelines for the designers of such products, but simultaneously to raise the designers' awareness of inclusivity concerns.

Finally, it should be noted that through this work, we expect to raise awareness and promote a more inclusive technological society, which will benefit not only the users but the developers, since increasing both offer and demand of ICT products will help foster the growth of the Silver Economy.

References

1. ICT4Silver: Project - ICT4silver. http://www.ict4silver.eu. Accessed 06 Aug 2019
2. Hawthorn, D.: Designing Effective Interfaces for Older Users (2006)
3. Nielson, J.: Usability for Senior Citizens. https://www.nngroup.com/articles/usability-for-senior-citizens. Accessed 06 Mar 2020
4. Pereira, L.: Princípios orientadores de design de interfaces para aplicações ITV orientadas para seniores portugueses. FBAUP (2014)
5. Czaja, S.J., Boot, W.R., Charness, N., Rogers, W.A.: Designing for Older Adults. (Boca, R., Ed.) (3rd ed.) CRC Press (2019). https://bookshelf.vitalsource.com/#/books/9781351682244. Accessed 27 Sep 2019
6. Fonseca, I., Amado, P., Costa, L.: Desenho de interfaces para seniores: desafios e oportunidades no projeto SEDUCE. Prisma.com (2014)
7. Material Design. https://material.io/design/usability/accessibility.html. Accessed 06 Oct 2019
8. Developer, A.: Make apps more accessible - Android developer. https://developer.android.com/guide/topics/ui/accessibility/apps, Accessed 03 Aug 2019
9. Developer Apple. https://developer.apple.com/design/human-interface-guidelines/accessibility/overview/introduction. Accessed 23 Aug 2019
10. W3C Web Accessibility Initiative: Older Users and Web Accessibility. https://www.w3.org/WAI/older-users/. Accessed 18 Apr 2020
11. Sousa, F., Martins, N.: Learning experience design:: instructional design applied to the onboarding of digital products. In: Martins, N., Brandão, D. (eds.) Advances in Design and Digital Communication: Proceedings of the 4th International Conference on Design and Digital Communication, Digicom 2020, November 5–7, 2020, Barcelos, Portugal, pp. 45–57. Springer International Publishing, Cham (2021). https://doi.org/10.1007/978-3-030-61671-7_5
12. Martins, N., Dominique-Ferreira, S., Lopes, C.: Design and development of a digital platform for seasonal jobs: improving the hiring process. J. Glob. Scholars Market. Sci. (2021). https://doi.org/10.1080/21639159.2020.1808851

13. Martins, N., Campos, J., Simoes, R.: Activerest: design of a graphical interface for the remote use of continuous and holistic care providers. Adv. Sci. Technol. Eng. Syst. J. **5**(2), 635–645 (2020). https://doi.org/10.25046/aj050279
14. Campos, J., Martins, N., Portela, D., Pereira, S., Simões, R.: Development of a graphical interface for continuous and holistic care providers. In: Proceedings of 14th Iberian Conference on Information Systems and Technologies (CISTI), Coimbra, 19–22 June 2019. IEEE, pp. 139–150 (2019). https://doi.org/10.23919/CISTI.2019.8760994
15. Oliveira, D., Neves, J., Raposo, D., Silva, J.: Research project management in communication design: design methodology applied to communication design research. In: Raposo, D., Neves, J., Silva, J., Castilho, L.C., Dias, R. (eds.) Advances in Design, Music and Arts: 7th Meeting of Research in Music, Arts and Design, EIMAD 2020, May 14–15, 2020, pp. 79–93. Springer International Publishing, Cham (2021). https://doi.org/10.1007/978-3-030-55700-3_6
16. Raposo, D.: Communicating Visually. In: Raposo, D. (Ed.), Communicating Visually: The Graphic Design of the Brand, pp. 20–41 (2018). Newcastle: Cambridge Scholars Publishing
17. Raposo, D., da Silva, F.M., Neves, J., Silva, J.: Clarifying the concept of corporate identity: from a collective vision to cultural interface. In: Rebelo, F., Soares, M. (eds.) AHFE 2017. AISC, vol. 588, pp. 600–609. Springer, Cham (2018). https://doi.org/10.1007/978-3-319-60582-1_60

Dynamic Analysis of a Robot Locomotion for an External Pipe Inspection and Monitoring

Bogdan Grămescu(✉), Adrian Cartal, Ahmed Sachit Hashim, and Constantin Nițu

Department of Mechatronics and Precision Mechanics, University "POLITEHNICA" of Bucharest, Bucharest, Romania
{bogdan.gramescu,adrian.cartal,constantin.nitu}@upb.ro

Abstract. The paper presents the dynamic analysis of a robot locomotion along the external oil and gas pipelines, for the purpose of their inspection and monitoring. The work is focused on both regular crawling motion and obstacles avoiding. During regular motion, the robot resembles to a crank-slider mechanism with equal lengths of the crank and rod, while for stepping over obstacles it becomes an open chain with double actuated leverage. For each of both motions, there were built models in 20-sim environment, for which the simulation results are presented.

Keywords: Mobile robot · External pipe inspection · Dynamic analysis

1 Introduction

The importance of the oil and gas industry for the world economy is obvious. The weather conditions of the areas where these resources are mainly exploited, and the personnel costs are reasons for use of mobile robotics within this industry [1]. Robotics is expected to help for monitoring, inspection and detection of pipelines defects as cracks, corrosion, and leakage, or for maintenance operations (e.g. deposition cleaning). For this purpose, the inspection mechatronic systems require movement along the pipe.

Most of the actual robots developed for this industry do the job inside the pipe [2–4]. There are different types of locomotion like wheels driven, inchworm or screw techniques. The main advantage of these inspection robots is the adjustment to variable pipe diameters and detection/cleaning of the internal depositions. The main disadvantage is the need for pipe dismantling for performing this operation.

An alternative solution is to use the outer surface of the pipe for locomotion and inspection tasks [5]. In this case, the inchworm technique or actuated wheels are regular solutions. The locomotion principle and mechanism should be also adequate to avoid obstacles like flanges, elbows turning left/right or up/down, etc. There is necessary to have at least two grasping mechanism, for locomotion over external surface of the pipes.

2 Developed Robotic Structure as Demonstrator

The robotic structure was developed [6] to travel along the pipelines, carrying sensorial equipment and communication devices, in order to perform the inspection of the pipes and data transmission.

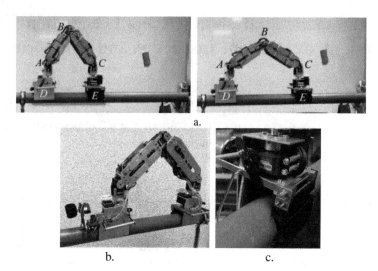

Fig. 1. Locomotion of the inspection robot along a pipe (a. - schematics, b. - small scale demonstrator, c. - clamping mechanism)

The complete system consists of a specific mechanism – electro-mechanical unit, which is the plant of the automatic system, integrated with sensors and actuators, two power supply units of different nominal voltages, with step down circuits, a control unit and a communication unit. This is a small-scale demonstrator (Fig. 1), able to travel along a pipeline of 50 mm diameter, and to avoid obstacles like elbows, flanges, etc. The analysis is also valid for a real scale robot, but the adequate components (mechanical structure, power supply and actuators) must be chosen when it is built.

An important function of the robot structure is its inchworm locomotion (see Fig. 1 a). This specific motion consists of alternating clamping of the worm's body ends, followed by body extensions and contractions alternatively. For clamping the ends of the robot body, if taken into consideration that its pathway is mostly a cylindrical surface (external surface of the pipe), a grasping mechanism seems to be the most appropriate solution. These devices are useful, during operation, mainly for two tasks: to ascertain their adequate position, in order to be aligned to the pipeline axis and to fasten the robot's ends on the pipe.

Using a rotational motor would diminish the size of the clamping mechanism, while letting fixed one of its jaws and the other mobile will lead to a cheaper design (Fig. 1 c). It is obvious that only a theoretical and experimental evaluation of the fixation grade could be the criterion for final decision making [7].

For the inspection of the surface, some scanning devices, camera and sensors must be moved around the pipe for inspecting its circumference. This requires a manipulator, attached to the robot body, which can position the camera and sensors (C&S) at equally distanced points on a circle, which is centered on the pipe axis, while the C&S incidence line is on radial direction of the pipe. The developed manipulator [8] consists of three bars open chain mechanism with active joints, as in Fig. 2.

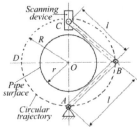

Fig. 2. Scanning manipulator

3 Modeling of the Robot Crawling Along the Pipe Using Actuator XYZ Robot A1–16

According to the inchworm locomotion presented in Fig. 1a, there are four motion stages for obtaining one step: two clamping, one extension and one retraction. The movement is accomplished by positioning of the active joints A, B, C, which are powered by servos XYZ A1–16.

3.1 Modeling of the Actuator XYZ Robot A1–16

The main advantages of this actuator are its low price, high torque and good resolution, in a small volume. The accompanying disadvantages are the lack of the data for building its model and the difficult access to the internal PID controller of the positioning closed loop, without its dedicated controller.

As concerns the poor datasheet of the servo, it was compensated by experimental research [9], which enabled measuring or identification of the servo's physical parameters, necessary for modeling. Generally, a DC motor is characterized by an electrical equation and a mechanical one:

$$u = L\frac{di}{dt} + Ri + k_\omega \cdot \omega \qquad (1)$$

$$T = k_t \cdot i = J_{eq}\frac{d\omega}{dt} + T_{fr} + T_L \qquad (2)$$

where: u – control voltage; i – current, L – winding inductance; R – winding resistance; ω - angular velocity of the output shaft; k_t - motor torque coefficient, multiplied by the gear ratio; k_ω - back EMF constant; T- motor torque, multiplied by the gear ratio; J_{eq} – equivalent inertia of the motor rotor and gears, at the output shaft; T_{fr} – internal friction torque of the servo; T_L – load torque.

The values determined by experimental research and processing of the device datasheet are: the control voltage is upper limited to $u = 12$ V; $R = 5.06$ Ω; $k_t=$ 1.034 N.m/A; $k_\omega=$ 1.034 V.s/rad, $J_{eq}= 5.7 \times 10^{-5}$kg.m^2.

The model of the servo actuator XYZ A1–16 is a simplified approach, where the actual PWM control of the DC motor is considered a current control, and the digital PID controller is replaced by a continuous one. The winding inductance was not measured/determined, because it has no significance, when the motor is current controlled.

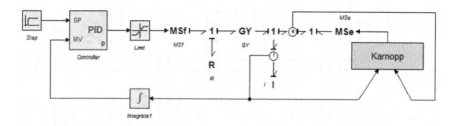

Fig. 3. Simplified model of the servo A1–16 in 20-sim

The model exploits the 20-sim facility of combining the Signal library with the Bond Graph one, which is based on power variables (Fig. 3). The internal friction of the actuator was experimentally determined and a modified Karnopp model was used to fit the measured data, because it is well known for solving the numerical drawback, when the velocity values are very small, almost zero.

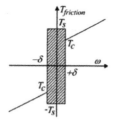

Fig. 4. Karnopp's modified model of friction

According to the Fig. 4, Karnopp introduced a small interval $[-\delta, +\delta]$ of the angular velocity, for which the friction torque is the minimum between $|T_S|$ and $|T_a|$, where T_S is the value of the stiction torque, experimentally determined [9] ($T_S = 0.109$ N.m) and T_a is the actuator torque. This period is called stiction, when the moving part is almost still. The model differs from the original one, by the linear variation with the angular velocity, instead of parallel lines to abscissa, at $+T_C$ and $-T_C$ ordinate values (Coulomb friction). Mathematical model of the modified Karnopp model is expressed by the following equations:

$$T_{slip} = \begin{cases} 0.0891 + 0.0082 \cdot \omega & \text{if } \omega \geq \delta \\ -0.0891 - 0.0082 \cdot \omega & \text{if } \omega \leq -\delta \end{cases} \quad (3)$$

$$T_{stick} = \begin{cases} \min(T_a, T_S) & \text{if } -\delta < \omega < \delta \text{ and } T_a \geq 0 \\ \min(T_a, -T_S) & \text{if } -\delta < \omega < \delta \text{ and } T_a \leq 0 \end{cases} \quad (4)$$

The new model is called modified because the original Karnopp model keeps the slip torque constant at Coulomb value, while the modified one introduces the viscous friction (linear dependence with angular velocity), as it was experimentally found. The equations are implemented into the new developed block Karnopp, as an equation submodel, shown in the model from Fig. 3.

This model does not contain a speed loop, because the Arduino board, used for control, cannot control both speed and position simultaneously, as the proprietary associated microcontroller unit and XYZ_Motor_Editor_Driver_v3 can do. The model from Fig. 3 uses a PID_p controller from the Signal library of 20-sim, which means a parallel form of the PID controller with the following transfer function:

$$C(s) = \frac{u}{\varepsilon} = K\left(1 + \frac{1}{T_i s} + \frac{T_d s}{1 + \frac{T_d}{N} s}\right) \tag{5}$$

where: u – output of the controller (command signal); $\varepsilon = SP\text{-}MV$ – error signal; K – proportional controller gain; T_i - integral time constant $(T_i > 0)$; T_d - derivative time constant $(T_d > 0)$; N – derivative gain limitation.

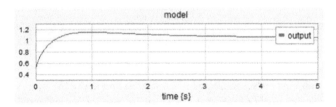

Fig. 5. A1–16 response for a 60° step reference (initial position 30°)

For a target (reference) position of $\pi/3$ (1.047) rad, when the following PID parameters are used: $K = 5$; $N = 100$; $T_i = 1.6$; $T_d = 0.5$, the result is the curve in Fig. 5. It is to be noticed that the target position is reached in about 5 s, after an overshoot of about 12%. The movement starts from an initial position, $\theta = \pi/6$, for a further comparison with the actuation of the robotic chain, during locomotion.

3.2 Modeling of the Inchworm Locomotion, Actuated by the Servos XYZ Robot A1–16

In this stage of the robot development, based on the mechanical design realized in a CAD environment, the values of the inertia and masses of the robot components, involved in the dynamic model, were determined, as well as the distances between the mass centers and the robot joints.

Due to the symmetrical built of the kinematic chain, the bodies attached to the joints A and C can be firstly fixed (A actuated and clamping mechanism E opened – the II stage of the complete inchworm cycle) and, then prismatic one (C actuated and clamping mechanism D opened – the IV stage of the complete inchworm cycle). From the geometric/kinematic point of view, there is only a sign difference between the two stages, but from the dynamic point of view, the difference is essential, because while in the first one the gravity forces help the motion, in the second one they oppose to it. That's why, both the kinematics and the dynamics of the stage IV are analyzed.

During the locomotion process, when one clamp, denoted by A or C, is activated, the arm attached to it performs a rotation around the joint with the same notation. In this case, for finding the moment of inertia of this arm, Steiner's theorem is applied. The arm BC rotates around the joint C (Fig. 6), and the moment of inertia is:

$$J_{BC(C)} = J_{BC(G1)} + m_{BC} \cdot l_{G1}^2 \qquad (6)$$

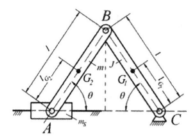

Fig. 6. Crank-slider mechanism in stage IV of locomotion

The distances between the mass centers of the arms BC, AB and the joints C and A, were measured in CAD environment: $l_{G1}= 112$ mm and $l_{G2}= 131$ mm.

The equivalent moment of inertia, which replaces the masses and moments of inertia of the entire kinematic chain, is obtained from the equivalence between the kinetic energy of the virtual substitute element and the sum of kinetic energies of each other components of the kinematic chain:

$$\frac{J_{eq} \cdot \omega^2}{2} = \frac{J_{BC(C)}\omega^2}{2} + \frac{J_{AB(G2)}\omega^2}{2} + \frac{m_{AB}v_{G2}^2}{2} + \frac{m_A v_{Ax}^2}{2} \qquad (7)$$

By taking into consideration the equations of the geometrical model [6], in respect with notations of Fig. 6, it results:

$$J_{eq} = J_{BC(G1)} + J_{AB(G2)} + m_{AB} \cdot l_{G2}^2 + 4[m_{AB} \cdot l(l - l_{G2}) + m_s l^2] \cdot \sin^2\theta \qquad (8)$$

The equivalent load torque can be evaluated from the mechanical work of the resistant forces, which are the gravity forces of the two arms and friction force of the slider:

$$T_{Leq} \cdot \omega = m_{BC}g \cdot v_{G1y} + m_{AB}g \cdot v_{G2y} + \mu m_s g \cdot v_{Ax} \qquad (9)$$

The Eq. (9) can be simplified:

$$T_{Leq} = (m_{AB}l_{G1} + m_{BC}l_{G2})g\cos\theta + 2\mu m_s \cdot gl\sin\theta \qquad (10)$$

The determined masses and moments of inertia involved in the Eqs. (8) and (10) are presented in the Table 1.

Table 1. Inertial parameters of the robotic chain components.

Parameter	m_s	m_{AB}	m_{BC}	$J_{AB(G2)}$	$J_{BC(G1)}$
Unit	kg	kg	kg	kg.m^2	kg.m^2
Value	0.503	0.331	0.376	6.93× 10^{-4}	7.53× 10^{-4}

Using the values from Table 1, the Eqs. (9) and (10) became:

$$J_{eq} = 0.0118 + 0.165 \cdot \sin^2\theta \tag{11}$$

$$T_{Leq} = 0.838 \cdot \cos\theta + 0.493 \cdot \sin\theta \tag{12}$$

where it was considered $\mu = 0.2$. The functions are defined for the values $\theta \in [30°, 60°]$. As in this range $\sin\theta$ is an increasing function, the maximum value of J_{eq} is reached for $\theta = 60°$, $J_{eq} = 0.135$ kg.m^2. For the entire range of θ, the equivalent moment of inertia varies between the limits $[0.053, 0.135]$ kg.m^2. On the other side, the equivalent load torque is practically decreasing on the same interval. As concerns the maximum value of T_{Leq}, the first derivative applied to (12) leads to: $\tan\theta = 0.493/0.838 = 0.588$. That means $\theta = 30.47°$. Practically, T_{Leq} is maximum at the interval left end and minimum at the right end of the range. T_{Leq} varies from 0.972 N.m for $\theta = 30°$, to 0.846 N.m for $\theta = 60°$. This fact can compensate the torque spent for the acceleration of the robot components, if the angular acceleration is kept under a certain limit. If an almost linear variation of J_{eq} and T_{Leq} is assumed, at the ends of the angular interval can be written:

$$J_{eqmin} \cdot \varepsilon + T_{Leqmax} = T_a \tag{13}$$

$$J_{eqmax} \cdot \varepsilon + T_{Leqmin} = T_a \tag{14}$$

where T_a is the actuator torque; ε - angular acceleration. By subtracting (13) from (14), it results a maximum angular acceleration:

$$\varepsilon = \frac{T_{Leqmax} - T_{Leqmin}}{J_{eqmax} - J_{eqmin}} = 1.54\,\text{s}^{-2} \tag{15}$$

The paper presents the first attempt to check the servo capability for the worst conditions, meaning that both J_{eq} and T_{Leq} will have maximum values.

As it can be seen in Fig. 7, a modulated effort source was added in order to introduce the load torque (0.972 N.m), while the equivalent moment of inertia for the crank-slider mechanism is added to the initial value of the inertance, I (equivalent inertia of the motor and gears).

The response is substantially improved just modifying the initial PID gain from $K = 5$ to $K = 10$ (see Fig. 8) but keeping the other parameters ($N = 100$; $T_i = 1.6$; $T_d = 0.5$). The deviation from the reference position is 4×10^{-3} rad, which very satisfying.

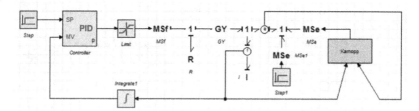

Fig. 7. Model of the loaded servo A1–16 with the maximum torque and inertia

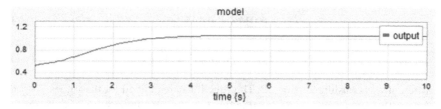

Fig. 8. Position response of the crank-slider mechanism for a 60° step reference (initial position 30°) with the increased gain to K = 10

These simulation results demonstrate the capability of the robotic chain to perform the proposed crawling along the pipeline, when it is actuated by only one servo. An alternative, and surely a reliable solution for performing the inchworm locomotion is the coordinated actuation of two servos, which can provide more input power, but tracking a certain trajectory.

3.3 Modeling of the Obstacles Overtaking

During locomotion, the robot can cross over flanges or fixing brackets, for which it needs to go over larger diameters than that of the pipe or should turn left-right or up-down, when an elbow is encountered. For the dynamic behavior verification, it was selected the movement stage, which is considered the most disadvantageous concerning the inertial load and the torque generated by the gravity forces.

From the strategy for flange/bracket, elbow up and elbow down overpassing, it was selected the transition state shown in Fig. 9, in which one actuator of the robotic structure is extremely loaded, both inertially and by torques of the gravity forces, because it has to raise vertically the arm BC, while the triangle ABC is kept equilateral.

The robot uses the actuator C to raise the equilateral triangle ABC, from initial position $\theta = \pi/3$ to the final position $\theta = \pi/2$ (see Fig. 9 b). For this transition, the equivalent moment of inertia and resistant torque must be evaluated, in order to perform the dynamics simulation. The model used is presented in Fig. 10.

For the state shown in the Fig. 9, b, using Fig. 10, the equivalent moment of inertia is derived from:

$$\frac{J_{eq} \cdot \omega_C^2}{2} = \frac{J_{BC(C)} \cdot \omega_C^2}{2} + \frac{J_{AB(G2)} \cdot \omega_C^2}{2} + \frac{m_{AB} \cdot v_{G2}^2}{2} + \frac{m_{A,CA} \cdot v_A^2}{2} \quad (16)$$

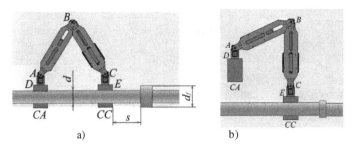

Fig. 9. Transition states verified for flange/bracket overpassing

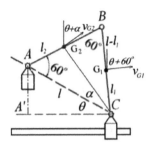

Fig. 10. For inertia and gravity torque calculation

where: ω_C – angular velocity around the joint C; $J_{BC(C)}$ – moment of inertia of the arm BC, around the axis C; $J_{AB(G2)}$ – moment of inertia of the arm AB, around its mass center; m_{AB} – mass of the arm AB; v_{G2} – linear velocity of the arm AB; $m_{A,CA}$ – mass of the clamping mechanism CA. The linear velocity of CA is the time derivative of the segment AA':

$$v_A = \frac{d}{dt}(l \cdot \sin\theta) = l \cdot \cos\theta \cdot \omega_C \qquad (17)$$

For calculating the linear velocity of G_2, the segment CG_2 must be firstly calculated. According to Fig. 10, this is:

$$CG_2^2 = l^2 + l_2^2 - 2l \cdot l_2 \cos 60° = l^2 + l_2^2 - l \cdot l_2 \qquad (18)$$

and
$$v_{G2}^2 = CG_2^2 \cdot \omega_C^2 = \left(l^2 + l_2^2 - l \cdot l_2\right) \cdot \omega_C^2 \qquad (19)$$

By substituting (17), (18), (19) in (16), it results:

$$J_{eq} = J_{BC(c)} + J_{AB(G2)} + m_{AB}\left(l^2 + l_2^2 - l \cdot l_2\right) + m_{A,CA} \cdot l^2 \cos^2\theta \qquad (20)$$

As concerns the equivalent resistant torque, generated by the robot gravity forces, it is calculated from the equivalent power conserving:

$$T_{req} \cdot \omega_C = m_{A,CA} \cdot g \cdot v_A + m_{AB} \cdot g \cdot v_{G2y} + m_{BC} \cdot g \cdot v_{G1y} \qquad (21)$$

Where:
$$v_{G2y} = v_{G2} \cdot \cos(\theta + \alpha) = \sqrt{l^2 + l_2^2 - l \cdot l_2} \cdot \omega_C \cdot \cos(\theta + \alpha) \qquad (22)$$

And
$$v_{G1y} = v_{G1} \cdot \cos(\theta + 60°) = l_1 \cdot \omega_C \cdot \cos(\theta + 60°) \qquad (23)$$

The unknown angle, α, can be derived from triangle ACG_2 (Fig. 10), by applying both sine and cosine theorems:

$$\frac{\sin \alpha}{l_2} = \frac{\sin 60°}{\sqrt{l^2 + l_2^2 - l \cdot l_2}} \qquad (24)$$

$$l_2^2 = l^2 + l^2 + l_2^2 - l \cdot l_2 - 2l\sqrt{l^2 + l_2^2 - l \cdot l_2}\cos \alpha \qquad (25)$$

From (24) and (25), it results:

$$\sin \alpha = \frac{l_2 \sqrt{3}}{2\sqrt{l^2 + l_2^2 - l \cdot l_2}} \qquad (26)$$

And
$$\cos \alpha = \frac{2l - l_2}{2\sqrt{l^2 + l_2^2 - l \cdot l_2}}. \qquad (27)$$

Firstly, there are introduced (17), (22) and (23) in (21) and it is obtained:

$$T_{req} = m_{A,CA} \cdot gl \cdot \cos\theta + m_{AB} \cdot g\sqrt{l^2 + l_2^2 - l \cdot l_2} \cdot \cos(\theta + \alpha) + m_{BC} \cdot g l_1 \cos(\theta + 60°) \qquad (28)$$

If (26) and (27) are introduced in (28), the equivalent resistant torque is expressed as:

$$\frac{T_{req}}{g} = \left[m_{A,CA} \cdot l + m_{AB}\left(l - \frac{l_2}{2}\right) + m_{BC} \cdot \frac{l_1}{2}\right]\cos\theta - \frac{\sqrt{3}}{2}(m_{AB}l_2 + m_{BC}l_1)\sin\theta \qquad (29)$$

Knowing that $l_1 = l_{G1} = 112$ mm, $l_2 = l_{G2} = 131$ mm and by use of the values from Table 1, the Eq. (6) gives $J_{BC(C)} = 5.47 \cdot 10^{-3}$ kg · m². The numerical result of (20), to be used for simulation is:

$$J_{eq} = 2.169 \cdot 10^{-2} + 1.562 \cdot 10^{-3} \cos^2\theta \quad \text{kg} \cdot \text{m}^2 \qquad (30)$$

where $\theta \in [0, 30°]$. Within this range, the maximum equivalent moment of inertia is for $\theta = 0°$

$$J_{eqmax} = 2.325 \cdot 10^{-2} \text{kg} \cdot \text{m}^2 \qquad (31)$$

The equivalent resistant torque can be calculated from (29), with the same values of masses and lengths and assuming g = 9.81 m/s². It results:

$$T_{req} = 2.1 \cdot \cos\theta - 0.7325 \cdot \sin\theta \quad \text{N} \cdot \text{m} \qquad (32)$$

The maximum value of the equivalent resistant torque is also for $\theta = 0°$, $T_{reqmax} = 2.1$ N.m.

It is to be noticed that both equivalent moment of inertia and the equivalent resistant torque have maximum values for the initial position, $\theta = 0°$. If in the model, the actuator can raise the structure with these maximum values, then, for the real actuation will be easier when the angle θ increases. The first attempt to simulate the model with the new values of J_{eq} and T_{req} was to check if the last PID parameters are adequate to perform the imposed task and the result was unsatisfying. The solution was to increase the PID gain from 10 to 25, for which the simulation result is presented in Fig. 11.

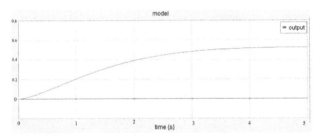

Fig. 11. Position response of the robotic chain for a 30° step reference (initial position 0°) with the torque coefficient diminished with 12% and $K = 25$ (PID)

Because it is not possible to change the PID parameters during operation, as having $K = 10$ for locomotion and $K = 25$ for obstacles avoiding, it has to be verified if the new PID parameters do not trouble the inchworm locomotion. Then, a simulation of the locomotion step, with the last gain, $K = 25$, was run, with the system response shown in Fig. 12.

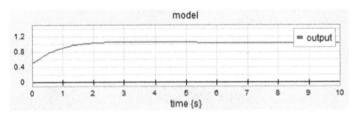

Fig. 12. Position response of the crank-slider mechanism for a 60° step reference (initial position 30°) with the increased gain to $K = 25$

It was also checked, with positive results, if a controller gain increase can compensate the negative deviation of the most important parameter of the DC motor – the torque coefficient. The lack of overshooting and oscillations at theoretical response is a satisfying result, but the variation of the equivalent inertia and resistant torque with the servo actuator position can introduce a velocity variation, which slightly modify the actual response, Anyway, for each case the reference is reached, because the simulations were run for the most disadvantageous circumstance, as it is the procedure in mechanical engineering.

4 Conclusions

When developing a mobile robot, as for any mechatronic system, the simulations performed on a model can save a lot of time of experimenting, because they approach to the actual parameters values, without a physical prototype, which is more expensive.

The key information for the robot developing is the actuators complete characterization, which can be done by help of simple experiments on them. This is a much stronger request, for the cheap components, due to the poor information provided by the producer. If possible, even the information from datasheet has to be checked.

The use of the servos in the development of mobile robots can solve effectively the problem of actuation. Their small size, high output torque, fairly good accuracy and affordable cost are convincing arguments for use at mini-robots and in this case for small scale demonstrator of external pipe inspection robot.

References

1. Shukla, A., Karki, H.: Application of robotics in onshore oil and gas industry-a review part I. Robot. Auton. Syst. **75**, 490–507 (2016)
2. Sharma, S.L., Qavi, A., Kumari, K.: Oil pipelines/water pipeline crawling robot for leakage detection/cleaning of pipes. Glob. J. Res. Eng. H Robot. Nano-Tech **14**(1), 31–37 (2014)
3. Schempf, H., Mutschler, E., Gavaert, A., Skoptsov, G., Crowley, W.: Visual and nondestructive evaluation inspection of live gas mains using the ExplorerTM family of pipe robots. J. Field Rob. **27**(3), 217–249 (2010)
4. Kotawad, A., Lad, K., Jadhav, S., Mandlik, R.: Identify the Deterioration in pipe by using wheel operated robot. Int. J. Res. Appl. Sci. Eng. Technol. **4**(II), 278–281 (2016)
5. Singh, P., Ananthasuresh, G.K.: A Compact and compliant external pipe-crawling robot. IEEE Trans. Rob. **29**(1), 251–260 (2013)
6. Nițu, C., Grămescu, B., Hashim, A.S., Avram, M.: Inchworm locomotion of an external pipe inspection and monitoring robot. In: Machado, J., Soares, F., Veiga, G. (eds.) HELIX 2018. LNEE, vol. 505, pp. 464–470. Springer, Cham (2019). https://doi.org/10.1007/978-3-319-91334-6_63
7. Grămescu, B., Cartal, L.A., Hashim, A.S., Nițu, C.: Clamping mechanisms of an inspection robot working on external pipe surface. In: Gheorghe, G.I. (ed.) ICOMECYME 2019. LNNS, vol. 85, pp. 218–230. Springer, Cham (2020). https://doi.org/10.1007/978-3-030-26991-3_20
8. Nițu, C., Grămescu, B., Cartal, L.A., Hashim, A.S.: Manipulator for Scanning of the Pipes External Surface, MATEC Web of Conferences, 290 08015 (2019). https://doi.org/10.1051/matecconf/201929008015
9. Hashim, A.S., Grămescu, B., Cartal, L.A., Nițu, C.: Modeling and Identification of a High Resolution Servo for Mobile Robotics, U.P.B. Scientific Bulletin, Series D Mechanical Engineering, ISSN 1454–2358, vol. 82 (2020)

The Choice of the Electric Energy Storage Device Type for the Hybrid Power Drive of Military Wheeled Vehicles

Dmitriy Volontsevich(✉), Sergii Strimovskyi, Ievgenii Veretennikov, Dmytro Sivykh, and Vadym Karpov

National Technical University "Kharkiv Polytechnic Institute", 2 Kyrpychova Street, Kharkiv 61002, Ukraine
vdo_khpi@ukr.net

Abstract. The purpose of the work is the formation of the scientifically based methodology for calculating the parameters and choosing the electric energy storage device criteria for the designed hybrid drive with the electromechanical transmission for promising military wheeled vehicles. It will be performed the analysis of the main characteristics of batteries and supercapacitors, which are used in vehicles, their advantages and disadvantages, as well as limitations and features of the application; the calculations of the required values of the electric power for the movement of the machine in various road conditions; calculations of the electric energy storage device capacity to ensure the power reserve during the autonomous driving with electric traction in various movement conditions; calculations of the maximum electric energy of the batteries and supercapacitors charge for the recuperation of the machine kinetic energy during emergency braking of the armored personnel carrier. Criteria are determined for choosing parameters of the electric energy storage device for the hybrid drive with the electromechanical transmission for military wheeled vehicles. All calculations are made on the example of the Ukrainian wheeled armored personnel carrier BTR-4.

Keywords: Electric energy storage device · Hybrid drive · Electromechanical transmission · Battery · Supercapacitor · Recuperated energy · Charge-discharge energy

1 Introduction

The analysis of modern military wheeled vehicles with the hybrid power drive and the electromechanical transmission showed that this type of the power drive and the transmission has already been used successfully on military technical support vehicles HEMTT-A3, Chevrolet Silverado (Colorado) ZH2; on reconnaissance vehicles Shadow RST-V, HE HMMWV, Panhard 4x4; on armored personnel carriers SEP, Panhard 6x6; on infantry fighting vehicles Marder, GCV, WARRIOR FV510 GENAIRCON [7] and on autonomous transport platforms of ground-based robotic complexes Tardec, Crusher,

MDARS [1–3]. The introduction of the hybrid power drive and electromechanical transmissions for various types of military wheeled vehicles is justified by the advantages, which are obtained in comparison with used diesel engines and mechanical or automated hydromechanical transmissions.

The main advantages of using the hybrid power drive and electromechanical transmissions for military wheeled vehicles are [8, 9]:

- the summary reduction of the fuel consumption and the increasing of the power reserve, due to the recovery of the braking energy [11, 12] and the body oscillation on the suspension [10];
- improved maneuverability and cross-country (terrain) capabilities;
- the ability to perform the covert movement (with lower noise and the heat dissipation in comparison with a running diesel engine);
- the ability to start driving with the electric energy storage instantly regardless of the weather conditions without starting the diesel engine;
- a simplified automation of the motion control, including the remote motion control;
- the ability to move back and forth at the same speed;
- the possibility of using the hybrid power drive as a diesel-electric station;
- the simplification of the main units layout on chassis of the machine without mechanical connections.

The obtaining of these advantages and the performance of the hybrid drive directly depends on the correct choice of parameters for the electric energy storage device, interacting with a diesel generator and traction motors. Therefore, it is important to consider, the scientific and technical task of choosing a type and a capacity of the electric energy storage device during designing the hybrid drive and electromechanical transmissions, taking into account the specifics of the military wheeled vehicles operation.

2 Literature Review

The wide variety of batteries are currently being manufactured for transport vehicles. They are divided into lead-acid, alkaline, iron-nickel (FeNi), nickel-cadmium (NiCd), nickel-metal hydride (NiMH), sodium-sulfur (NaS), lithium-sulfur (LiS) [15] and lithium-ion (Li-ion), depending on the chemical system of elements [4]. All of these types of batteries in terms of the resistance to operating conditions can be used to build the energy storage in hybrid drives on military wheeled vehicles. However, it is necessary to choose batteries with the highest capacity, the energy density, the shortest charge time and long life cycle to obtain the maximum efficiency of the military wheeled vehicles hybrid drive, even at the expense of their cost. The lithium-ion battery system is the best suited to these requirements [16]. It provides the specific energy density in the range of 100…240 W·h/kg and includes six main types of Li-ion batteries. These are lithium-cobalt ($LiCoO_2$), lithium-manganese ($LiMn_2O_4$ or LMO), lithium-nickel-manganese-cobalt ($LiNiMn_2O_4$ or NMC), lithium-iron-phosphate ($LiFePO_4$), lithium-nickel-cobalt-aluminum ($LiNiCoAlO_2$ or NCA), lithium-titanium ($Li_4Ti_5O_2$) batteries. The first type of $LiCoO_2$ has the highest specific energy intensity, but it can only work at low discharge-charge currents. Therefore, it is used for portable electronic devices with the low power consumption of the electric energy. The remaining 5 types can operate at higher currents, which are sufficient for using in power units, power tools, electric vehicles, hybrid cars, but at the same time they have lower specific energy consumption in comparison with $LiCoO_2$. The greatest energy intensity among chemical battery systems, which are used for powerful consumers of electric energy, is the NCA system. This technology is used by Tesla to make traction batteries for electric vehicles. The $Li_4Ti_5O_2$ system is also promising. It provides the fast battery charging, the wide range of operating temperature, long life cycle, but at the same time, the $Li_4Ti_5O_2$ system has low specific energy intensity, compared to the NCA system and the higher cost.

The characteristics of the batteries were considered, which are used in mass-produced electric vehicles and cars with the hybrid drive and the electromechanical transmission. The characteristics of the batteries are shown in the Table 1, which are installed in the most common electric cars, trucks and cars, trucks with the hybrid drive and the electromechanical transmission.

Table 1. Characteristics of batteries, which are installed in electric vehicles and cars with the hybrid drive and the electromechanical transmission.

Car model	Car mass, kg	Battery capacity, kW·h	Battery type	Battery mass, kg	Energy density, W·h/kg	Type of cooling	Mileage on storage energy, km
Tesla model S	2208	85/90/100	NCA	540/571/627	157,4…159,5	Liquid	320…424
Nissan leaf	1680	24/30	LMO	218/272	110	Aerial	130…160
BMW i3	1440	22	LMO	204	107,8	Aerial	135
Toyota prius	1530	2,15	NiMH	39	55,1	Aerial	20
Hino dutro hybrid	6500	2,15	NiMH	39	55,1	Aerial	–
Volvo FE hybrid	26000	5	LMO	46	108,7	Aerial	–
Renault hybrid	25000	5	LMO	46	108,7	Aerial	–
Volvo FL electric	16000	50×2(6)	LMO	455×2(6)	110	Aerial	160…350
Tesla semi	36000	100×6(10)	NCA	627×6(10)	159,5	Liquid	480…800

The Table 1 shows that the highest energy density is had by batteries, which are used in Tesla cars and trucks. The rated voltage of the battery is 400 V. The battery has the liquid cooling, the NCA chemical system and produces a current of up to 850 A for a battery with a capacity of 85 kW·h and up to 1000 A for a battery with a capacity of 100 kW·h. It is obtained, as a result of calculating the electric power, the supplied or the received maximum power of the battery is 340 kW and 400 kW, respectively, for batteries 85 kW·h and 100 kW·h.

It should be also noted there is a possibility of using supercapacitors to build electric energy storage devices [13, 14]. Maxwell, Yunasko, Nesscap, ELTON (ESMA) companies suggest to use supercapacitors not only in electric systems for starting internal combustion engines, but also in hybrid drives and electric transmissions for the accumulation of the recuperated electric energy during the vehicle braking [5]. Supercapacitors have the high specific power (W/kg), a faster charge-discharge process and a longer service life compared to Li-ion batteries, but they also have the much lower specific energy density (W·h/kg) and larger overall dimensions compared to Li-ion batteries. Therefore, the power reserve of the vehicle with energy storage devices in supercapacitors will not be large.

Currently, supercapacitors are used in hybrid power drives of some models of buses, trucks and locomotives. For example, Oshkosh Corporation applied four Maxwell BMOD0063-P125 modules in series in the hybrid drive of the HEMTT-A3 technical support truck. The modules provide the total accumulation of the electrical energy of up to 1,9 MJ. Also, on the experimental "Krymsk" armored personnel carrier, ELTON modules 30EK405 were tested with the total stored electric energy of up to 3 MJ [6].

3 Research Methodology

The characteristics of the selected battery or super-capacitor to build the energy storage device must provide:

- The efficient acceleration on electric traction subject to the maintaining of the diesel engine in the economy mode;
- the accumulation with minimal losses of the recuperable electric energy during braking;
- the required stand-alone power reserve of the military wheeled vehicle with the electric traction;
- the fulfillment of the conditions for limiting the overall and mass parameters of the electric energy storage device by the internal space and the maximum permissible mass of the designed machine.

It is necessary to calculate the following required parameters to make a choice of the electric energy storage device:

- the maximum instantaneous values of the discharge and charge power of the electric energy storage in vehicle acceleration and braking modes;

- the minimum value of the electric energy storage device capacity for the ensuring of the effective operation of the hybrid power drive in vehicle acceleration and braking modes;
- the power reserve of the vehicle during driving with the electric energy storage device.

The calculations are performed for the armored personnel carrier with the 8x8 wheel formula with the hybrid power drive and the electromechanical transmission, are made according to the sequential construction scheme, with power transfer from the traction motor to the drive wheel. The simplified power transmission scheme is shown in Fig. 1.

In the diagram of Fig. 1 the 7, 8, 9, 10 positions contain eight elements each. This provides the construction of eight branches for transmitting of power flows to each wheel separately. In these branches, in the acceleration mode, the power is converted and transmitted from the traction electric generator 3 and the electric energy storage 6 to the drive wheels 10 of the machine. In the braking mode, the power is converted and transmitted from the drive wheels 10 to the electrical energy storage 6.

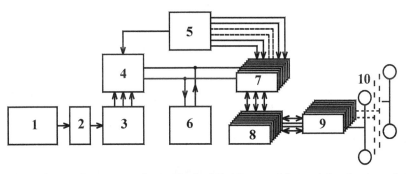

Fig.1. The scheme of power transferring in the hybrid power drive and the electromechanical transmission in the sequential construction scheme, where: 1 – the diesel engine; 2 – the matching gear; 3 – the traction electric generator; 4 – the power control unit for the traction electric generator; 5 – the top level controller; 6 – the electric energy storage device; 7 – the inverter; 8 – the traction electric motor; 9 – the gearbox; 10 – the driving wheel.

The acceleration mode of the wheeled armored personnel carrier was examined on the straight road. The required parameters for the electric energy storage device can be determined. For this, the current value of the required electric power is calculated by the formula from the source [18]:

$$N_{EP} = \frac{F_D v_v}{1000 \eta_{GB} \eta_{TM} \eta_{inv}}, \qquad (1)$$

N_{iEP} is the current value of the electric power for the movement of the vehicle (kW); F_{iT} is the current value of the vehicle traction force (N); v_{iv} is the current value of the vehicle speed (m/s); $\eta_{GB} = 0,9235$ is the gearbox efficiency; $\eta_{TM} = 0,94$ is the traction motor efficiency; $\eta_{inv} = 0,98$ is the inverter efficiency.

The current value of the traction force, which is necessary for the movement of the vehicle, can be calculated by the formula [18]:

$$F_{iT} = fmg + K_V F v_{iv}^2 + m\delta a_i, \qquad (2)$$

f is the coefficient of the resistance to the movement, $f = 0,015$ is for the asphalt concrete, $f = 0,045$ is for the dirt road; m is the vehicle mass (kg); $g = 9,81$ (m/s^2) is for the gravity acceleration; $K_V = 0,65$ (N·s^2/m^4) is the aerodynamic flow coefficient of the machine body; F (m^2) is the area of the vehicle frontal projection; $\delta = 1,246$ is the coefficient of accounting for rotating masses; a_i– is the current value of the vehicle acceleration (m/s^2).

The current value of the discharge and charge power of the electric energy storage device is determined by the formula:

$$N_{iEES} = N_{Gnom} - N_{iEP}, \qquad (3)$$

N_{iEES} is the current value of the electric energy storage power (kW); N_{Gnom} is the generator power of the diesel engine in the economy mode (kW).

The power of the generator is calculated by the formula [18]:

$$N_{Gnom} = N_{Dnom} \eta_{red} \eta_{Gen} \eta_{inv}, \qquad (4)$$

N_{Dnom} is the diesel engine power in a case of the minimum specific fuel consumption (kW); $\eta_{red} = 0,985$ is the efficiency matching gear; $\eta_{Gen} = 0,94$ is the efficiency of the traction generator; $\eta_{inv} = 0,98$ is the inverter efficiency.

If the value N_{iEES} is positive, it is the charge power of the electric energy storage device, if the value is negative, it is the discharge power of the electric energy storage device.

The required capacity of the electric energy storage in the acceleration mode of the wheeled armored personnel carrier is determined by the formula:

$$E_a = \frac{\sum_{i=1}^{t} N_{iEES}}{3600}, \qquad (5)$$

E_a is the required capacity of the electric energy storage in the acceleration mode (kW·h); i is a calculation step is 1 s; t is the vehicle acceleration time, s.

Next, the regenerative braking mode of the wheeled vehicle was studied. During the performing uniform braking, the maximum value of the recuperated electric power is determined by the formula [19]:

$$N_{ER} = \frac{m v_M^2}{2t} \eta_{GB} \eta_{TM} \eta_{inv} \eta_{CB}, \qquad (6)$$

N_{ER} is the recuperated electrical power (kW); t is the braking time (s); $\eta_{GB} = 0,95$ is the charge battery efficiency; v_M is the vehicle speed (m/s);

The braking time of the vehicle was calculated from the condition of equally slow motion to a complete stop $t = v_M/a$, where a is the deceleration of the vehicle. In a case of the emergency braking, the deceleration is approximately 4 m/s^2.

The amount of the recuperated electric energy was calculated at the performing emergency braking of the armored personnel carrier from the initial speed of 16,7 m/s (60 km/h) to the complete stop. As a result, the braking time is $t = 16,7/4 = 4,2$ s. The maximum value of the recuperated electric power is:

$$N_{ER} = \frac{25000 \cdot 16,7^2}{2 \cdot 4,2} 0,9235 \cdot 0,94 \cdot 0,98 \cdot 0,95 = 670,8 \ (kW).$$

During 4,2 s, eight traction electric motors of the armored personnel carrier will generated 2817,3 kJ (670,8 kW·4,2 s) of the electrical energy.

The uniform movement of the armored personnel carrier on the straight road was studied and the power reserve of the armored personnel carrier in the electric vehicle mode at the average speed of 12,5 m/s (45 km/h) on asphalt-concrete and dirt roads was determined.

The required electric power for the movement of the military wheeled vehicle can be calculated by the formula [18]:

$$N_{EP} = \frac{F_T v_M}{1000 \eta_{GB} \eta_{TM} \eta_{inv}}, \qquad (7)$$

N_{EP} is the electric power for the movement of the vehicle (kW); F_T is the vehicle traction force (N); v_M is the vehicle speed (m/s).

The traction force, which is necessary for the movement of the vehicle, can be calculated by the formula [18]:

$$F_T = fmg + K_V F v_M^2, \qquad (8)$$

f is the coefficient of the resistance to the movement, $f = 0,015$ is for the asphalt concrete, $f = 0,045$ is for the dirt road; m is the vehicle mass; $K_V = 0,65$ (kg/m³) is the aerodynamic flow coefficient of the machine body; F (m²) is the area of the vehicle frontal projection; $g = 9,81$ (m/s²) is the gravity acceleration.

The required battery capacity to ensure the movement of the vehicle for the given distance is calculated by the formula [19]:

$$E = N_{EP} \frac{S}{3,6 v_M}, \qquad (9)$$

E is the battery capacity (kW·h); S is a distance (km).

From the formula (9) the power reserve equation can be derived:

$$S = E \frac{3,6 v_M}{N_{EP}}. \qquad (10)$$

After performing calculations according to formulas (1) – (10), the highest values of the discharge and charge power of the electric energy storage device in the acceleration and braking modes of the vehicle are selected. And also the largest value of the required capacity of the electric energy storage device is selected for choosing its type during the designing of the hybrid power drive for the military wheeled vehicle.

4 Results

The Simulation of the acceleration mode, calculations of the recovered electric power during performing uniform braking and determining the power reserve of the machine were performed for the Ukrainian wheeled armored personnel carrier BTR-4. The calculated acceleration characteristics of the promising BTR-4 armored personnel carrier with the designed hybrid power drive and the electromechanical transmission (Fig. 1) were obtained by calculating formulas (1) – (5) in the source [9] and they are presented in Fig. 2.

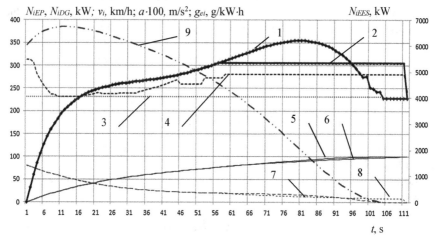

Fig. 2. The calculated acceleration characteristics of the promising BTR-4 with the designed hybrid power drive and the electromechanical transmission, where: 1 – the current value of the required electric power NiEP (kW) for the movement of the wheeled armored personnel carrier; 2 – the current value of the generated electric power NiDG (kW) by the diesel generator of the 3TD-3 engine for the movement of the wheeled armored personnel carrier with the deactivated electric energy storage devices; 3 – the minimum value of the specific fuel consumption g_e (g/kW·h) for the 3TD-3 engine; 4 – the current value of the specific fuel consumption g_{ei} (g/kW·h) of the 3TD-3 engine during accelerating of the armored personnel carrier with deactivated electric energy storage devices; 5 – the experimental characteristic of the change in the speed of the movement v_i (km/h) of the BTR-4 during acceleration; 6 – the calculated characteristic of the change in the speed of the movement Vi (km/h) of the armored personnel carrier with the hybrid power drive; 7 – the change in acceleration on $a \cdot 100$ scale, m/s^2 is for the characteristic 5; 8 – the change in acceleration on $a \cdot 100$ scale, m/s^2 is for the characteristic 6; 9 is the change in the electrical power of NiEES (kW) in the electric energy storage device during accelerating of the armored personnel carrier with the hybrid power drive to a speed of 100 km/h.

According to the results from Fig. 2, the short-term consumption of the electric energy due to traction electric motors with the power of up to 352 kW for 10 s is required for ensuring the dynamic acceleration of the armored personnel carrier BTR-4 with the weight 25 tons to a speed of 27,8 m/s (100 km/h). The long-term consumption of electric energy with a capacity of 306 kW is necessary for maintaining the movement at the maximum speed of 27,8 m/s (100 km/h). At least 5 kW of electric energy should be added to these values to ensure the operation of all electrical devices (consumers) on the armored personnel carrier. If the acceleration of the armored personnel carrier is performed in the electric vehicle mode (the battery is the energy source), the required acceleration dynamics is provided by a battery with a capacity of 100 kW·h or two parallel-connected batteries with a capacity of 85 kW·h. When the hybrid drive with the diesel engine is in the minimum fuel consumption mode, the generator makes 192 kW of electricity. In this case, to ensure the required acceleration dynamics of the armored personnel carrier from the battery, the short-term output of 160 kW of the electric energy and the long-term output of 114 kW of electric energy are required for maintaining the maximum speed.

Also, the initial charge of the electric energy storage device must be at least 6030 kW. From the graph 9 it is seen that during the acceleration of the machine, firstly the charge of the drive increases to 6735 kW for 12 s and after that, it decreases to zero by 104 s.

As a result, during the acceleration of the armored personnel carrier with the hybrid drive with the worked diesel engine (the sources of energy are a diesel generator and a battery), the using of a battery with a capacity of 85 kW·h will be excessive. Therefore, the battery capacity can be reduced to 50 kW·h with the maximum short-term energy output of up to 200 kW.

The recuperated electric power, which is equal to 670,8 kW, was calculated by the formula (6) in regenerative braking mode. If one battery with a capacity of 100 kW·h, which can accumulate not more than 400 kW per second, is used for the energy storage, it can accumulate not more than 1680 kJ per 4,2 s. As a result, about 1137,3 kJ will be lost. It is enough to connect two 85 kW·h batteries in parallel to save the recuperated electrical energy without losses completely. They will be able to accumulate up to 680 kW per second instantly. As a result, the total capacity of the two batteries will be 170 kW·h, and the mass is 1080 kg.

The capabilities of the supercapacitor were studied as an energy storage device for the hybrid drive of the armored personnel carrier. The characteristics of one module of the BMDC0063-P125 supercapacitor from Maxwell were set for analyzing the operating modes of the hybrid drive [5]. The charge-discharge operating voltage is 125 V, the maximum charge voltage is 135 V, the capacity is 63 F, the specific energy intensity is 2,4 W·h/kg, the specific maximum power is 3,7 kW/kg, and the mass is 58 kg. As a result it is determined the number of required modules for building the energy storage device, which is based on the required amount of the electrical energy reserve. The maximum value occurs during the performing emergency braking with the deceleration of 4 m/s². Eight traction electric motors of the armored personnel carrier will generated 2817,3 kJ (670,8 kW·4,2 s) of the electrical energy per 4,2 s. The accumulation of the electrical energy in the drive, which consists of supercapacitors, is calculated by the formula [19]:

$$W_C = \frac{CU^2}{2000}, \tag{11}$$

where W_C is the total charge energy of the energy storage device, which consists of supercapacitors (kJ); C is the total capacity of the energy storage device, (F); U is the charge voltage in the U–$U/2$ (V) window.

The maximum value of the charge voltage ≥ 400 V was set, it is close to the operating voltage of the considered characteristics of a Li-ion battery with a capacity of 100 W·h. Three modules of BMD0063-P125 supercapacitors must be connected in series to fulfill this condition. As a result the maximum value of the charge voltage can be got, it is equal to $3 \cdot 135$ V $= 405$ V. The working charge voltage will be 375 V, and the total capacity of 3 series-connected modules will be 21 F. The charge energy, which is calculated by formula (11), will be equal to $W_C = 21 \cdot 375^2/2000 = 1476$ kJ, which is approximately half less than the required value.

It is necessary to connect two groups of series-connected batteries in parallel (3 modules in each group) to double the capacity of the energy storage device. The total number of modules will be 6 pieces. In this case, the total capacity of the energy storage device will be 42 F, and the total charge energy is 2953 kJ at an operating voltage of 375 V. This totally fits into the condition for the full secondary using of the recuperated electrical energy, which is received during the emergency braking of the armored personnel carrier. If all 6 modules are connected in series, the value of the operating voltage of the energy storage device can be increased to 750 V. In this case, the total capacity of the energy storage device will be 10,5 F, and the total charge energy will be unchanged.

As a result, the total mass of 6 modules will be equal to 348 kg. This is the mass of the supercapacitors energy storage device that provides the complete recuperation of the electric energy from the emergency braking of the wheeled armored personnel carrier with the mass of 25 tons. If the same energy storage device will be built on Li-ion batteries, its mass will be 1080 kg.

During the performing power reserve calculations, according to formulas (7) – (10), it was obtained that when the armored personnel carrier with the mass of 25 tons moves along the straight asphalt-concrete road at a speed of 12,5 m/s (45 km/h), the traction force should be equal to 4,2 kN, and the required electric power for the movement should be 65,4 kW. Accordingly, 11,6 kN and 178,8 kW are for the dirt road.

If a battery with a capacity of 100 kW·h is used for the electric energy storage device, the armored personnel carrier will be able to move autonomously along the straight asphalt road with 68,8 km or along the straight dirt road with 25 km. It is necessary to increase the number of batteries due to connecting them in parallel to increase the power reserve, as well as to use the automatic systems for optimally electricity consume during the vehicle movement [17].

For comparison, it was noted that the mass of a Li-ion battery with a capacity of 100 kW·h is 627 kg. This is 1,8 times more than the mass of the energy storage device of supercapacitors 6 modules. The number of modules will be increased 2 times more to 12 pieces to approximate the mass of the electric energy storage device on supercapacitors to the mass of Li-ion batteries with a capacity of 100 kW·h. Then the mass of the energy storage on supercapacitors will be 696 kg. The cruising range of the armored personnel carrier weighing 25 tons on energy storage devices from 12 BMD0063-P125 modules will be calculated. It is supposed that there are 4 groups of modules connected in parallel in the energy storage device. Each group has 3 series-connected modules. Then, having the performed calculations according to the formula (6), it is obtained the total charge energy of the energy storage device is 5906 kJ.

Previously, it was found that the required electric power for the driving at a speed of 12,5 m/s (45 km/h) on the straight asphalt-concrete road is 65,4 kW, and on the dirt road is 178,8 kW. If these power values are taken every second from the energy storage device, the armored personnel carrier will be able to move 90,3 s on the asphalt-concrete road and 33 s on a dirt road. As a result, the armored personnel carrier will move 90,3 s·12,5 m/s = 1128,7 m along the smooth asphalt-concrete road, and 33 s·12,5 m/s = 412,5 m along the dirt road. These values are significantly less than the power reserve, which is performed by the armored personnel carrier with a Li-ion battery with a capacity of 100 kW·h.

5 Conclusions

The choice of the electric energy storage device methodology, which was presented, allows to analyze the efficiency of the designed hybrid power drive with the electromechanical transmission of the wheeled armored personnel carrier in acceleration, braking and uniform motion modes.

The chosen electric energy storage device for the hybrid drive with the electromechanical transmission must meet the required instantaneous maximum values of discharge and charge powers in the acceleration and braking modes of the wheeled armored personnel carrier.

The capacity of the electric energy storage device must ensure the continuous operation of the diesel engine in the economical mode at accelerating of the wheeled armored personnel carrier and accumulating of the recuperated power without losses in performing emergency braking of the wheeled armored personnel carrier at the current speed of 16,7 m/s (60 km/h).

Li-ion batteries should be chosen, which have the NSA chemical system, to ensure the maximum power reserve during the motion from the electric energy storage device. The energy storage devices in Li-ion batteries will have lower mass in comparison with the using of supercapacitors, considering the energy density (W·h/kg) of Li-ion batteries is significantly higher than that of supercapacitors.

References

1. Sivakumar, P., Reginald, R., Venkatesan, G., Viswanath, H., Selvathai, T.: Configuration study of hybrid electric power pack for tracked combat vehicles. Def. Sci. J. **67**(4), 354–359 (2017). https://doi.org/10.14429/dsj.67.11454
2. Walentynowicz, J.: Hybrid and electric power drive combat vehicles. J. KONES Powertrain Transp. **18**(1), 471–478 (2011)
3. Military Use of Hybrid Electric Drives. Army Guide Monthly, vol. **12**(15), 16–18 (2005)
4. Robert Bosch. Bosch Automotive Handbook. 9th edn, Robert Bosch GmbH (2014)
5. Schepmann, S.: Ultracapacitor Heavy Hybrid Vehicle: Model Predictive Control Using Future Information to Improve Fuel Consumption. All Theses TigerPrints Clemson University 168 (2010)
6. Kozachenko, V.F., Ostrirov, V.N., Lashkevich, M.M.: Transmission on the basis of a valve-inductor motor with independent excitation. Electr. Eng. **2**, 54–60 (2014)
7. Galvagno, E., Rondinelli, E., Velardocchia, M.: Electro-mechanical transmission modeling for series-hybrid tracked tanks. Int. J. Heavy Veh. Syst. **19**(3), 256–280 (2012). https://doi.org/10.1504/ijhvs.2012.047916

8. Volontsevich, D.O., Veretennikov, E.A., Kostianik, I.V., Iaremchenko, A.S., Efremova, A.I., Karpov, V.O.: Determination of the electric drive power for lightly armored caterpillar and wheeled vehicles using single- or two-stage mechanical gearboxes. Electr. Eng. Electr. Mech. **1**, 29–35 (2019). https://doi.org/10.20998/2074-272X.2019.1.05
9. Volontsevich, D.O., Klyuchka, R.V., Sobko, A.P., Strimovskiy, S.V.: Analysis of operating modes of a hybrid drive with an electromechanical transmission on a promising wheeled armored personnel carrier. Integr. Technol. Energy Saving [Integrovani texnologiyi ta energozberezhennya] **4**, 34–47 (2018)
10. Liu, J., Li, X., Wang, Z., Zhang, Y.: Modelling and experimental study on active energy-regenerative suspension structure with variable universe fuzzy PD control. Shock. Vib. **2016**, 1–11 (2016). https://doi.org/10.1155/2016/6170275
11. Kotsur, M.I., Kotsur, I.M., Blizniakov, A.V.: Increase effectiveness of reversible braking mode realization of the wound-rotor induction motor. Eastern-Eur. J. Enterp. Technol. **8**(73), 27–30 (2015). https://doi.org/10.15587/1729-4061.2015.36670
12. Sinchuk, O., Kozakevich, I., Kalmus, D., Siyanko, R.: Examining energy-efficient recuperative braking modes of traction asynchronous frequency-controlled electric drives. Eastern-Eur. J. Enterp. Technol. **1**(1), 50–56 (2017). https://doi.org/10.15587/1729-4061.2017.91912
13. Sevilla, M., Mokaya, R.: Energy storage applications of activated carbons: supercapacitors and hydrogen storage. Energy Environ. Sci. **7**(4), 1250–1280 (2014). https://doi.org/10.1039/c3ee43525c
14. Fomin, O., Sulym, A., Kulbovskyi, I., Khozia, P., Ishchenko, V.: Determining rational parameters of the capacitive energy storage system for the underground railway rolling stock. Eastern-Eur. J. Enterp. Technol. **2**(1), 63–71 (2018). https://doi.org/10.15587/1729-4061.2018.126080
15. Maurer, C., Commerell, W., Hintennach, A., Jossen, A.: Capacity recovery effect in lithium sulfur batteries for electric vehicles. World Electr. Veh. J. **9**(2), 34 (2018). https://doi.org/10.3390/wevj9020034
16. Martinez-Laserna, E., Herrera, V.I., Gandiaga, I., Sarasketa-Zabala, E., Gaztañaga, H.: Li-ion battery lifetime model's influence on the economic assessment of a hybrid electric bus's operation. World Electr. Veh. J. **9**(2), 28 (2018). https://doi.org/10.3390/wevj9020028
17. Chen, Z., Liu, W., Yang, Y., Chen, W.: Online energy management of plug-in hybrid electric vehicles for prolongation of all-electric range based on dynamic programming. Math. Probl. Eng. **2015**, 1–11 (2015). https://doi.org/10.1155/2015/368769
18. Aleksandrov, E.E., Epifanov, V.V., Medvedev, N.G., Ustinenko, A.V.: Tjagovo-skorostnye harakteristiki bystrohodnyh gusenichnyh i polnoprivodnyh kolesnyh mashin [Trailer-speed characteristics of high-speed track and full-wheel drive wheeled vehicles]. Kharkiv, NTU "KhPI" publ. (2007)
19. Gnatov, A.V., Argun, Shh.V., Trunova, I.S.: Teoriya elektropryvodu transportnyh zasobiv: pidruchnyk [The theory of electric vehicles: a textbook]/Kharkiv, Kharkiv National Automobile and Highway University publ. (2015)

Machinery Retrofiting for Industry 4.0

Pedro Torres[1,4](✉), Rogério Dionísio[1,3,5], Sérgio Malhão[1], Luís Neto[2,4], and Gil Gonçalves[2,4]

[1] Instituto Politécnico de Castelo Branco, Castelo Branco, Portugal
{pedrotorres,rdionisio}@ipcb.pt, smalhao@ipcbcampus.pt
[2] FEUP, Faculdade de Engenharia, Universidade Do Porto, Porto, Portugal
{lcneto,gil}@fe.up.pt
[3] DiSAC – R&D Unit, Avenida Do Empresário, 6000-767 Castelo Branco, Portugal
[4] SYSTEC, Research Center for Systems and Technologies, Porto, Portugal
[5] INESC TEC, Porto, Portugal

Abstract. The paper presents an approach for the retrofitting of industrial looms on the shop floor of a textile industry. This is a real case study, where there was a need to update the equipment, providing the machines with communication features aligned with the concept of Industry 4.0. The work was developed within the scope of the research project PRODUTECH-SIF: Solutions for the Industry of the Future. Temperature, Inductive, Acoustic and 3-axis Accelerometers sensors were installed in different parts of the machines for monitorization. Data acquisition and processing is done by a *SmarBox* developed on a cRIO 9040 from *National Instruments*. A *SmartBox* processes data from one to four looms, allowing these old machines to have communication capacity and to be monitored remotely through the factory plant's MES/ERP. Communication can be done through the OPC UA or MQTT architecture, both protocols aligned with the new trends for industrial communications. The sensor data will be used to feed production and manufacturing KPIs and for predictive maintenance. The approach presented in this paper allows industries with legacy equipment to renew and adapt to new market trends, improving productivity rates and reduced maintenance costs.

Keywords: Retrofitting · Cyber physical system · Industrial IoT · Predictive maintenance · Industry 4.0

1 Introduction

The exponential growth of technology and the growing requirements for personalized products mean that manufacturers need to invest in automation to be aligned with this industrial revolution that is Industry 4.0. However, old infrastructure and outdated processes can make it difficult to reach a better productivity and competitiveness. Communications are one of the main keys of Industry 4.0, allowing sensors, machines and people to communicate and cooperate with each other. This is the fundamental basis for Internet of Things (IoT) and Industrial Internet of Things (IIoT). Manufacturers invest in smarter equipment, with a greater ability to acquire and process data in real time, without compromising communication requirements, preferably choosing IP-based protocols to ensure global connectivity.

Any industrial equipment is designed to be robust and have the longest possible life. Industrial machinery typically has high acquisition costs, reason enough to consider if a replacement machine is the better choice, unless there is no other alternative in order to obtain the maximum return on investment. Upgrading is an economical way to improve machine interoperability and productivity by renewing legacy equipment for the Industry 4.0 era. The implementation of a monitoring and control system, such as Supervision Control and Data Acquisition (SCADA) or Distributed Control System (DCS), can help in the upgrading, as it allows the system to operate as a whole and not as an isolated component. Integration of Human Machine Interfaces (HMI) to act, visualize and report processes is also important in the retrofitting process.

An accurate insight of the production process is obtained by installing new sensors and improving connectivity, connecting old machines to the intranet or the internet. The new sensors can be connected to the machine's legacy controllers, a generally difficult process, or new controllers can be installed. The controller collects data and converts it into a usable format [1], such as Open Platform Communications - Unified Architecture (OPC UA) or Message Queuing Telemetry Transport (MQTT), so that the information can be made available to the rest of the factory network.

In the context of the Portuguese textile industry, legacy industrial loom machines have been used for decades. Currently, they are sharing the shop floor along with modern machines ready for the Industry 4.0 challenges, that causes disparities between different sections of the manufacturing process and shop floor management. The objective of this paper is to present an approach for the retrofitting of legacy industrial looms. This is a real case study, where there was the need to update the equipment, providing the machines with communication features aligned with the concept of Industry 4.0. The remaining of this paper is organized as follows. The first chapter presents the introduction to the concept and need of industrial loom machines retrofitting, followed by a section dealing with the choice of and industrial communication protocol. A case study is then described in detail in Sect. 3. The lessons learned from the retrofitting process are presented in Sect. 4, followed by the conclusions.

2 Industrial Communication Architectures

One of the problems of industrial automation is the lack of uniformity in communications, especially when there is a multitude of equipment from different manufacturers. Although today there are several communication standards, many of the factories, especially older ones, have machines without the ability to communicate. The use of standards-based data connectivity is essential to ensure that the machines may "talk" the same language. Of the various market protocols, OPC UA is the one that truly serves as a data connectivity standard to make Industry 4.0 a reality. Instead of replacing standards, OPC UA complements them, creating a common layer for exchanging information.

This chapter describes OPC UA and MQTT, the two standards implemented for communication between the *SmartBox* installed on the shop floor and the different automation systems in the factory.

2.1 OPC UA

OPC UA is the standard for communications in Industry 4.0. The German initiative RAMI 4.0 (Reference Architectural Model for Industry 4.0) [5], under the *Plattform Industrie 4.0*, released several specification reports for communication between Industry 4.0 components, using OPC UA [6–8]. A fundamental condition for RAMI 4.0 is that all actors involved must be able to understand each other, regardless of domain, vendor and technology. The OPC UA architecture provides a data model and several information models, these define modulation rules and several types that can be used to model information and data. With this, all actors involved can understand each other by sharing a common syntax and semantics. During the development of this use case the SOSD (Smart Object Self-Description) information model [4] was developed on top of the OPC UA Data Access information model. All aspects of the newly developed model were based in the *Plattform Industrie 4.0* recommendations for Industry 4.0 components. Figure 1 shows the OPC UA architecture supported on the communication model between client and server.

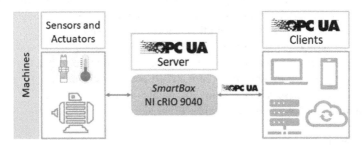

Fig. 1. OPC UA based network architecture.

2.2 MQTT

MQTT is a publish/subscribe communication protocol developed by IBM with a special emphasis in achieving a lightweight footprint. This protocol is widely used in IoT and IIoT applications due to its simplicity and robustness [10]. MQTT communication relies in a central broker that allows information publishers and subscribers to fill/consume message queues identified by topics. Despite lacking a standard information model as OPC UA, this protocol can be used in less capable controllers and is very simple to setup, which allows for time and cost savings. The popularity of MQTT caught the attention of OPC Foundation, resulting in the Part 14 of the OPC UA standard [9]. This bridging between both architectures enables the use of OPC UA over MQTT, allowing for: transmission of OPC UA standardized information over the internet, using MQTT as a transport mean; easy integration of IoT gateways and devices based on MQTT with OPC UA clients and servers. In this use case, IoT devices and low-cost sensors were integrated through MQTT. This was a successful cost saving method to add environmental and other non-critical sensor data to the use case. Figure 2 shows the MQTT communication model implemented in this work.

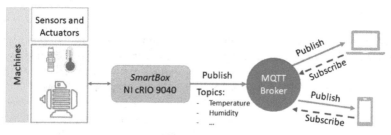

Fig. 2. MQTT communication model.

3 Case Study – Preparation of Legacy Looms

The case study is implemented on a textile factory that produces clothing tags. In the manufacturing plant, it has looms from different generations. This work focuses on the oldest equipment, as these are the ones that need intervention to maintain their usefulness and desired production levels. The preparation of these legacy looms was carried out in 5 stages: gathering of requirements, installation of sensors, data acquisition and processing by a *SmartBox* [2, 3], integration with the factory network and, finally, data analysis.

During the requirements assessment, the company identified the main problems the old looms were suffering. All electrical and mechanical properties of the machines were summarized, so the set of sensors to be installed could be correctly selected. Another important step was to assess the company communication infrastructure and technologies used to communicate with the local MES/ERP.

Installed Sensors:

- 1 Temperature sensor in the machine electrical box for monitoring the temperature inside the box due to overheating and prevent fault situations;
- 1 Temperature sensor on the Jacquard loom for monitoring the temperature inside the equipment;
- 1 Temperature sensor on the loom calender for local temperature monitoring;
- 1 Inductive sensor to measure the motor rotations and consequently calculate the length of produced tissue;
- 1 Acoustic sensor to gather mechanical noise of the machine during production, in the context of fault detection;
- 1 3-axis accelerometer to measure mechanical vibrations during production. Useful in a context of predictive maintenance by identifying altered spectrum frequencies.

All sensors are wired to a multiple input/output board on the NI cRIO 9040, as well as signals from the tower light to identify different levels of machine operability as summarized on Fig. 3.

The calibration of the sensors and the real-time processing of the data is done in the *SmartBox* that works as an OPC UA server. OPC UA clients running on smartphones, tablets or computers can easily access to the information remotely at different parts of the shop floor. All data are stored in a database accessed by the company's MES/ERP. The developed SOSD model, allows for other components developed in the context of the PRODUTECH-SIF project to communicate and understand context and production data of the old machines [4]. The case study components, according with the SOSD model, illustrated in Fig. 4., are divided in four data types: *SmartObjectType* describes the *SmartBox* properties and relations; *DeviceType* describes machines, sensors and actuators and its relations; *ServiceInstanceType* describes services available in the OPC UA Server, these can be provided by software modules, machines, *SmartBoxes* and their relations; *PointDescriptionType* describes endpoints in the network with which the OPC UA server can communicate.

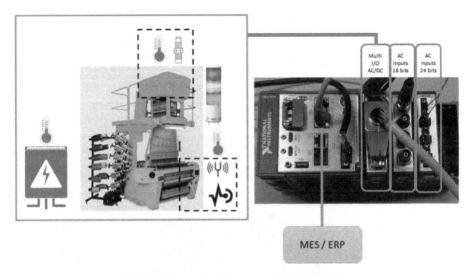

Fig. 3. Physical architecture of the installation.

The intervention on the shop floor will be done in different phases. Until the moment was prepared a testbed with 4 machines. After testing and adjustments all legacy looms on the shop floor will be retrofitted. Figure 5 illustrates the intervention in one machine and the location of 2 installed sensors.

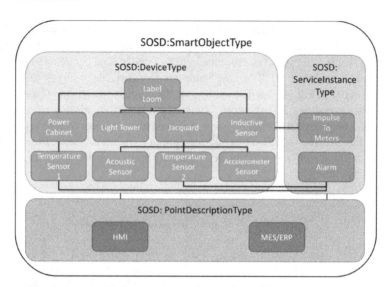

Fig. 4. Case study elements mapped according with SOSD data model.

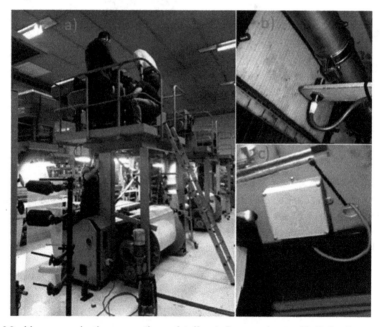

Fig. 5. Machine sensorization operations detail. a) Legacy loom, b) Inductive sensor, c) Temperature sensor protected by box.

4 Conclusions

The Industry 4.0 is coming with Internet of Things, Cyber-Physical Systems, Cloud Computing and intelligent networking based on wireless, mobile and sensor devices autonomously exchanging information, triggering actions and controlling each other independently. Smart gateways for I-IoT are implemented in the 4th industrial revolution due to their capabilities for data analytics, diagnostics, prognostics and self-awareness, and simplify greatly the process to export data and functionality of the physical devices to end users on the web.

Two important lessons learned during the first phase are related with the requirements assessment: 1) The *SmartBox* was sampling at 1 Hz for some time, this revealed as a problem because the vibration and acoustic signals could not be accurately reconstructed, and therefore, be correctly analyzed. 2) The logic levels of the machine signals, low cost sensors, industrial sensors and the I/O board on the NI cRIO 9040 were not compatible. This required to use an optocoupler to match the machine and industrial sensor signals with the I/O board.

The work presented in this paper is important in the process of retrofiting old machines. Sensorization and implementation of standardized communications is essential in for the concept of connected factories. As future work will be implemented techniques of data analytics and machine learning in scenarios of predictive maintenance.

Acknowledgments. This research was supported by the project PRODUTECH SIF - Solutions for the Industry of the Future, financed by the Portuguese National program COMPETE 2020.

References

1. Torres, P., et al.: Cyber-Physical production systems supported by intelligent devices (smart-boxes) for industrial processes digitalization. In: 5th International forum on Research and Technology for Society and Industry (RTSI), pp. 73–78, IEEE, Florence, Italy (2019). https://doi.org/10.1109/RTSI.2019.8895553
2. Malhão, S., Dionísio, R., Torres, P.: Industrial IoT smartbox for the shop floor. In: 5th Experiment International Conference (exp.at 2019), pp. 258–259, IEEE, Funchal (Madeira Island), Portugal (2019). https://doi.org/10.1109/EXPAT.2019.8876562
3. Rubio, E.M., Torres, P.M., Dionísio, R.P.: Smart gateways for IOT-factory integration: trends and use case. In: Ferreira, L., Lopes, N., Silva, J., Putnik, G., Cruz-Cunha, M., Ávila, P. (Eds.) Technological Developments in Industry 4.0 for Business Applications, pp. 149–170. IGI Global (2019). https://doi.org/10.4018/978-1-5225-4936-9.ch007
4. Neto, L., et al.: An industry 4.0 self-description information model for software components contained in the administration shell. In: International Conference on Intelligent Systems and Applications, 8, pp. 1–6. IARIA, Rome, Italy. ISBN: 978–1–61208–723–8
5. Hankel, M., Rexroth, B.: The reference architectural model industrie 4.0 (rami 4.0). ZVEI, April 410 (2015)
6. Rauchhaupt, L., Kadel, G.: Network-based Communication for Industrie 4.0–Proposal for an Administration Shell. Plattform Industrie 4.0 (2016)
7. P. I. 4.0: Interaction Model for Industrie 4.0 Components, 3 (2016). https://www.plattform-i40.de/I40/Redaktion/EN/Downloads/Publikation/interaction-model-I40-components.pdf?__blob=publicationFile&v=9. Accessed 02 Feb 2020

8. VDMA. Industrie 4.0 Communication Guideline Based on OPC UA. 10 12 2017. https://industrie40.vdma.org/en/viewer/-/v2article/render/20625194. Accessed 02 Feb 2020
9. OPC Foundation. OPC UA Part 14 - PubSub Specification. OPC Foundation, 6 2 2018. https://opcfoundation.org/developer-tools/specifications-unified-architecture/part-14-pubsub. Accessed 02 Feb 2020
10. Peralta, G., Iglesias-Urkia, M., Barcelo, M., Gomez, R., Moran, A., Bilbao, J.: Fog computing based efficient IoT scheme for the Industry 4.0. In: Electronics, Control, Measurement, Signals and their Application to Mechatronics (ECMSM). 2017 IEEE International Workshop of (2017)

Conceptual Design of a Positioning System for Systematic Production of Needle Beds

Luis Freitas[1(✉)], Rui Oliveira[1], Teresa Malheiro[2], A. Manuela Gonçalves[2], José Vicente[3], Paula Monteiro[4], and Pedro Ribeiro[5]

[1] School of Engineering, MEtRICs Research Centre, University of Minho, Guimarães, Portugal
b7245@dep.uminho.pt
[2] School of Sciences, CMAT Research Centre, University of Minho, Guimarães, Portugal
[3] InsideLimits, Canidelo, Vila Nova de Gaia, Portugal
[4] Center of Computation Graphics (CCG), Guimarães, Portugal
[5] School of Engineering, Algoritmi Research Centre, University of Minho, Guimarães, Portugal

Abstract. In-Circuit-Test Machines are nowadays one of the most important systems on the productions lines in the electronic industry. Printed Circuit Board quality standards have become increasingly higher and all processes involved in its production must be checked and controlled with a high degree of reliability. One particular aspect that clearly did not follow the technological breakthroughs is the manufacturing process of the In-Circuit-Test machines, and more particularly its needle beds. This process is still a manual process that results in critical quality problems that are going to be reflected on the final production lines. This paper proposes a new manufacturing process for needle bed production that aims at reducing the production process constraints and therefore improving all respective quality parameters. The problem was analyzed and several project parameters were defined, such as project goals, engineering requirements and functions analysis. A conceptual solution is developed and proposed to solve the identified problem.

Keywords: In-Circuit-Test machines · Mechatronic design

1 Introduction

In the new era of industrial automation, characterized by concepts on the domain of Industry 4.0, a new paradigm is driving the development of industrial equipment for industrial companies. The so-called Cyber-Physical Systems (CPS) equipment must be developed and adapted for the new industrial challenges [1], and prepared to connect and communicate [2–8] with other systems by collecting data and making available useful information.

Technological production is nowadays increasing at breakneck pace, forcing producers of electronic equipment into a supply race of electronic components at an increasingly higher pace and quality level. Printed Circuit Boards (PCB) are one of these electronics components and they are especially important because they are widely used [9].

With the increasing complexity and the need for lot size reduction, one of the main problems still affecting the PCB production chain is its electrical quality testing phase

(In-Circuit Test systems - ICT systems), often leading to PCB stagnation (production line stops to PCB rework) and/or rejection. A part of these problems is not related to the PCB itself, but to manufacturing failures of ICT systems arising from the fact that the manufacturing process is still mostly handmade [10]. Since the number of test points is always increasing and the ICT system is getting more and more complex, the present manufacturing process of the ICT system struggles to keep up the high quality standard of the ever-demanding automotive industry. The key to avoid a part of the problems during the testing phase is to prevent manufacturing failures of the ICT systems.

This paper will focus on the development of an automatized system for the most critical production phase of the ICT system production – needle receptacle insertion.

To achieve the proposed goals, the paper is organized as follows: Sect. 2 presents a brief ICT description and its current manufacturing process by highlighting the goals proposed for the system to be developed and the proposed methodology to do it; Sect. 3 describes all steps of the design tasks development to achieve the proposed conceptual solution to solve this challenge; Sect. 4 presents the main results by focusing on the more important aspects. Finally, the last section presents the conclusions by summarizing the goals achieved with this conceptual solution and the steps to be taken next.

2 State of the Art

As an article aiming at presenting a conceptual solution to solve a well-known problem, this chapter presents on 2.1 subchapter, a description of the current systems and manufacturing process used. More specifically, it covers ICT and fixtures concepts and its constitution, as well as the current manufacturing process of the fixtures and the problems it entails.

The proposed goals are also presented, on 2.2 subchapter, based on the problems previously explained, as well as the proposed design process methodology to achieve these goals.

2.1 ICT Systems Manufacturing

ICT systems include both software and hardware parts. Focusing on the hardware, Fixture it is an electromechanical device roughly consisting of a mobile plate where the Device Under Test (DUT) – this paper only addresses PCBs as DUT – and a Needle Bed ("Bed-of-nails") are placed. A PCB ICT provides a quick contact with hundreds of access points (test points) between the PCB and the needles by pressing the mobile plate down against the Needle Bed. The Needle Beds are the most critical parts in terms of quality problems due to their importance in the entire system and its handmade manufacturing process [9].

The needle bed consists of a test plate custom manufactured with several through-holes where the receptacles and the needles will be inserted. On the other side of the test plate, the receptacles-needle set is connected to a wiring system. This wiring system allows measuring electrical parameters in the PCB, connecting the PCB and software part [11]. Figure 1 shows both sides of the test plate, bottom and top: the Needle Bed and the Wiring system, respectively.

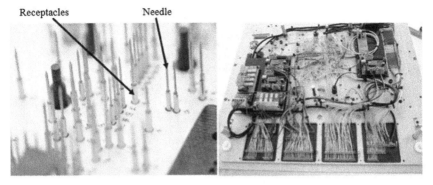

Fig. 1. Needle beds [10]

The manufacturing process consists in two main parts: the wiring process, which creates an entire network of electrical connections, and the insertion process of the receptacles and needles. This paper will focus be on the receptacles insertion process.

The receptacles are manually inserted in the test plate and then the needles are also manually inserted into these receptacles, that serves as guidance for the needles. The two following figures present a technical drawing of a receptacle (Fig. 2) and photos of the receptacles, needles and receptacles-needle set sizes compared to an ordinary pen (Fig. 3).

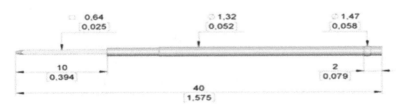

Fig. 2. Receptacle technical drawing (KS-075 47 E03 Ingun) [13]

Note that receptacle diameter and size are, in themselves, a problem since precision levels are around hundredths of a millimeter.

Receptacle insertion is done using a press-in tool (Fig. 4), which helps the technician to press the receptacles to their final position.

Therefore, the angle and the force applied by the technician on the receptacles may deform them and cause the needles to be uneven at the end of the process. Since it is a handmade process, it may also be difficult to ensure perpendicularity between the receptacle's axis and the test plate. This manufacturing process can take up to more than a week of work depending on the PCB to be tested, and it is a truly repetitive process, requiring high concentration and sensitivity levels from the technicians throughout the whole process. This manufacturing process is highly dependent on the operator's skill and requires years of training and experience, which is a competitive disadvantage.

It is also important to note that if the receptacle is misplaced, consequently the respective needle, and the needle beds aren't fully validated internally, since it is almost

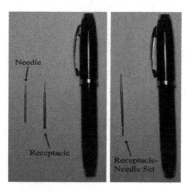

Fig. 3. Receptacles/needle size

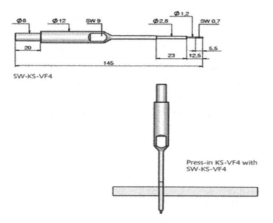

Fig. 4. Press-in Tool (SW-KS-VF4, Ingun) [14]

undetectable by the human eye, the error detection is postponed to the production line phase.

All the conditions inherent to the abovementioned process resemble more a handcrafted process than an industrial one. Therefore, automation is a viable path both to ensure tolerances, perpendicularity and accuracy in the receptacle placement – and so the needle position - and to carry out a quality control of these parameters, and consequently to reduce both time and cost of production.

This upgrade is expected to create a competitive advantage in this manufacturing process and consequently achieve some significant goals, such as those presented in Table 1.

Table 1. Project goals

Features	Current Market Situation	Project Goals
Accuracy increase for access point	0,8 mm minimum diameter recommended for the access points	Automated system capable of 0,6 mm minimum diameter of access point
Accuracy increase in centering the needles' contact points in regard to the access points	70% average of accurate contacts	Automated system capable of 95% average of accurate contacts
Fixtures production process optimization	Handmade receptacles insertion (high operator dependency; error-prone process)	Automated system of receptacles insertion and quality checking
Fixtures production time reduction	100% = 104 h	50% = 40 h
Receptacle insertion repeatability process	Handmade process makes repeatability impossible	Automated system that allows receptacle insertion repeatability process

2.2 Methodology

Throughout the years, several authors tried to explain and systematize the design process. From Descriptive Models, that simply describe the steps and phases of the design process, such as the ones proposed by French [15], for example, to models like the ones proposed by Jones, Archer, or even the VDI 2221 [16] guideline, named Prescriptive Models, that tries to define and encourage designers to adopt a more analytical and systematized way of working [17].

Given the innovative nature of this project, it was chosen a methodology that does not set a definitive line between analysis of the problem and development of solution concepts, to allow designers a more flexible approach to the problem. This is the methodology proposed by Cross [17].

3 Project Development

This chapter presents all the necessary phases that will lead to the development of system. In the first Subsect. 3.1, the project is dismembered into its objectives, functions and specifications, that are absolutely crucial to develop all the conceptual system alternatives presented in Subsect. 3.2. It is also presented the evaluation of each alternatives, and how it compares with each other regarding the system requirements.

3.1 Exploration Step

The goal of this step is to understand and define, with some level of detail, all the aspects to enable designing a system to match all the project needs.

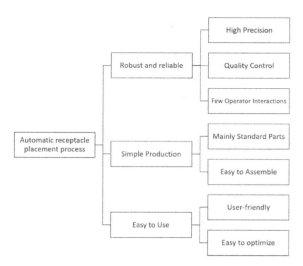

Fig. 5. Objectives tree

After a careful and thoroughly analysis, an objectives tree [15] was built (see Fig. 5).

The main objective for this project is the design of a system to automate the receptacle placement process.

To enable the use of this system in an industrial context, and taking into account all the demands of a highly competitive industry like the automotive industry, other sub-objectives were added.

The first one is that, whatever system is chosen, it will have to be a robust and reliable system in terms of precision, quality control of the needle beds and operator dependency. Moreover, the system needs to be of simple production to allow good reproducibility, cost-efficiency and easy maintenance. The majority of the parts should be standard and bought-out and easy to assemble. Finally, and since this system will be used with an operator, the user aspect is highly important and must include a user-friendly interaction, with easy optimization and modification of the process parameters.

After all the objectives are stated, it is necessary to break them down into functions. The overall function diagram is simple (see Fig. 6). [17].

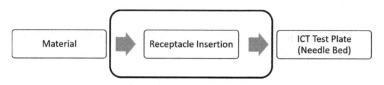

Fig. 6. Main function diagram

But to do a more systematic analysis, it is necessary to break the main function down into its respective sub-functions.

Figure 7 presents all the tier 2 functions that allow the system to fulfill all the objectives previously presented.

Conceptual Design of a Positioning System for Systematic Production

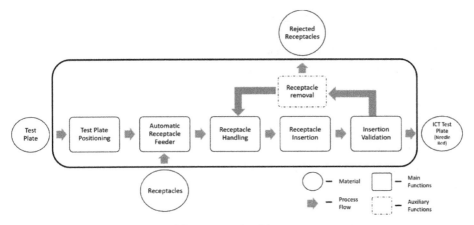

Fig. 7. Function Diagram

These main sub-functions are described below:

- **Test Plate Positioning–** The system must ensure that the plate receiving the receptacles stands in a known position and within the specified tolerances.
- **Automatic Receptacle Feeder –** Since one objective is to reduce operator interaction as far as possible, it is necessary to ensure that the receptacles are fed into the system in a way that guarantees their correct positioning.
- **ReceptacleHandling –** This sub-function describes the sub-system that will position the receptacles received via the automatic receptacle feeder into their correct positions on the test plate.
- **ReceptacleInsertion –** This sub-function relates to the task of placing the receptacles into their final position on the test plate.
- **Insertion Validation–** The final sub-function defines the sub-system that will guarantee that all receptacles (one by one or all at the same time) are placed correctly on the test plate.

Finally, an auxiliary function was added that is not directly related to the objectives but adds value to the system:

- **Receptacle removal–** This sub-function will allow the system, within its own process, to correct possible bad receptacle insertions, thus reducing the time and therefore adding value to the entire system.

The final step of this first phase is the listing of the requirements and/or engineering characteristics extracted from the problem analysis. Table 2 presents a list of all engineering characteristics and requirements for all the sub-functions:

Starting with all these assumptions, all the necessary information is thus gathered to start developing and comparing system alternatives.

Table 2. Engineering requirements

Sub-functions and engineering requirements	Tolerances/Dimensions
Test Plate Positioning	
Positioning Tolerance of the test plate relative to the system structure	0.05
Poke-Yoke system for the test plate	
Maximum test plate dimension	500 mm × 500 mm
Automatic Receptacle Feeder	
Minimum number of receptacles per cycle	200
Two types of receptacles	
Receptacle positioning	
Positioning Tolerance of the receptacles relative to the test plate	0.05 mm
Repeatability for receptacle positioning	0.01 mm
Receptacle Insertion	
Positioning Tolerance of the receptacles relative to the test plate	0.05 mm
Insertion Validation	
Verify receptacle presence	
Verify receptacle position	0.01 mm
Verify receptacle perpendicularity	0.05 mm
Verify receptacle height in relation to the test plane	0.05 mm

3.2 System Alternatives

Matching the system objectives with the requirements/characteristics, a set of alternatives solutions were developed for each specific sub-function described in the previous section. Table 3 presents a morphological chart with the all the sub-solutions for all the sub-functions.

As previously mentioned, for the purpose of this paper only the designing process for the Receptacle Positioning and Receptacle Handling/Insertion sub-functions will be addressed. These two functions are directly connected since the system for receptacle positioning, handling and inserting could be the same. The next tables present the selection matrix for these two sub-functions and their respective sub-solutions.

In Table 2, some parameters were defined to will serve as evaluation parameters for the sub-function Receptacle Positioning. These parameters are directly connected to the system's requirements and objectives. The first two parameters are quantitative, since a real value can be defined, and self-explanatory.

"Flexibility" refers to the sub-solution's ability to embrace new sub-functions or auxiliary functions that are not mentioned here, and the more flexible this solution is, the easier it is to integrate this sub-function with the Receptacle Handling/Insertion. Regarding the space that this system needs to ensure the plate's maximum dimension's requirement, the name chosen for the parameter was "overall dimensions".

Conceptual Design of a Positioning System for Systematic Production

Table 3. Sub-solutions alternative per sub-function

	Sub-solution 1	Sub-solution 2	Sub-solution 3	Sub-solution 4	Sub-solution 5
Plate Load/Positioning	Guiding pins	Optical alignment			
Receptacle Feeder	Vibration System	Circular Cube	Manual Load in predefined positions		
Receptacle Positioning	3 axis System	6 axis Robot	Scara Robot	Delta Robot	Cylindrical Robot
Receptacle Handling/Insertion	Gripper Fingers System	Pneumatic-actuated System	Vacuum System		
Insertion validation	Industrial Vision	Mechanical Verification System	Operator Validation		

It is important to note that the values and the score presented on Table 4 do not refer to the state-of-art of the current technology but compare systems within the same price-range. Circles are used to represent the sub-solution scores, regarding the "flexibility" and the "overall dimensions", considering a scoring range from 1 to 3, where 1 circle is the minimum score and 3 is the maximum score.

Table 4. Receptacle positioning sub-solution score

	Positioning Tolerance	Repeatability Tolerance	Flexibility	Overall Dimensions
3 axis System	0.02	0.02	●	●
6 axis Robot	0.05	0.05	●●●	●●
Scara Robot	0.02	0.02	●	●●●
Delta Robot	0.05	0.05	●●	●●●
Cylindrical Robot	0.05	0.05	●●	●●

As demonstrated in the previous table, the 6 axis robot is the option that fulfils the requirements and achieves a higher scoring. Like the others, this robot fulfils the positioning and repeatability tolerances, but it also provides a better flexibility due to its 6th axis, thus providing an easier integration with the handling system and the ability to absorb possible tolerance and positioning problems in the feeding system. In terms of overall dimensions, it also scores positively.

The same process was applied to the handling sub-function. Table 5 presents the scoring of three conceptual sub-solutions regarding three evaluation parameters. The first parameter refers to the possibility of each concept ensuring the positioning tolerance required for the system and has a preponderance of 0.5. The second parameter addresses the system's easiness of producing and assembly, with a preponderance of 0.25. In this case, flexibility relates to the concept's ability to absorb and interact with the different sub-functions alternatives and has the same preponderance as the machinability.

Note that the notification and the scoring range to each sub-solution is the same as in Table 4, where 1 circle is the minimum score and 3 is the maximum score.

Table 5. Receptacle handling sub-solution score

	Scheme	Positioning Tolerance (0.5)	Machinability (0.25)	Flexibility (0.25)
Gripper Fingers System		●	●●●	●●●
Pneumatic-actuated system		●●●	●●	●
Vacuum System		●●●	●●●	●

Considering the relative weight of the parameters, the solution that best suits this project is the pneumatic system because it provides a better positioning tolerance while maintaining a relatively easy machinability. This derives from the way the solution is designed, thus making it easier to ensure much-needed close tolerances; but to do so, the machinability and the assembly process are more complex. Finally, it scored worse in the flexibility since it needs that the receptacles are fed in a very strict position.

4 Results

The main proposal of this paper is to develop a conceptual solution to solve the problem identified. The study shown in the previous chapter allowed, among several theoretical solutions, to achieve a conceptual solution.

This conceptual solution is detailed presented in this chapter and manages to carry out the sub functions that best fulfills the requirements previously identified as necessary to solve the problem.

This conceptual solution in presented and explained in this chapter.

4.1 Conceptual Solution

As the development of the concept was taking place, it was observed that there were two main aspects that the gripper needed to ensure in the first place: a gripping function and a positioning system that guarantee that the receptacle is in a known position.

The concept presented here (see Fig. 8), can be divided in two parts. The central part, the "guiding pin", will enter the receptacle and will guide the part into its correct position, and the gripping part will grab the receptacle with the help of a pneumatically actuated mechanism.

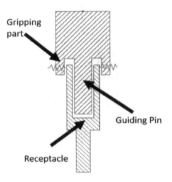

Fig. 8. Conceptual Solution

After this conceptual description, the team started to 3D-modeling the system, as presented in Fig. 9.

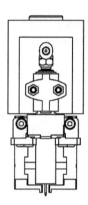

Fig. 9. Front View of the 3D-modeled solution

For a better understanding, the system and all of its sub-components are presented in an exploded view in Fig. 10.

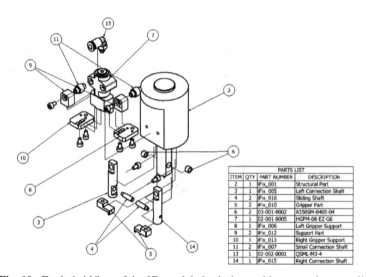

ITEM	QTY	PART NUMBER	DESCRIPTION
2	1	iFix_001	Structural Part
3	1	iFix_005	Left Connection Shaft
4	2	iFix_016	Sliding Shaft
5	2	iFix_010	Gripper Part
6	2	03-001-0002	A350SM-0405-04
7	1	02-001-0005	HGPM-08-EZ-G6
8	1	iFix_006	Left Gripper Support
9	2	iFix_012	Support Part
10	1	iFix_013	Right Gripper Support
11	2	iFix_007	Small Connection Shaft
13	1	02-002-0001	QSML-M3-4
14	1	iFix_015	Right Connection Shaft

Fig. 10. Exploded View of the 3D-modeled solution and its respective parts list

To ensure the needed precision, the part that will guide the receptacle needs to have very good concentricity with the part that will connect to the robot. The best way to achieve this is to aggregate these two functions in the same part and the result is a structural part that is responsible to position the receptacle and will be directly assembled into the robot flange. This structural part is shown in Fig. 11.

The robot will position the handling system in such a way that the guiding pin will enter in the receptacle, ensuring that the receptacle is fully guided in all directions. This part that needs to have tighter tolerances and it will be dimensionally controlled for guiding pin concentricity, relative to the robot flange part, and its diameter.

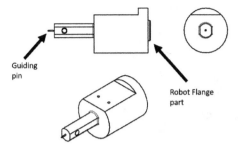

Fig. 11. Views of the structural part

After receptacle positioning is guaranteed, the system grabs it and holds it in the same positioning throughout the insertion process. For this, a system was developed, based on a pneumatic finger gripping system that is presented in Fig. 12.

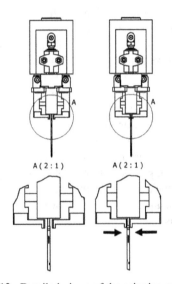

Fig. 12. Detailed views of the gripping system

The gripper parts, that are solidary with the parallel fingers gripper through the connection shafts, will press the receptacle against the guiding pin with little but enough force to hold it in place.

This system will be integrated into a 6-axis Robot. This robot allows a precise positioning of the receptacles and also provides flexibility that can help when interacting with the automatic feed system that will be developed in the future.

Figure 13 presents the handling system assembled into the robot.

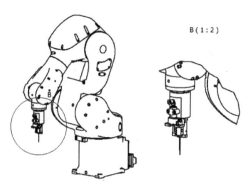

Fig. 13. Overview of the system attached to the robot

5 Conclusions and Future Developments

This paper addressed the specific problem of finding an accurate solution for the automation of a manual manufacturing process.

The conceptual solution here presented highlights the need for automation of this process and the consequent reduction of all parameters that cause the final product's current quality failures. The achieved precision, ensured by the system, is 0.05 mm – thus accomplishing the main goal – and, more importantly and so hard to achieve manually, the repeatability was also set to 0.05 mm. The solution also allows a considerable time reduction in the receptacle insertion and wiring process of at least 50%, from the previous mentioned week (40 h) to an estimative of 20 h.

This paper stresses the importance of a well-established methodology to ensure solution development according to design objectives and engineering requirements.

The next steps will include the detailed design of all sub-solutions and their integration with each other. Finally, a proof-of-concept prototype will be developed to test and validate all the sub-solutions .

References

1. Canadas, N., Machado, J., Soares, F., Barros, C., Varela, L.: Simulation of Cyber Physical Systems Behavior Using Timed Plant Models. Mechatronics **54**, 175–185 (2018). https://doi.org/10.1016/j.mechatronics.2017.10.009
2. Kunz, G., Machado, J., Perondi, E., Vyatkin, V.: A formal methodology for accomplishing iec 61850 real-time communication requirements. IEEE Trans. Ind. Electron. **64**(8), art. no. 7878522, 6582–6590 (2017). https://doi.org/10.1109/TIE.2017.2682042
3. Kunz, G., Perondi, E., Machado, J.: Modeling and simulating the controller behavior of an automated people mover using IEC 61850 communication requirements. In: IEEE International Conference on Industrial Informatics (INDIN), art. no. 6034947, pp. 603–608 (2011). https://doi.org/10.1109/INDIN.2011.6034947
4. Silva, M., Pereira, F., Soares, F., Leão, C.P., Machado, J., Carvalho, V.: An Overview of Industrial Communication Networks. In: Flores, P., Viadero, F. (eds.) New Trends in Mechanism and Machine Science. MMS, vol. 24, pp. 933–940. Springer, Cham (2015). https://doi.org/10.1007/978-3-319-09411-3_97

5. Leão, C.P., Soares, F.O., Machado, J.M., Seabra, E., Rodrigues, H.: Design and development of an industrial network laboratory. Int. J. Emerg. Technol. Learn. **6**(SPECIAL ISSUE.2), 21–26 (2011). https://doi.org/10.3991/ijet.v6iS1.1615
6. Campos, J.C., Machado, J.: Pattern-based analysis of automated production systems. In: IFAC Proceedings Volumes (IFAC-PapersOnline), 13 (PART 1), pp. 972–977 (2009). https://doi.org/10.3182/20090603-3-RU-2001.0425
7. Campos, J.C., Machado, J., Seabra, E.: Property patterns for the formal verification of automated production systems. In: IFAC Proceedings Volumes (IFAC-PapersOnline), 17 (1 PART 1) (2008). https://doi.org/10.3182/20080706-5-KR-1001.4192
8. Leão, C.P., et al.: Web-Assisted Laboratory for Control Education: Remote and Virtual Environments. In: Uckelmann, D., Scholz-Reiter, B., Rügge, I., Hong, B., Rizzi, A. (eds.) ImViReLL 2012. CCIS, vol. 282, pp. 62–72. Springer, Heidelberg (2012). https://doi.org/10.1007/978-3-642-28816-6_7
9. Ancău, M.: The optimization of printed circuit board manufacturing by improving the drilling process productivity. Comput. Ind. Eng. **55**(2), 279–294 (2008). https://doi.org/10.1016/j.cie.2007.12.008
10. Bateson, J.T.: In-Circuit Testing. 1st edition Springer, Dordrecht (1994)
11. Professional Plastic. https://www.professionalplastics.com/professionalplastics/BedofNailsTestFixturePCBOverview.pdf. Accessed 02 Feb 2021
12. I.-T. Solutions. http://www.insidelimits.pt/. Accessed 25 Jan 2021
13. Ingun. https://ingun.com/en/Products/Productfinder/GKS_KS/KS-075/KS-07547E03. Accessed 15 Jan 2021
14. Ingun. https://ingun.com/ingun-media/Global/Downloads/Print/PI_GKS_SW-KS-xxx.pdf?la=de-DE&hash=1EC2BBC6AAD3E4FDDA0D2D14F04295120EB3DEA7. Accessed 15 Jan 2021
15. French, M.J., Gravdahl, J.T.: Conceptual Design for Engineers. 3rd edition. Springer, Heidelberg (1985)
16. Konzipieren technischer Produkte. VDI-Richtlinie 2221. VDI-Verlag, Düsseldorf (1973)
17. N. Cross., T.: Engineering Design Methods: Strategies for Product Design. 3rd edition. John Wiley & Sons, Ltd, Hoboken (2000)

Selection and Development of Technologies for the Education of Engineers in the Context of Industry 4.0

Pedro José Gabriel Ferreira[✉], Silvia Helena Bonilla, and José Benedito Sacomano

Universidade Paulista, Doutor Bacelar, São Paulo - SP 1212, Brazil
pedro.ferreira@docente.unip.br

Abstract. The fourth industrial revolution has been changing the way new products are produced, global competition and labor relations. More than that, it alters social relationships and educational methodologies. Educational administrators are responsible for adjusting the curricula of engineering courses in order to prepare new professionals for the needs of the industry. In this work an exploratory research is carried out in the literature for the selection and development of educational technologies, which can assist in the new teaching methodologies for the education of Engineers, in the context of Industry 4.0. A market research is carried out to evaluate the technologies available in Brazil with good cost-benefit and partnerships are established for the development of an industrial automation kit. The selected and developed equipment is presented, as well as the design of a laboratory for innovation and entrepreneurship called Fablab and implemented by the Universidade Paulista - UNIP.

Keywords: Industry 4.0 · Fablab · Virtual reality · Engineering education

1 Introduction

The fourth industrial revolution has spread across the world and makes industrial processes increasingly competitive. The concept of I4.0 is based on the integration of information and communication technologies and industrial technology, and is mainly dependent on building a cyberphysical system to realize a digital and intelligent factory, to promote manufacturing to become more digital, information led, tailored, and green [1].

Nine technologies are reshaping production processes [2], they are: Robotics, big data, virtual reality (VR), integration of vertical and horizontal systems, internet of things (IOT), cyber security, cloud computing, additive manufacturing and augmented reality. As a reflection, the need for universities to adapt their curricula arises, in order to provide the market with professionals trained to work with these new concepts. Education must be aligned with industry to increase competitiveness and generate the talents required (interdisciplinary) for Industry 4.0 [3]. In order for these new professionals to be properly trained, we will need to invest in the development of engineering courses and innovation.

It is understood, therefore, that the concept of Industry 4.0 or Advanced Manufacturing requires strong investment in education and innovation so that it can contribute to the dissemination of knowledge that make it possible to understand its real impact on production processes and changes in the Brazilian economy [4].

For some time, we have been discussing what skills would be needed for the training of Engineers. Professional competence means the personal ability to mobilize, articulate and put into action knowledge, skills, attitudes and values necessary for the efficient and effective performance of activities required by the nature of the work and the technological development [5]. The latest revisions made in the Brazilian curriculum guidelines for engineering courses have already addressed the issue of competencies, a deficiency in the development of non-technical competencies during graduation is demonstrated [6]. It is up to the educational administrators to remodel engineering courses not only with these competencies provided for in the resolution cited, but also to understand which skills are necessary for the training of the Engineer capable of acting in Industry 4.0.

In order to stimulate new skills, educational tools can be used. UNIP uses in its Engineering courses what is known as, Supervised Practical Activity - SPA, also known as challenge activity. SPA activity enriches physics teaching at a higher level, preparing and training students and reinforcing their cognitive processes through experimentation and social and interpersonal relationships [7].

Other tools can help SPA and encourage student innovation and entrepreneurship. The need for people able to develop solutions to problems of the technologically oriented modern world has steadily increased [8]. One example is the Maker Space or FabLabs, which enable students to develop and create their products without necessarily having complex means of manufacturing them. Maker Spaces are environments where apprentices, designers, engineers and anyone with an idea can exercise their creativity safely and easily, with the help of technical facilitators and/or technology in the development of creative work [9]. 3D printing helps engineering students learn via rapid prototyping and enhances CAD drawing skills, troubleshooting, and optimization [10].

Tools such as Virtual Reality and Augmented Reality, in addition to being pointed out as technologies that are reshaping production, also become increasingly easy and cheaper to use. The use of Augmented Reality and Virtual Reality devices is on the rise and these devices are increasingly realistic and less costly financially [11].

Another very important item for the teaching of industrial automation and consequently the concepts of Industry 4.0 is the Programmable Logical Controller or PLC. It is a vital component in any automation process and considered as one of the technologies that will be present in this type of concept for a long time, even with the new technologic advances. Even in the age of Industry 4.0 and industrial internet, it can be assumed that these controllers will continue to be required to a large extent for tomorrow's production [12].

Based on the industry 4.0 development scenario and exploratory and market research, this work aims to identify, select and develop tools that support educational methodologies that can provide new professionals with the competencies required by the Brazilian national curriculum guidelines and those necessary to work in Industry 4.0.

2 Methodology

The first stage of the work was to conduct an exploratory research in the literature, evaluating which technologies could be used to assist in teaching methodologies aimed at training Engineers, with the skills needed by the industry. Educational specialists of our university were consulted to validate the technologies selected.

Subsequently, a market research was conducted with suppliers of potential industrial and educational equipment to identify which technologies could be provided immediately and which ones could be developed together. Two of the selected suppliers presented technologies already developed and that could be tested immediately, such as 3D printers and 3D virtual commissioning software for factory automation and equipment. The third supplier suggested a partnership to develop kits that integrated industrial processes through a suitcase of experiments with a programmable logic controller (PLC) and a Human Machine Interface (HMI), along with an exercise book.

After establishing the technologies that would be employed, the equipment was provided for testing and the PLC /HMI kit was developed together. Training was conducted to train teachers and technicians in their use.

Finally, a space called FabLab UNIP was developed to accommodate 3D printers, through operational tests with teachers, technicians and students to assess needs and subsequent use throughout the community.

3 Discussion and Results

As a result of the exploratory research and market research conducted, some equipment was acquired by the university aimed at assisting in learning by challenge. For students to start their Reverse Engineering work, a scanner has been selected for creating 3D images and later printing, presented in Fig. 1. It is possible to position an object on the turntable, performing a full scanning automatically.

Fig. 1. Scanner used for reverse engineering.

For students to perform various operations in the manufacture of a prototype, such as laser cutting and engraving, milling and 3D printing, a hybrid equipment has been

Fig. 2. Hybrid equipment with its milling and printing heads.

selected, which is presented in Fig. 2. It is possible to change the heads and perform different operations on the same equipment.

For prints that require greater definition or require special materials, such as biocompatible scans, the University acquired the equipment with stereolithography printing technology, presented in Fig. 3.

Fig. 3. 3D Printer - digital light processing (DLP).

For the installation of printers and creation of an environment where students can be creative and innovative, designing, manufacturing and testing their prototypes, the University has designed a laboratory. This laboratory was certified by the FabLab network, which means it belongs to the worldwide network of laboratories with common characteristics. This type of space will allow an exchange of knowledge between students, communities and businesses from around the world. The laboratory plant is presented in Fig. 4 and the laboratory in Fig. 5.

The 3D virtual commissioning software for factory automation chosen was Ciros, presented in Fig. 6. Students and teachers can simulate and program industrial processes, see the manufacture of products, interfere in the process, cause errors and see the corrections made by the system.

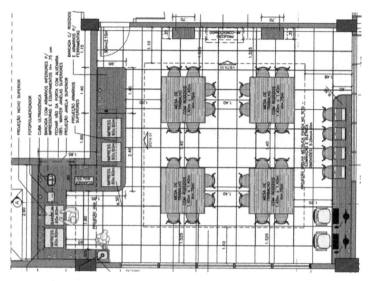

Fig. 4. Fablab UNIP plant, located on the Bacelar campus.

Fig. 5. FabLab UNIP pictures

Fig. 6. Ciros software screen - Industrial production line

The kit presented in Fig. 7 was developed by one of the authors in conjunction with an automation company.

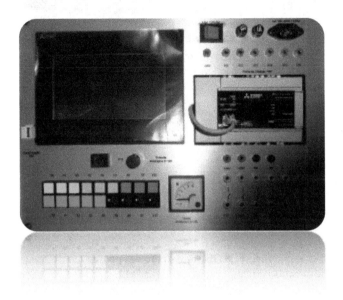

Fig. 7. PLC and HMI kit developed in partnership with automation company.

It is important to notice that although HMI and PLC cannot be considered as innovative technologies, the integration of them will bring to students, the possibility of monitoring the results of their programming inputs simulated on screen as shown in Fig. 8. The adoption of the suitcases offers a cheaper opportunity (unique in Brazil) to the university since real processes need space and are expensive to be mounted. In a suitcase, HMI and PLC are packaged together and a book with 12 exercises has been developed, where the student will learn how to program a PLC, with data recording and movement, the use of the timer on delay and timer off delay, how to use their digital and analog inputs and outputs and program the HMI. Real industrial processes were chosen to elaborate the exercises, which are visualized on the HMI screen. Answers to the exercises are provided to assist teachers during simulations. Two examples of exercises are cited as follows:

The first one involves the control of a tank level (low and high level) by triggering a pump. In this way, the student can program the PLC to trigger the pump as a function of the tank level provided by the level sensors (Fig. 8).

Fig. 8. First example: control of a tank level (screen).

The second example involves a thermic treatment process control where the number of samples as well as the treatment time can be established by the student. A conveyor belt is triggered to direct the samples to the oven and a control of opening/closing gates can be programmed, as shown in Fig. 9.

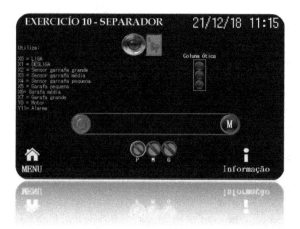

Fig. 9. Second example: thermic treatment process control (screen)

Table 1 shows the selected technologies, the learning outcomes and the justification related to curriculum improvement for each one of them.

Table 1. Technologies, learning outcomes and justification.

Technology	Outcomes	Curriculum improvement
3D Printers	Innovation Entrepreneurship	Enable the students to design, produce, test and modify the prototype in a fast and cheap way Students can do the whole process at the university (not depending on other factories, without spending more time and money) Enable the exchange of prototypes between our university and the worldwide network (FabLab concept)
Virtual reality	Problem solving ability Technical knowledge Global vision	Allows the testing of the industrial processes that currently could be expensive to the university Vision of system capacity of correction
Kit PLC + HMI (suitcase)	Technical knowledge Process vision	Enable the students to visualize the industrial process operating and the results of programming change (Screen of HMA)

4 Conclusions

Exploratory research has contributed to the selection of educational tools that can be used in teaching methodologies aimed at developing specific capabilities for new engineers to act in the context of industry 4.0. The market research helped to select equipment that has good acquisition cost and capability. Fablab is being implemented and will serve to support challenge activities and final course work. Its efficiency will be measured with questionnaires that will be applied to teachers and students, after the use of new technologies. Since this is the first year that students are subjected to the new technologies the learning outputs are difficult to evaluate and thus the research shows this limitation. Even so, learning outputs will be evaluated after the first year of implementation through performance tests and surveys. Simultaneously, the research team of Industry 4.0 is gathering insight through an industrial experts' survey in order to gain feedback of the selected techniques and broaden the outlook on other possibilities adopted by industrial network. To evaluate the technologies and proposed methodology a survey is now been conducted involving professors of different countries (Poland, Japan, United States, Brazil and Portugal).

References

1. Mendoza, R.A.R., Menendez, R.M., Saldivar, R.P.: Engineering education 4.0: proposal for a new curriculum. In: IEEE Global Engineering Education Conference 2018 (EDUCON), pp. 1273–1282. Curran Associates, Red Hook (2018)

2. BCG Homepage. https://www.bcg.com/capabilities/operations/embracing-industry-4.0-rediscovering-growth.aspx. Accessed 28 Feb 2020
3. Chou, C.M., Shen, C.H., Hsiao, H.C., Shen, T.C.: Industry 4.0 manpower and its teaching connotation in technical and vocational education: adjust 107 curriculum reform. Int. J. Psychol. Educ. Stud. **5**(1), 9–14 (2018)
4. Alarcon, D., Da Rosa, L.Q., da Silva, R.S., Muller, F.M. De Souza, M.V.: Os desafios da educação em rede no contexto da indústria 4.0. In: Congresso Internacional de Conhecimento e Inovação 2018, vol. 1, pp. 53–61 (2018)
5. CNE Homepage. http://portal.mec.gov.br/index.php?option=com_docman&view=download&alias=15766-rces011-02&category_slug=junho-2014-pdf&Itemid=30192. Accessed 25 Feb 2020
6. Carvalho, L.D.A., Tonini, A.M.: Uma análise comparativa entre as competências requeridas na atuação profissional do engenheiro contemporâneo e aquelas previstas nas diretrizes curriculares nacionais dos cursos de Engenharia. Gestão Produção **24**(4), 829–841 (2017)
7. Ferreira, P.J.G., et al.: Ensino de Física em Cursos de Engenharia e Atividades Práticas Supervisionadas: Uma proposta de ensino baseada na aprendizagem por desafio. In: Brito, C. da R., Ciampi, M.M. (eds.) International Conference on Engineering and Technology Education 2014, vol.13, pp. 261–265 (2014)
8. Verner, I., Merksamer, A.: Digital design and 3D printing in technology teacher education. In: Shpitalni, M., Fischer, A., Molcho, G. (eds.) CIRP 25th DESIGN CONFERENCE INNOVATIVE PRODUCT CREATION 2015, vol. 36, pp. 182–186. Elsevier, Amsterdam (2015)
9. Brockveld, V.V.M., Teixeira, C.S., Da Silva, M.R.: A Cultura Maker em prol da inovação: boas práticas voltadas a sistemas educacionais. In: ANPROTEC 2017. Anprotec, Brasília (2017)
10. Chong, S., Pan, G., Chin, J., Show, P.L., Yang, C.K., Huang, C.: Integration of 3d printing and industry 4.0 into engineering teaching. Sustainability **10**(11), 3960 (2018)
11. Morenilla, A.J., Romero, J.L.S., Mora, H.M., Miralles, R.C.: Using virtual reality for industrial design learning: a methodological proposal. Behav. Inf. Technol. **35**(11), 897–906 (2016)
12. Langmann, R., Pena, L.R.: PLCs as industry 4.0 components in laboratory applications. Int. J. Online Eng. **12**(7), 37–44 (2016)

An Exploratory Approach with EEG – Electroencephalography in Design as a Research and Development Tool

Bernardo Providência[1] and Rute Silva[2(✉)]

[1] Teacher at School of Architecture, University of Minho, Guimarães, Portugal
`providencia@arquitetura.uminho.pt`
[2] Master in Product and Service Design, University of Minho, Guimarães, Portugal

Abstract. This article discusses methodologies based on EEG - electroencephalography, applied to ED - emotional design and consequently, to the relationship between the user and the product. The objective is to understand whether, based on the brain response, it is possible to identify the individual's emotional factors when interacting with the products. The process by which users enjoy experiences when interacting with products is the result of a context of use, fundamental for companies as well as for designers in the design and development of more positive new products. In this way, it is crucial to approach different areas of design and engineering on multidisciplinary teams. EEG technology and ED are approached, based on practical cases as well as the evaluations of experiments carried out in the Psychological Neuroscience laboratory that validate the application of these concepts. Through this, it is expected that the contribution of the EEG and the self-report tools, will validate Wellington's (2007) ethnographic assessment assumptions about motivation and engagement. As a methodology, they are expected to contribute to the creation tools that meet the new trends more focused on semantic values - ideological and playful, than on pragmatic values - practical and critical. Thus, it becomes relevant to answer questions about how the incorporation of this technology - EEG - can contribute to the strengthening and complementarity of the interaction between engineering and design, in the construction of products and services that, in addition to satisfying basic needs, contribute to your wellbeing [1].

Keywords: Electroencephalography (EEG) · Product design · Engineering · Design and emotion · Neuro design

1 Introduction

Based on the literature review, it is intended to present a theoretical model for a practical intervention in the evaluation and development of experiences with the product through the EEG. In particular, the intersection of neurosciences and emotional design as a contribution to engineering. This investigation emerges as a response to the existent gap in the semantic evaluation of emotional design [2], since the responses given in the interaction with the product refer more commonly to functional and usability assessment. In specific, understand if the communication process at semantic level can be validate using technologies, through the EEG, which allows us to perceive the brain's response to stimuli. This study is part of a partnership with the Psychological Neuroscience Laboratory of the Faculty of Psychology of the University of Minho, with extensive experience and publications in this area, which allowed the development of the necessary protocols and investigations with the EEG evaluation technology.

A good understanding of neural and emotional response between users and products is essential for success in developing new products. In the design process – product evaluation/validation – self-report scales are applied to assess user's feelings and preferences. Recently, to improve these assessments, EEG has been applied to measure brain activity in order to understand sensory perceptions of consumers' emotional responses. The reason why this technology has been useful to support self-report scales is the fact that it directly measures users' physiological and emotional responses [3].

The intervention of the EEG in the emotional issue of the human being comes from the need to identify emotions [4]. However, studies report that human beings can control their facial expressions or vocal intonation, manipulating the expressions they intend to transmit to others. Thus, the focus is on the recognition of internal emotions through the EEG, witch can be visible in real-time through the intensity, magnitude, and speed of the brain response, as well as the correlation between the left and right hemisphere in response to food stimulus [3]. This approach leads us to question the possibility of correlating EEG data with António Damásio's somatic markets.

> *"Damásio (1996), with the formulation of the somatic marker hypothesis, proposes a mechanism through which the emotional processes can guide the behavior (or influence a tendency for a determined behavior), particularly the decision making. This hypothesis declares that emotion acts automatically as a positive and negative marker in cognitive decisions, to maintain the system necessary for survival in a healthy society. Thus, the markers work as guides in the problem-solving processes and we resort to them when it is necessary to decide."* [5].

Emotional design is one of the main considerations when it comes to sensory perceptions. This concept emerges in the late twentieth century and works the user's interaction with objects from sensory and cognitive experiences, underlining perception, and stimuli:

> *"Cagan and Vogel (2002), identified a group of values to be considered by designers, named Value Opportunities (VO). The OV set includes emotion, aesthetics, product identity, impact, ergonomics, core technology, and quality. (...) This show, that it is remarkable that emotion is the first VO that adds value to a product and improves the experience of users."* [2].

Through these authors, it is possible to identify that emotion adds value to a product, improving the experience and can affect the purchase decision. From the scientific point of view, the triangulation resource is important because it allows for understanding a research question from multiple perspectives. Patton (2002) points out that the goal of triangulation isn't to get consistency across data sources or approaches [6]. In Patton's view, these divergences should be an opportunity for deeper meaning in the data and not be something that weakens the research purpose.

Briefly, the investigation included:

The bibliographic review on theoretical concepts as well as the relationship between them, thus allowing to support the purpose of this investigation.

Understand the operation of EEG technology and its potential, starting the fieldwork that involved: observation, experimentation and graphic recording in the laboratory. This step also helped to understand how these methodologies are applied in this investigation. Then, methods that can be complementary were used, such as self-report scales - which are already used in this type of investigation.

The experimental phase involved the development of a practical case - a case study. The design of the experience was defined based on the protocols used in the EEG technique and with the concepts of emotional experience, namely the approach by (Norman, 2014) which refers to three levels of assessment: visceral, behavioral, and reflective [7].The participant is evaluated using the EEG technique and the Likert scales that allow the evaluation of semantic differentials [8].

Following the previous steps, it is concluded that this technique can be an asset in the evaluation phase in the development of new products, allowing to make the emotional characterization of the response to objects.

2 Methodologies

In this study, two different methods are approached that allow understanding of how the human being feels about a certain product - brain processing evaluation scale, EEG and self-report scales. In the design of the investigation, it is important to consider the methodological triangulation, for a validation of the questions to be answered. Thus, from the methodology, it is interesting to compare the different studies and methods - quantitative and/or qualitative, thus allowing to create a sustained model to increase the veracity of the study.

One of the considerations of the two methods is that, if, on the one hand, the data collected through the EEG with the electrodes allows to directly record the participant's brain activity, on the other, the information collected through the questionnaire with self-report method allows to obtain the opinion of each participant. Songsamoe (2019), argues that:

> "However, these measurements are subject to cognitive bias from consumers because all of the measurement data are obtained from consumers' reported thoughts or questionnaires." [3]

Still, on the perspective of self-report assessment based on semantic analysis, Krippendorff says:

> "The study of the symbolic qualities of man-made forms in the cognitive and social contexts of their use and the application of the knowledge gained to objects of industrial design. By this definition, product semantics is not a style, program, or movement. Rather, it is a concern for the sense artefacts make to users, for how technical objects are symbolically embedded in the fabric of society." [2]

Krippendorff also mentions, contrary to other authors, that *"the form does not follow the function, but the meanings"* [2]. The principle of product semantics is not primarily concerned with issues of functionality, the semantics of the product is based on what people tend to understand when interacting with the product. The human being sees beyond the characteristics of a product, these serve as a means of message, through which the user unleashes meanings. This question leads us to consider that if an object is purely utilitarian, it means that it lacks presence and is only visible in the act of use. The emotional component - emotional design, is another dimension that goes beyond what is functional, thus correlating the concept of functionality and symbolism in the same product. For a product to be understood from symbology, the user needs to be interconnected with the product, that is, there is interaction with it. Only in this way, it is possible to know what the user says because the meanings emerge from the interaction process. From the research point of view, the symbolism inherent in a given product is expressed by the user's attitude and what he says, that is, the user is exposed to a certain stimulus - product, in a given context, where the interaction results.

Regarding the methodology in its practical component, the investigation follows five stages:

First, the design of the experiment - based on EEG technology and self-report scales. Human resources - participants, material and space - are also considered at this stage. When? How? which makes it possible to gather the necessary conditions to start this phase of the investigation.

Second, data collection - This phase considers a universe between 12 to 20 participants, with adulthood.

The experiments are carried out in the facilities in the EEG collection room - Psychological Neuroscience laboratory. Also, all equipment and material are made available by the same. About the experience, the focus of the evaluation is based on two variables: visual evaluation through photographs and the real object; tactile evaluation through the object. The technique of recording brain activity is prepared, in the period in which the EEG is registered, the participant is exposed to a certain stimulus - in this case, a product (kitchen utensil).

The realization of the experiment, after preparing the laboratory and the participant, involves three tasks for data collection. In the presentation, each photograph is shown to the participant for 1 s with a repetition of 50 times. The repetitions of the images are interspersed with the Likert scale for the participant to classify according to the categories being evaluated. In the physical product, in the visual variable, the participant observes each different type of object for 5 s with a repetition of 35 times. In this case, the Likert scale is only answered by the participant at the end of the task. In the tactile variable, the considerations are the same. At the end of the three tasks, the participant responds to a questionnaire with open answers that complete the categorization through the categories exposed on the Likert scale.

The third phase involves the treatment and analysis of data from the methods applied in the evaluation. The data are obtained through the following methods: registration of the EEG technique; self-report scales - are answered by the participant during the recording of brain activity; questionnaire - answered at the end of the activity registration. In the treatment of data from information collected through self-report scales, categories are defined based on the evaluated concepts, for example, interesting and uninteresting concept. Subsequently, the EEG data are separated based on these same categories, thus allowing a comparison between both dimensions. This comparison is made at the statistical level, parameters of amplitude and latency of the components of the potential evoked by the product.

Finally, relate the theoretical foundation and practical results.

The methods and tools applied in the study were defined through a literature review. The next section describes the emotion assessment tools considered to apply in the practical case - self-report scales and brain function measures.

2.1 Brain Processing Evaluation Measures

The use of technology to measure brain processing using magnetic resonance images has allowed to deepen the knowledge about brain processing, namely using fMRI (Functional magnetic resonance images) as in the case of studies on the emotional stimuli associated with the music of Oliver Sacks, where from the magnetic resonance records it allows to perceive the mapping of certain areas of the brain during the listening of music scores [9]. At the same time, the use of EEG has become more prominent in this type of mapping from the point of view of the clinical forum and recently, with the application of this technique in the design area, making it possible to identify several advantages of using EEG in the emotional evaluation of the user with a particular product.

> *"The EEG technique is currently the most commonly used method for measuring brain waves with high temporal resolution. It can measure the brain waves in real time and provide much useful data. EEG data obtained from brain wave measurements can be analysed in terms of mood, acceptance tendency and brain functioning* (Andersen et al., 2018). *Therefore, it is not surprising that the brain activity of consumers is very important, and the EEG technique is emphasised for application in sensory research and consumer behaviour evaluations."* (Shaw & Bagozzi, 2018) [3].

The EEG is the study of how the cognitive functions (including perception, memory, language, emotion, behavior monitoring/control, and social cognition) are supported or implemented by brain the electrical activity [10]. This technique of hightemporal-resolution is an exceptional tool for study of cognitive processes and one of the reasons is the ability to capture cognitive dynamics in the period in which cognition occurs [10]. The collection of data from the EEG, in addition to be a non-invasive technique, allows us to obtain data-rich information – depending on the number of electrodes that are used. The author also says that cognition is fast, brain rhythms are fast, and brain imaging technologies, such as EEG, can capture this fast dynamic by directly measuring brain activity.

2.2 Self-report Measures

In the field of design, the use of data collection methods for self-report evaluation is frequent, which consists of applying a set of tools that go through observation, questionnaire or semi-structured interviews and that allow the evaluation of quantitative and qualitative parameters/responses [11]. In conclusion, these self-report evaluation measures, in addition to being simple to access, allow information to be obtained quickly and objectively. Thus, in this investigation aimed at applying semantic analysis, the observation tool and questionnaires were applied - completed with self-report scales.

Questionnaire
The questionnaire is a means of communication between the researcher and the participant, that is, it is an investigation method with a set of questions (open, closed or mixed) that aims to achieve a certain knowledge [11].

SAM
The Self-Assessment Manikin (SAM), developed by Peter Lang – 1980, is a technique of image evaluation of the affective response associated with an object or context. This combines three types of image: Fig. 1 - to measure pleasure - happy or unhappy; motivation - active or relaxed; dominance - greater or less control in each situation, when faced with a wide variety of stimuli. This method is easy to apply and non-verbal, which allows you to assess the reports of affective experiences quickly [12].

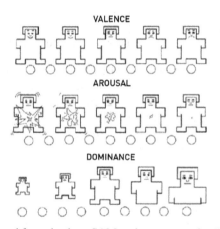

Fig. 1. Adaptad from the three SAM scale representation (Lang, 1980)

PANAS
The Positive and Negative Affect Schedule or (PANAS), developed by psychologists David Watson, Lee Anna Clarck and Auke Tellegen – 1988, is a scale with different words that describe feelings and emotions. PANAS measures the positive and negative affect of the user, realizing how he feels at a given moment and how those same emotions influence him to act and make decisions. On the one hand, positive affect – positive emotions and expressions, such as joy or satisfaction; on the other hand, negative affect – negative emotions and expressions, such as anger fear, or sadness. This scale is applied through a 5-point Likert scale questionnaire for scoring. PANAS is a reliable and consistent

scale, with scores ranging from 0.86 to 0.90 for PA and from 0.84 to 0.87 for NA (Magyar-Moe,2009) [13].

POMS

The Profile of Mood States or (POMS), developed by McNair, Lorr, and Droppleman-1971, is a scale that measures the six factors of the mood state with a 5-point Likert scale - the basis for answering a questionnaire with 65 questions. Several studies prove its suitability for sensitively, accurately and validly measuring the moods of individuals, whether in a psychiatric context or with a non-psychiatric population (Boyle, 1987; McNair et al., 1971; Norcross, Guadagnoli & Prochaska, 1984; Weckowizc, 1978) [14].

3 Study Cases

This section aims to demonstrate and validate the application of EEG concepts as an evaluation method in the product development process. First, the equipment and the data collection process with EEG used in the laboratory and this investigation - pilot case, are mentioned. Then, the methodology applied in the experiment is approached, which is based on the investigations that have been carried out in the Neuroscience laboratory, taking into account the methods and results analyzed in the case studies. These case studies are referred to in the article, to demonstrate how the EEG technique has been applied in the product area about the participant's evaluation.

3.1 Equipment and Conditions for Data Collection with EEG

Regarding the collections in the area of Psychology, the method used in the Psychological Neuroscience Laboratory of the University of Minho is considered, where the fieldwork is carried out to understand the technology, observe and developing experiences of the collecting EEG data. The data collection includes a laboratory coordinator, the researcher, the participant, equipment for recording brain waves, hardware, and software for collecting and analyzing data. Briefly, the procedure follows the following order: 1. Preparation of the laboratory - computers and all the necessary material to carry out the collection in an isolated space of any type of radiation; 2. Preparation of the participant – introduce and explain the session; cap placement (perimeter measurement to choose the appropriate cap for the participant); skin preparation; placing the electrodes on the skin of the face; placement of electrodes in the cap – Fig. 2 – layout applied in the experiments. To start the collection, the participant moves to another room compartment, where he is accommodated to make the smallest possible movements, and the electrodes are connected to the registered electrical signal amplifier. Through a computer, after explaining what is wanted on the part of the participant, a sequence of images is presented that contains the exercise through which the registration is made in real-time. After the collection, all the electrodes of the participant are removed, and later the analysis of the records is made.

The EEG signal are recorded using an ActiveTwo Biosemi system (Biosemi, Amsterdam, the Netherlands) with 64 pin-type active electrodes inserted in an elastic cap according to the international 10-10 system. Electrooculogram is also recorded by two electrodes placed next to the outer canthi of the eyes (HEOG), and by one electrode placed below the participants' left eye (VEOG), using a Common Mode Sense (CMS) and Driven Right Leg (DRL) montage of two electrodes located over the vertex as reference. The electrode offset is always kept below 30 mV. EEG data is filtered online between 0.01 and 100 Hz, and digitized at a sampling rate of 512 Hz. EEG signals are stored for posterior off-line analyses using the Matlab toolboxes EEGLab (Delorme & Makeig, 2004) and ERPLab (Lopez-Calderon & Luck 2014).

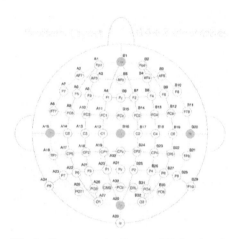

Fig. 2. Representation Layout (BioSemi, n.d).

In addition to this, the fieldwork, allowed to participate in the collections regarding the preparation of the participant; identify the methods of experiments that are applied to obtain a certain type of result and identify possibilities and limitations. Two investigations were observed, in which the participant's experience was performed through the representation of a computer image. In one of the studies, the participant had to observe the images on the screen and say what the same image corresponded to - using only the voice. In the other study, the participant had to interact with the computer and make decisions on how he wanted to organize the objects. These questions allow us to consider some problems in the data collection phase and to design the design of the experiments.

3.2 Application of Methodologies in Design – Case Study

Article - Understanding consumer physiological and emotional responses to food products using electroencephalography

In this article, the author focuses on the principles and applications of EEG in food research, with food being the factor under evaluation. According to Tammela, a consumer's brain activity, in terms of physiological response, is important for the emotional response related to food and food preference [3]. Linforth reinforces these issues by referring that the consumer's brain waves can sensitively change when facing stimuli, names, food appearance, odors, and food flavors [3].

As an evaluation method, consumers are stimulated by different factors: food properties, environmental factors, and internal consumer factors. The assessment is done in two ways: first, "biofeedback" response; second, changes in brain activity and emotional response, for example, like, dislike. In these types of products, brain waves are usually measured before, during, and after exposure to stimuli. The use of the EEG technique included: the participant, the stimuli (food, image, sound, environment, among others), the recording equipment - the electrodes, and the analysis software. In the general procedure, the electrodes are placed on the scalp and then connected to the brain wave amplifier. After preparation, the participant is exposed to stimuli, and brain waves are recorded in real-time and subsequently analyzed.

The author's final considerations for the study, are that although self-report scales are used to study consumers' emotional responses to food products, these measures are obtained from thoughts reported by consumers or through questionnaires. With this, the application of EEG is seen as an opportunity because it allows measuring the consumer's physiological response directly. They say that changes in alpha and beta waves clearly illustrate changes in consumers' emotions and brain states with food stimulation [3].

Article - *Assessing Product Design Using Photos and Real Products*

The article investigates the validity of using photos of the product instead of real products in the user's responses to the design with the evaluation of the design of a car. The authors of this article used explicit and implicit methods such as EEG, eyetracking, and questionnaires [15]. In the evaluation - comparing the users' perceptive responses when using photos and real cars - they used three distinct perceptual aspects: explicit responses - using a questionnaire; visual responses - patterns of looking; neural responses - use of the EEG technique.

In the experiment procedure, two-car models were considered, named A and B of different brands. Two experiments were carried out, one with photos and the other with real cars. Car A was used in both experiments, while B was a different color in each experiment. In the first experiment, the participants evaluated the photos of the product, and in the second, the physical product (the participants in each experiment were different so as not to induce the memory recovery process). In the questionnaire, with a scoring system from 1 to 7, three aspects were considered: preference - indicates whether to like or dislike the car design, which is crucial for the decisionmaking process; luxury - often considered in this type of product design; harmony - consistency between the design of the different components of the car. Subsequently, the considerations were made from an average and standard deviation [15]. The EEG technique was used only in car A since the question of color variation in car B could be an unwanted influence on neural responses. Regarding the data recording, a gray screen was displayed for five seconds to record the basic signals, for twenty seconds a photograph of the car and between the images, five seconds of a gray screen. This evaluation showed that the use of real products is desirable to understand users' responses more precisely, as the photo and real car responses show to be considerably different [15].

3.3 Deduction for Study Cases

Intending to legitimize this investigation, the work that is carried out in the neuroscience laboratory, the experiments carried out so far and the bibliographic references analyzed - having been presented in the article two practical cases allow to understand the following aspects: in both investigations the EEG method following the common collection process, that is, similar to what is done in the laboratory of the University of Minho; the collection of EEG data in both is done before, during and after the experiment - even if different times can be considered; in the first article, it considers self-report scales as a traditional method and to improve data the EEG, while the second considers questionnaires as a traditional method and as a complement to the EEG and eye-tracking technique. All the information collected so far, and the methodologies have been validated and are common in different sources, allow to emphasize the need for methodological triangulation between the EEG technique and the self-report scales. In addition to a larger number of data, it allows to compare them and understand what is possible to obtain from each one. If, on the one hand, through EEG it is possible to perceive the participant's reaction to the stimulus, the self-report scale allows us to define this response. This pertinent investigation illustrates whether we can characterize the product through this technology and whether it is possible to define the participant's motivation with the product. but as mentioned above, the emotion factor is one of the main factors for humans, and it is increasingly important to consider its importance as functionality.

4 Conclusion

This article intends to demonstrate the purpose of an ongoing investigation, through the contents covered: bibliographical research, laboratory experiments, methodology design and analysis of practical cases that demonstrate its results. During this article, the purpose of ongoing research is addressed through the analysis of key concepts and case studies. Based on the work accomplished so far, it is intended, through brain mapping, to understand the user's motivation and involvement in each product. So, adapting to the product development process in the evaluation component - answering questions such as: Does this form motivate me? Is it functional? Do I feel comfortable? Therefore, understanding the emotional responses by consumers' biofeedback is essential for success in product development because self-report scales are subject to cognitive changes. After all, all measurement data is obtained from the user's reported thoughts or through questionnaires.

The article focuses on its purpose through the analysis of key concepts and practical cases, but although engineering is not addressed explicitly, it is also related to these issues, such as design, when it comes to product development. This intervention by multidisciplinary teams - designers and engineers - in the evaluation of users' responses is important in both areas.

The intervention of this investigation with the EEG technique can be seen as a tool for working with emotional design, which can, in turn, influence the design of the product. Although these issues arose from a design gap, they are implicit in the new engineering processes. The application of the EEG technique has been an asset in the application in engineering projects that have become successful from functionality and usability.

However, from meaning, motivation, and engagement, there is still a considerable gap as demonstrated in this text. In the case of practical cases, the value of this approach to engineering is noticeable, for example, in the article Assessing Product Design Using Photos and Real Products, the investigation was supported by the Hyundai Motor Company brand. In engineering, the environment is merely functional, but for example, the factor of shapes and textures can influence the motivation of users who consequently influence the purchase decision. Once again, the importance of the bridge between engineering, and the design of the intervention of this investigation is reinforced. In a completely different area, the food area, the article Understanding consumer physiological and emotional responses to food products using electroencephalography (EEG), demonstrates that through emotions, it is possible to identify what consumers like or dislike and create a model, which influences work of engineers.

Acknowledgments. This work was financed by the Project Lab2PT - Landscapes, Heritage and Territory laboratory - UIDB/04509/2020 through FCT - Fundação para a Ciência e a Tecnologia

References

1. Desmet, P., Jimenez, S., Pohlmeyer, A.: Positive Design, Reference Guide. Delft, The Nerthelands (2015)
2. Medeiros, W.: Meaningful Interaction A Proposition for the Identification of Semantic. Staffordshire University, England, Pragmatic and Emotional Dimensions of Interaction with Products (2007)
3. Songsamoe, S., Saengwong-ngam, R., Koomhin, P., Matan, N.: Understanding consumer physiological and emotional responses to food products using electroencephalography (EEG). Trends Food Sci.Technol. **93**, 167–173 (2019)
4. Providência, B., Ciurana, J., Cunha, J.: Connecting emotion to product customization an integration model system. In: Proceedings of 8th International Design and Emotion Conference London (2012)
5. Carvalho, M.: O design emocional na construção de interfaces. Universidade de Aveiro, Aveiro (2013)
6. Patton, M.Q.: Two decades of developments in qualitative inquiry: a personal, experiential perspective. Qual. Soc. Work Res. Pract. **1**(3), 261–283 (2002)
7. Norman, D.: Why we love (or hate) everyday things. Perseus Books Group, New York (2004)
8. Ribeiro, I.M.: Mapeamento da hedonomia e das experiências emocionais. UFPE, Recife (2020)
9. Langleben, D.D., Moriarty, J.C.: Using brain imaging for lie detection: Where science, law, and policy collide. Psychol. Public Policy Law. **19**(2), 222–234 (2013)
10. Cohen, M.X.: Analying Neutral Time Series Data: Theory and Practice. MIT Press, The Nertherlands (2014)
11. Brace, I.: Questionnaire Design: How to plan, Structure and Write Survey Material for Effective Market Research, 2nd edn. Kongan Page, London (2008)
12. Bradley, M.M., Lang, P.J.: Measuring emotion: the self-assessment manikin and the semantic differential. J. Behav. Ther. Exp. Psychiatry **25**(1), 4959 (1994)
13. Positive Psychology, PANAS. https://positivepsychology.com/positive-and-negativeaffect-schedule-panas/. Accessed 24 Jan 2020

14. Faro Viana, M., Almeida, P., Santos, R.C.: Adaptação portuguesa da versão reduzida do Perfil de Estados de Humor – POMS. Análise Psicológica. **19**(1), 77–92 (2012)
15. Moon, S.-E., Kim, J.-H., Kim, S.-W., Lee, J.-S.: Assessing product design using photos and real products. In: Proceedings of the 2017 CHI Conference Extended Abstracts on Human Factors in Computing Systems, 17, pp. 1100–1107 (2017)

Modelling IT Specialists Competency in the Era of Industry 4.0

Maciej Szafrański, Selma Gütmen, Magdalena Graczyk-Kucharska(✉), and Gerhard Wilhelm Weber

Faculty of Engineering Management, Poznan University of Technology, ul. Rychlewskiego 2, 60-965 Poznan, Poland
magdalena.graczyk-kucharska@put.poznan.pl

Abstract. The topic of the future and development of human capital in the context of issues related to the digitization of the economy, including Industry 4.0, matching IT competencies to the needs of enterprises and the labor market is a growing challenge for public and training institutions in making decisions as to its development in the region. The availability of competency resources of future employees depends on many factors, including the region, matching the candidate to the job profile, date of publication of the offer or completion of the candidate's education. The literature review was done in this paper in the context of the issue of professional competencies of potential candidates in the IT profession in the era of automation and digitization of enterprises. The authors focus in particular on emphasizing the matching of professional competencies to the needs of enterprises in the era of the fourth industrial revolution. An important and innovative element of the study is the mathematical model which describes the impact of significant independent variables on the examined dependent variable which is the group of professional competencies of the IT technician. The research sample is based on 2,173 competency profiles of students studying in the profession of an IT technician, described in total by 3,874 skills and 619 job offers describing over 3,133 skills. The aim of the article is to create a mathematical model supporting decisions, taking into account significant dependent variables for matching professional competencies in the IT area to the needs of the labor market.

Keywords: Professional competence · Competence management · Industry 4.0 · Data mining · Multivariate adaptive regression splines · Labor market

1 Introduction

The accelerating world and the economy require innovation. In the second decade of the twenty-first century, the system of organizational and technological solutions referred to as Industry 4.0. became one of them. Among other things, this name encompasses the development of the Internet, including the Internet of things, communication between humans and machines and cloud computing processes [1]. Job creation and virtual industrialization [2] are also characteristic of Industry 4.0. The specifics of that issue were described in more detail in an article devoted to a bit different subject, i.e. competence management in a network of production companies [3]. The article quoted works of authors engaged in the issues addressed by Industry 4.0 [4–6]. The expectations regarding the direction of the development of companies encompassed under the phrase Industry 4.0 led to a growth in the demand for work provided by IT specialists, including programmers. At the moment, the expected acceleration in the improvement of efficiency and effectiveness in companies is hampered by shortages of staff in that professional group. Therefore, the following issues can be seen:

- shortage of IT specialists, including programmers,
- insufficient level of competencies among IT specialists relative to expectations on the labor market.

Competencies are defined in different ways. Those issues were addressed in the article [7]. With respect to an individual, e.g. a single employee, they are very often simply presented as a set of knowledge, skills and attitudes [8]. From the point of view of decision-makers in companies, they are perceived as an input resource to labor processes [9]. Companies undertake various activities in order to either acquire competencies from the market or train employees so that they acquire or develop them [10]. Looking for competencies on the market requires communicating with it. With that respect, it is quite difficult to name competencies properly so that the communicating parties (businesses, job candidates, educating institutions) [11] understood the same thing under different names of competencies [12]. Other challenges include improvement of competency selection methods so that to ensure their expected bundles on job positions. In a turbulent reality, those bundles often include competencies which in traditional, still dominating education systems, are taught as part of different professions and thus it is additionally difficult to find employees meeting the competency requirements. Job candidates often have competency gaps or sets of excessive competencies at the same time. From the point of view of the labor market and companies, every solution which will accelerate the shaping of competency bundles expected on job positions is interesting. For instance, it would be very useful if a method was developed thanks to which on the basis of identification of one competency in a candidate, it could be concluded with a high degree of probability that they will also have a bundle of other needed competencies. It is exactly such a method that the Poznań University of Technology team is working on together with its international partners. This article is going to present the outcomes of works carried out to date in reference to the profession of an IT specialist which is very important in the global dimension.

2 Problem Definition

In solving social problems, and works on looking for relations between separate competencies, it is very useful to have a big amount of data, however, it is often difficult to collect them. It is not enough just to have those data, they should also be up to date and of high quality. Whereas the results of student examinations are collected on an ongoing basis in education systems, the evaluation criteria are often ill-adjusted to the requirements pursued on the market. In reference to the needs of the labor market, employers' needs are sometimes analyzed, for instance on the basis of data from jobcentre websites but firstly, it is difficult to compare them with data from education systems and secondly, companies do not publish them or do so against a significant fee and thirdly, there are no standardized competency glossaries, a fact which makes the flow of knowledge about the needs of the labor market more difficult [12]. However, when it is eventually possible to collect the data, as it was the case in the project "Time of professionals - Wielkopolska professional education" (2012–2015) where system.zawodowcy.org was used [3, 13], a shortage of methods with which the data could be analyzed is noticeable. Most often, the analyses boil down to the publication of simple reports with circle or bar charts. There are some advantages to that since such charts are understandable to most recipients of the information. The lack of in-depth studies of dependencies between variables such as competencies, features of the job or training or internships candidates, or variables describing the environment in which the companies or the job candidates function, makes it impossible to draw conclusions on the dependencies among them. The absence of methods supporting inference about relations among the needs of companies, the condition of competencies and factors which affect the shaping of all those variables makes it more difficult to make decisions in enterprises, education systems, regional, and domestic and international decision-making centres. The limited knowledge about the nature and changeability of competencies treated as a resource makes it more difficult to manage them both on the microeconomic level (e.g. in companies) as well as the macroeconomic level (e.g. in the region or the state).

It was mentioned that data on the competencies of future employees, including IT employees, were and are still being collected via system.zawodowcy.org. It would be advantageous if before the data are collected, research assumptions were made about dependencies among different variables. Unfortunately, it is not possible. Firstly, there are so many categories of variables listed in the system that the number of assumptions would be close to infinity. Secondly, it is the purpose of the system to collect and present information about job candidates and employers and thus research purposes must give way to the information needs of users of the system.

The methodology gap in that research area might be filled by the use of advanced statistical methods such as *MARS*, Linear Regression or *ANN* [14].

This article is going to focus on *MARS*. The method is known in mathematics and already used in some areas of finance [15], the environment [16], natural resources [17]. Its insufficient use, given the benefits it can generate, is noticeable in problems concerning the functioning of social systems or human development. The use of *MARS* in the works whose outcomes will be presented in the following points stemmed from the identification of very interesting research problems which can be described as follows:

- are there any dependencies between the professional competencies of IT job candidates and social competencies and other features of the candidates or the environment they function in?
- how to notice dependencies between a big number of categories of variables which are collected without primary assumptions of dependencies among them?
- are there tools, methods or statistical techniques which can be used to look for dependencies among the features, including competencies of potential IT employees, in big sets of unordered data?

- how can sets of IT specialists' competencies be modelled on the basis of identified dependencies among the features of potential IT employees, including their competencies, and on the basis of data concerning the needs of employers?

3 Methodology

3.1 Conditions of the Research and Assumptions Made

The formulated research problems became the basis for the development of a research method and carrying out studies.

The following conditions for the studies have been assumed.

The research used data from system.zawodowcy.org. In connection with the fact that the system collects data on competency profiles of students of technical schools who learn different professions in Wielkopolskie province (Poland), the scope of the data was limited only to competency profiles of students learning the profession of IT specialists. The data were collected for the period of October 2012 to November 2015. They continue to be collected in the system, however after a change was introduced to their collection method, including the competency glossary, it is no longer possible to combine data from the studied period with data from later periods. The historical period also stems from the fact that the data from said period had been anonymized and thus the system administrator was able to disclose them for the purposes of the studies. The historical period is sufficient for the development of a method for studying dependencies between the features of potential IT employees which also include competencies. It can be argued that the application of example data would be possible for the purposes of testing research methods but actual data make it possible to verify the method by coming up with conclusions which refer to the reality. The system also presents job offers posted by employers in the same period. They were also created based on the dictionary on the basis of which the students of technical schools developed their competency profiles. Thanks to the use of the same dictionary, it is possible to compare how the student's competencies from their profile match the expectations of the employer presented in the offer. A 1:1 comparison is simple and automated. The comparison of aggregate data is also simple. Unfortunately, it is impossible in the system to study dependencies among competencies at the same time taking into account the influence exerted over those dependencies by the needs reported by employers, i.e. representatives of the labor market. Neither does the system have any mechanism for studying how the occurrence of competencies depends on other features of the student than their competencies and the features of the environment they live in. Among other things for those reasons, it is impossible to model competency sets in the profession or answer questions such as: will the occurrence of one competency be accompanied by the occurrence of other or others? Is the occurrence of a given competency or competency bundle affected by the variable characterizing the environment or a variable other than competency which characterizes the student or a variable which characterizes the company which formulated the job offer? In the competency dictionary, they were divided into three groups, according to the Polish "Vocational education core curriculum" [18] which will be in effect until 2021. Therefore, professional (P), common (C) and general (G) skills were distinguished in the dictionary. In the dictionary, the competencies were described with more detailed skills.

The following research assumptions were made:

1. there are dependencies between professional and common and general competencies of future IT specialists (i.e. students learning the IT profession),
2. there are dependencies between professional competencies and other features of the students who are potential IT employees,
3. there are dependencies between the features of the environment in which the students learning to be IT specialists live and their professional competencies.

3.2 Preparation of Data for the Research

Taking the aforementioned research conditions into account, 3346 students were identified in system.zawodowcy.org in the profession of the IT technician. Moreover, students learning other professions who have at least 3 skills attributed to the IT technician profession, i.e. an extra 566 people, were taken into account. Those students were included because quite often it is not the profession which they learnt that is the search criterion on the labor market but their actual skills or, in wider terms, competencies.

In the studied period, 423 (small, midsize and big) companies described their competency needs in 619 job offers.

3807 skills were indicated in the student's competency profile or the job offer at least once in the period in question. 3324 of them were (P), 335 were (C), and 148 were (G). All the skills became input data for the studies. Moreover, variables such as student features other than skills, jobs offers from employers and the environment the students live in were taken into account for the input.

Table 1 presents all the variables used in the studies.

System.zawodowcy.org contains many other variables which were not taken into account in the presented studies. Therefore, the variables presented in Table 1 should be treated as representatives of separate categories of variables. Some of the variables could not be taken into account due to the fact that they were only descriptions, not quantified variables. It made their use impossible, given the statistical technique applied.

The number of skills and students identified in the system was so high that the data had to be reduced. The available statistical tools would not have coped with such a big number of columns and records. Therefore, the data were synthesized with special mathematical procedures which are not going to be presented here in detail and 3807 skills were reduced to 3 variables, i.e. Y_Z, X_{14} or X_G, X_{15} or X_C. It must be pointed out that those three variables encompass the complete information about all the skills of the students (in the 3 separate groups) and at the same time about all the competency needs included by the employers in the job offers published in the system. For instance, in order to create the variable Y, analyzed were all competency profiles developed by the students in the system and all the job offers posted there by employers. Both parties used the same dictionary of competencies. Therefore, every student used the entire dictionary of competencies, choosing some of them as those they possess. Employers did a similar thing when describing their competency needs in the job offers. All the records in the offers' data set and all the records in the student profile data set were compared with

Table 1. Variables used in the studies.

Variable	Symbol
Dependent variable concerning professional competencies	
The student's total evaluation value of professional skills for a particular offer	Y_P
Independent variables (predictors)	
Concerning the student's features	
Student's gender	X_1
Student's birthday	X_2
Student's profile creation date	X_3
Concerning the job offer	
Job offer visible from	X_4
Job offer visible to	X_5
Job offer's date of work start	X_6
Job offer's date of creation	X_7
Job offer's type of employment	X_8
Job offer's working time	X_9
Job offer's no. of job positions	X_{10}
Job offer's shift work	X_{11}
Job offer's non-stationary work	X_{12}
Concerning the environment's features	
The job offer and the student are located in the same county	X_{13}
Concerning competencies other than professional	
The combined evaluation value of the student's common skills for a particular offer	X_{14} or X_G
The combined evaluation value of the student's general skills for a particular offer	X_{15} or X_C

one another, subsequently taking into account only those pairs of students and job offers which had at least one skill from the skill dictionary in common (in the case of the variable Y, the comparison was made only for professional skills). Subsequently, the generated data were aggregated to one number for every offer-student pair described in the profile. A detailed depiction of the data aggregation procedure for the variable Y and analogously developed variables X_C and X_G would require a separate article in the scope of mathematics or operational research.

3.3 The Research Technique

MARS or Multivariate Adaptive Regression Splines method was used in the study. It is a statistical technique which addresses one of the key defined problems, i.e. the fact that researchers have a set of unordered data and cannot specify research hypotheses at the input for the studies. Therefore, it is necessary to apply a non-parametric technique [19].

MARS is such a technique. The basics of the application of MARS were described by authors such as [20–23].

In MARS, the starting point is the following general mathematical formula which shows the connection between the response variable (output variable) and the input variables (predictors):

$$Y = f(X) + \varepsilon \qquad (1)$$

where:

Y – the response variable,
$X = (X_1, X_2, \ldots, X_m)^T$ – the vector representing each input variable,
ε – noise with zero mean and limited variance.

The model can be developed to the following form:

$$Y = \gamma_0 + \sum_{n=1}^{N} \gamma_n F_n(X) + \varepsilon \qquad (2)$$

(Y) can be calculated as a function of the predictors X (and their interaction). The elements of the equation include the input ordinate (γ_0) and weight-weighted (γ_n) sum total of many base functions $F_n(X)$ which can be expressed with the symbol (BF_n). The functions are often called splines.

The base function (BF_n) can be graphically depicted as in Fig. 1.

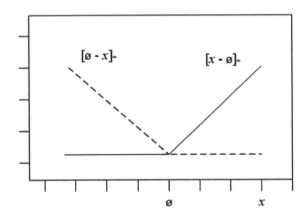

Fig. 1. Graphic expression of a single base function used in MARS. Where: $\varphi, x \in \mathbb{R}$.

Figure 1 presents two component functions of the base function. They are: $[\varphi - x]_+$ and $[x - \varphi]_+$. The parameter φ is the knot of the base function and divides it into two sections of linear regression (it must be remembered that MARS is an advanced linear regression technique). The location of the knot stems from the data. With more accuracy, the functions can be presented as follows:

$$[x - \varphi]_+ = \begin{cases} x - \varphi, & \text{if } x > \varphi \\ 0, & \text{otherwise} \end{cases} \quad (3)$$

$$[x - \varphi]_- = \begin{cases} \varphi - x, & \text{if } \varphi > x \\ 0, & \text{otherwise} \end{cases} \quad (4)$$

As it can be seen in Fig. 1, only the sections following the φ knot are taken into account in the analyses. The value of the function before the knot is zero.

As it was presented, the use of *MARS* involves searching the space of all input and output values and connections between them. More base functions, chosen from the set of all admissible functions, are added to the model of the reality described with the variables so that to maximize the overall level of the model's fit to the described reality. This practice leads to finding the most important independent variables and the most important interactions among them.

MARS is a very accurate tool, therefore it makes it possible to identify a lot, even very little, dependencies and develop an excessively fitted model. Taking such a big number of connections into account would make very difficult to draw conclusions which independent variables affect the dependent variable to the most extent, therefore, the omission of those of the base functions which present the lowest fit of the model to the reality is one of the stages in the procedure of identifying dependencies. For that purpose, *MARS* applies the lack of fit criterion (LOF) which is described by the Generalized Cross Validation (GCV). It makes it possible to arrive at the optimum number of base functions (BF_n) which ensure the best predictive fit. The GCV formula was developed by Friedman in 1991 [21] and looks as follows:

$$LOF\left(\hat{f}_b\right) = GCV(b) := \frac{\sum_{i=1}^{l}(y_i - \hat{f}_b(x_i))^2}{(1 - \frac{P(b)}{l})^2} \quad (5)$$

where:

b – optimal number of terms.
$\hat{f}_b$ – optimal number of effective terms b,
l – number of observations,
x_i – corresponding data vector.
y_i – the y value which is related to data vector xi.

In formula (5):

$$P(b) := v + dK \quad (6)$$

which is read as follows: P(b) is the efficient number of parameters.

where:

v – the number of linearly independent functions,
d – the price of "penalty" for one unit of complexity (most often designated as 2 or 3); 2 is preferred in additive models [15],
K – number of knots.

In order to describe the connections between the variables listed in Table 1, the following function was eventually adopted in the presented studies:

$$Y = \gamma_0 + \sum_{n=1}^{N} \gamma_n B_n + \varepsilon \qquad (7)$$

4 Model Development

SPM 2018 software was used to develop the model of dependencies among the variables defined in Table 1. Following the aforementioned procedure of the application of *MARS*, it was possible to arrive at the following 23 base functions (formula 8):

$$BF_1 = \max\{0, X_{14} - 2\};$$
$$BF_2 = \max\{0, 2 - X_{14}\};$$
$$BF_3 = \max\{0, X_8 - 1\} \cdot BF_1;$$
$$BF_5 = \max\{0, 1 - X_{15}\} \cdot BF_2;$$
$$BF_6 = \max\{0, X_4 - 5281\} \cdot BF_5;$$
$$BF_7 = \max\{0, 5281 - X_4\} \cdot BF_5;$$
$$BF_8 = \max\{0, X_4 - 5269\} \cdot BF_5;$$
$$BF_{10} = \max\{0, X_4 - 5297\} \cdot BF_5;$$
$$BF_{12} = \max\{0, X_4 - 5163\} \cdot BF_5;$$
$$BF_{14} = \max\{0, X_2 + 1695\} \cdot BF_3;$$
$$BF_{16} = \max\{0, X_{15} - 31\} \cdot BF_1;$$
$$BF_{17} = \max\{0, 31 - X_{15}\} \cdot BF_1;$$
$$BF_{19} = \max\{0, 4864 - X_3\} \cdot BF_{16};$$
$$BF_{20} = \max\{0, X_5 - 5426\} \cdot BF_5;$$
$$BF_{21} = \max\{0, 5426 - X_5\} \cdot BF_5;$$
$$BF_{22} = \max\{0, X_4 - 5115\} \cdot BF_5;$$
$$BF_{24} = \max\{0, X_4 - 5178\} \cdot BF_5;$$
$$BF_{26} = \max\{0, X_4 - 5233\} \cdot BF_5;$$
$$BF_{28} = \max\{0, X_5 - 5426\} \cdot BF_3;$$
$$BF_{29} = \max\{0, 5426 - X_5\} \cdot BF_3;$$
$$BF_{30} = \max\{0, X_3 - 5373\} \cdot BF_5;$$

$$BF_{33} = \max\{0, 32 - X_{15}\} \cdot BF_3;$$
$$BF_{34} = \max\{0, X_5 - 6575\} \cdot BF_5; \tag{8}$$

On the basis of formula 7, the model presented in formula 9 with 21 base functions was developed for the research problem in question.

$$\begin{aligned} Y = {}& 12.41 + 4.21 \cdot BF_1 + 10.26 \cdot BF_3 - 81.92 \cdot BF_5 \\ & - 29.62 \cdot BF_6 + 0.369 \cdot BF_7 + 20.81 \cdot BF_8 \\ & + 11.22 \cdot BF_{10} - 9.461 \cdot BF_{12} - 0.005 \cdot BF_{14} \\ & - 0.1405 \cdot BF_{17} - 0.00172 \cdot BF_{19} - 0.0531 \cdot BF_{20} \\ & - 0.317 \cdot BF_{21} + 3.040 \cdot BF_{22} + 8.216 \cdot BF_{24} \\ & - 4.307 \cdot BF_{26} - 0.0315 \cdot BF_{28} + 0.098 \cdot BF_{29} \\ & + 0.263 \cdot BF_{30} - 0.374 \cdot BF_{33} + 0.116 \cdot BF_{34} + \varepsilon. \end{aligned} \tag{9}$$

The symbol "−" means the related BF has a negative effect on the response variable while " +" means it has a positive effect.

1 Observation

Therefore, 21 BF_n were included in the model described with formula 8. The analysis of those functions indicates that they take into account 7 input variables (X) out of all 15 listed in Table 1. Those variables need to be considered the most important ones when it comes to their influence over Y, i.e. the student's total evaluation value of professional skills for a particular offer. They are:

X_2 – 1 instance, student's birthday.

X_3 – 2 instances, student's profile creation date.

X_4 – 8 instances, Job offer visible from...

X_5 – 5 instances, Job offer visible to...

X_8 – 1 instance, Job offer's type of employment.

$X_{14} = X_G$ – 1 instance, total value of the evaluation of the student's common skills for a particular offer,

$X_{14} = X_G$ – 3 instances, total value of the evaluation of the student's general skills for a particular offer.

Some of the remaining input variables might also be important but statistically "covered" by the dependency through the 7 variables that were already taken into account.

2 Observation

6 out of the 15 variables included in the studies were time-related. 4 of them were included among the 7 which were considered the most important for Y. That shows that the variable Y is significantly dependent on time. For instance, in the period in question, the match of the professional skills of students declaring that they have the skills required from IT specialists to the expectations included in job offers for such specialists largely depended on the time at which those offers were posted in system.zawodowcy.org (X_4).

3 Observation

Table 2 shows all the estimated values of coefficients γ_n, corresponding to selected BF developed through *MARS* in the model: *of the influence of* different factors (variables X) *on the match between:* the competency profiles of students who have the professional skills of IT specialists and the professional skills expected in the job offer.

Table 2. Coefficient values in the model prepared with the *MARS* technique.

y_0	y_1	y_2	y_3	y_4	y_5	y_6	y_7
12.41	4.21	10.26	−81.92	−29.62	0.369	20.81	11.22
y_8	y_9	y_{10}	y_{12}	y_{13}	y_{14}	y_{15}	y_{16}
−9.461	−0.005	−0.1405	−0.00172	−0.0531	−0.317	3.040	8.216
y_{17}	y_{18}	y_{19}	y_{20}	y_{21}	y_{22}		
−4.307	−0.0315	0.098	0.263	−0.374	0.116		

MARS assigns these coefficients for each BFs in the model. The ones with minus have a negative effect on response variable Y, the others with plus have positive effects on it, as it has been explained in Fig. 1 and in Formula 3 and 4.

5 Conclusions and Outlook

The use of *MARS* in the research turned out to be very promising. Data collected in system.zawodowcy.org are just one of many examples where, due to their complexity and amount, it is difficult to see specific dependencies while making hypotheses and verifying them could take almost an eternity. The technique used makes it possible to identify dependencies among the variables very fast. It would be difficult to apply *MARS* without an IT tool. Having it makes it possible to quickly change the variable sets and repeat the study many times over a short period of time.

During the studies, several challenges which will have to be faced in the future have also been identified. Firstly, deciding which data to use in the research and further works connected with their preparation for the analysis is a very time-consuming process. Secondly, whereas *MARS* does provide immediate information about dependencies among the variables, it does not provide detailed information about the nature of those dependencies. That nature has to be studied with the use of other statistical methods. Thirdly, if a decision is made to aggregate the data, there appears the problem of how extensive the aggregation should be. Excessive aggregation does make it possible to take a big amount of data into account, however, the output information is very general.

It was the main goal of the studies to create a mathematical model supporting decisions, taking into account significant dependent variables for matching professional competencies in the IT area to the needs of the labor market. It was an ambitious and realistic goal which was fully achieved.

The studies should be deepened in order to get a more accurate grasp of the nature of the identified dependencies among the 7 variables X and the variable Y with the use of additional research methods, including statistical ones and those conducted in interdisciplinary teams. Changing the position of the dependent variable might prove interesting. For instance, current variable $Y = X_P$ can be treated as an input variable while other variables, e.g. X_C or X_G can assume the role of Y in order to test whether and how X_P and other X variables affect them. The X data set can be enlarged by adding e.g. macroeconomic data or additional features of the students to test whether other factors affect the match between the competency profiles of students who have the professional skills of IT specialists and the professional skills expected in the job offer. It might also be interesting to use the *MARS* tool in relation to other professions and job positions as system.zawodowcy.org contains much bigger data resources than those used in the presented research. Finally, it might also be interesting to compare the effectiveness of *MARS* with other parametric or non-parametric methods such as the basic Linear Regression [21, 24] or Artificial Neutral Networks [25, 26].

The results of the studies are promising enough to justify the need for their continuation as well as intensification. On the basis of the generated results, it is not yet possible to recommend what kind of decisions should be made by business owners, schools or representatives of local governments so that the competencies of graduates learning IT professions adapted faster to the working conditions in the companies. Such fast adaptation and then requalification and fast learning are necessary in the conditions of Industry 4.0. In the opinion of the authors, the studies should continue for instance as part of projects dedicated to basic research.

References

1. Lee, J., Behrad, B., Kao, H.A.: A cyber-physical systems architecture for industry 4.0-based manufacturing systems. Manufact. Lett. **3**, 18–23 (2015)
2. Magruk, A.: Uncertainty in the sphere of the industry 4.0-potential areas to research. Bus. Manage. Educ. **14**(2), 275–291 (2016)
3. Graczyk-Kucharska, M., Szafrański, M., Goliński, M., Spychała, M., Borsekova, K.: Model of competency management in the network of production enterprises in Industry 4.0 – assumptions. In: Hamrol, A., Ciszak, O., Legutko, P., Jurczyk, M. (eds.) Advances in Manufacturing, pp. 195–204. Springer, Heidelberg (2018). https://doi.org/10.1007/978-3-319-68619-6_19
4. Lu, Y.: Industry 4.0: a survey on technologies, applications and open research issues. J. Ind. Inf. Integr. **6**, 1–10 (2017)
5. Brettel, M., Friederichsen, N., Keller, M., Rosenberg, M.: How virtualization, decentralization and network building change the manufacturing landscape: an industry 4.0 perspective. Int. J. Mech. Ind. Sci. Eng. **8**(1), 37–44 (2014)
6. Thames, L., Schaefer, D.: Software-defined cloud manufacturing for industry 4.0. Proc. CIRP **52**, 12–17 (2016)
7. Szafrański, M.: Threefold nature of competences in enterprise management: a qualitative model. In: Proceedings of the 20th European Conference on Knowledge Management, Universidade Europea de Lisboa, Lisbon, Portugal, vol. 2, pp. 1006–1015 (2019)
8. Boyatzis, R.E.: The Competent Manager: A Model for Effective Performance. Wiley, New York (1982)
9. Russ-Eft, D.F.: Human resource development, evaluation, and sustainability: what are the relationships? Hum. Resour. Dev. Int. **17**(5), 545–559 (2014)

10. Spychała, M., Goliński, M., Szafrański, M., Graczyk-Kucharska, M.: Competency models as modern tools in the recruitment process of employees. In: Proceedings of the 10th European Conference on Intangibles and Intellectual Capital ECIIC 2019, Published by Academic Conferences and Publishing International Limited, Chieti-Pescara, Italy, pp. 282–291 (2019)
11. Szafrański, M., Goliński, M., Graczyk-Kucharska, M., Spychała, M.: Cooperation of education and enterprises in improving professional competences - analysis of needs. In: Hamrol, A., Grabowska, M., Maletic, D., Woll, R. (eds.) MANUFACTURING 2019. LNME, pp. 155–168. Springer, Cham (2019). https://doi.org/10.1007/978-3-030-17269-5_11
12. Szafrański, M., Bogurska-Matys, K., Goliński, M.: Problems in communication between businesses and technical education system. Manage. Prod. Eng. Rev. **8**(2), 9–18 (2017)
13. Spychała, M., Szafrański, M., Graczyk-Kucharska, M., Goliński, M.: The method of designing reference models of workstations. In: Proceedings of the 18th European Conference on Knowledge Management ECKM 2017, Academic Conferences and Publishing International Limited, Barcelona, pp. 930–939 (2017)
14. Graczyk-Kucharska, M., Özmen, A., Szafrański, M., Weber, G.W., Goliński, M., Spychała, M.: Knowledge accelerator by transversal competences and multivariate adaptive regression splines. Cent. Eur. J. Oper. Res. **28**(2), 645–669 (2019). https://doi.org/10.1007/s10100-019-00636-x
15. Köksal, E.: Modeling of exchange rates by multivariate adaptive regression splines and comparison with classical statistical methods. Master's thesis. Institute of Applied Mathematics, Middle East Technical University (2017)
16. Taylan, P., Özkurt Yerlikaya, F., Weber, G.W.: Precipitation modeling by polyhedral RCMARS and comparison with MARS and CMARS. Environ. Model. Assess. **19**, 425–435 (2014)
17. Özmen, A., Yılmaz, Y., Weber, G.W.: Natural gas consumption forecast with MARS and CMARS models for residential users. Energy Econ. **70**, 357–381 (2018)
18. Regulation of the Minister of National Education of 31 March 2017 on the core curriculum for vocational learning, item 860 as amended. Accessed 22 Dec 2019
19. Ghasemzadeh, A., Ahmed, M.M.: Utilizing naturalistic driving data for in-depth analysis of driver lane-keeping behavior in rain: non-parametric MARS and parametric logistic regression modeling approaches. Transp. Res. Part C: Emerging Technol. **90**, 379–392 (2018)
20. Friedman, J.H.: Multivariate adaptive regression splines. Ann. Stat. **19**(1), 1–65 (1991)
21. Hastie, T., Tibshirani, R., Friedman, J.: The Elements of Statistical Learning. Springer, Heidelberg (2009). https://doi.org/10.1007/978-0-387-84858-7
22. Salford predictive modeler. https://www.salford-systems.com/products/spm. Accessed 15 Feb 2019
23. Multivariate Adaptive Regression Splines (MARSplines). https://www.statsoft.pl/textbook/stathome_stat.html?https%3A%2F%2Fwww.statsoft.pl%2Ftextbook%2Fstmars.html. StatSoft Electronic Statostic TextBook. Accessed 22 Dec 2019
24. Aster, R., Borchers, B., Thurber, C.: Parameter Estimation and Inverse Problems. Academic Press (2003)
25. Lippmann, R.: An introduction to computing with neural nets. IEEE ASSP Mag. **4**, 4–22 (1987)
26. Mass, J., Flores, J.: The application of artificial neural networks to the analysis of remotely sensed data. Int. J. Remote Sens. **29**, 617–663 (2008)

Original Constructive Solutions for the Development of Industry 4.0 in Romania

Gheorghe Gheorghe[1(✉)], Badea Sorin-Ionut[1], Iulian Ilie[1], and Despa Veronica[2]

[1] INCDMTM Bucharest, Pantelimon Road, 6-8, 2nd District, 021631 Bucharest, Romania
[2] Valahia University of Targoviste, 130105 Targoviste, Romania

Abstract. The scientific paper includes the original constructive solutions provided for the development of Industry 4.0 in Romania.

The scientific paper will focus next on examples of contributions and elaborations of original concepts and constructions of intelligent systems, technological platforms and COBOT networks that support the digital enterprise and the Intelligent Industry (4.0) in Romania.

Keywords: COBOT networks · COBOT technology platforms · Medical applications · Intelligent control · Production lines control · Checking tightness automotive parts · Car body assembly · Electronics pads control · Ecological agriculture

1 Introduction

Examples of COBOT Networks of COBOT platforms with immediate applications in Digital Enterprise and Intelligent Industry (4.0).

Some works [1–3] have been developed in the context of Industry 4.0 and this issue has an upmost importance in several countries, including Romania. Since several pillars that are part of Industry 4.0 context, Industry 4.0 is based, mainly, in new approaches for designing and developing cyber-physical systems. Those systems are based on a perfect combination of physical and virtual worlds [4–6]. The communication and exchanging of information is an issue that has a lot of work for being developed [7–11].

2 COBOT Network of Technology Platforms for Integrated Control, Sorting and Storage, Servicing and Operational Assistance

At European level, a complex and ambitious program called Industry 4.0 has been elaborated, with the aim focused on a new industrial revolution understood more towards a much improved or modernized industry, by using and integrating cyber-physical and cyber-mixmechatronics systems in intelligent manufacturing. According to Fig. 1, the cobot platform network - for intelligent control, sorting & storage, servicing and operational assistance, has an organizational and functional structure, like any Web network, consist of:

COBOT technology platform - for intelligent control processes in the field of Electronics, structured on three important sectors: the whole sector of Intelligent Mechatronic Control Equipment for the electronic domain (for different electronic components and subsystems in the digital electronic manufacturing process); the data of the intelligent measurement and control are taken over and protocol and then transmitted, through the cyber environment/system, to the user, these being provided through the cyber protection systems; after the presumptive appearance of the first errors, through the telemonitoring/telecontrol/teleconfiguration and telementenance center, contact the applicant beneficiary, to whom it is transmitted by e-mail, telephone, etc., all the operations to be undertaken, in order to remove the identified errors in order to reintroduce the intelligent mechatronic control equipment, only if the applicant beneficiary agrees to be interrogated through the INTERNET and the INTRANET of the beneficiary entity. By applying remote monitoring, the interventions for Good Functioning are cheaper! Also, the Center for telemonitoring, telecontrol, teleconfiguration and telementenance [12], is the one that ensures the smooth operation of the mechatronic equipment of intelligent control, based on the complex structure of advanced software programs.

COBOT technology platform [13] - for sorting and storage processes in the industrial environment, also structured on three important sectors - the whole sector of Intelligent Mechatronic Equipment for sorting and storage processes in the industrial environment,

– the cyber environment sector and - the telemonitoring center sector.

The COBOT type technological process comprises all the robot-operator technological operations, from the robot and operator input sizes, to their transformation into processed data and protocol, to their transmission on the supplier-beneficiary communication bus, from the cyber space, to the beneficiary who performs the specific activities of the robotic equipment, finally pursuing the fulfillment of the output sizes, through the remote monitoring and remote control center of the COBOT technology platform for technological processes, it is structured on three levels: the level of intelligent robotic equipment, the level of the cyber space and the level of the telemonitoring center.

The technological processes served by this COBOT technology platform are very many, very diverse and very complex.

Depending on these technological processes, the intelligent robotic equipment specific to the type of process is integrated and adapted, with its adaptability, to the cyber space and the telemonitoring center and to the specially used and applied software.

COBOT technology platform for machine tool assistance (CNC), is structured on three sectors/levels such as: intelligent robot equipment, cyber space and telemonitoring center.

The technological assistance process of the MU with the CNC, includes all the technological operations carried out, from the detection of the elements in the working space of the robot - such as materials, devices, operator, etc., to the assistance through active or passive operational service, to the communication assistance with the operator, to the feedback of each activity, to the OK of each activity.

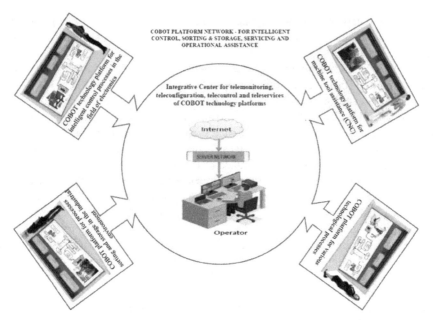

Fig. 1. COBOT network of technology platforms for integrated control, sorting & storage, servicing and operational assistance

3 COBOT Network of COBOT Platforms for Intelligent Handling and Control, Processes, Positioning, Measuring and Intelligent Control Processes in Metrology and Industry

According to Fig. 2, the cobotic network of COBOT-type platforms - for intelligent handling and control processes, positioning processes, integrated verification and control processes and intelligent measurement and control processes in industry and metrology, comprises COBOT platforms, so:

- COBOT platform for integrated verification and control processes in the MixMechatronic domain;
- COBOT platform for intelligent measurement and control processes for technological equipment;
- COBOT platform for intelligent manipulation and control processes in the field of Mechatronics and Cyber-MixMechatronics;
- COBOT platform for positioning processes in metrology laboratories.

The COBOT platforms mentioned above with each typed and corresponding structure, technological platforms specific to the services performed and specific to the related industrial fields [14, 15].

Each COBOT type technological platform has, in its turn, a typical structure, namely: intelligent mechatronic equipment/system, built to perform the functions of measurement/verification/control/positioning and manipulation, which transmits the data flow processed in the processes that they carries out, to the communication bus to the beneficiary and/or the supplier of the equipment, in order to remotely monitor its quality but also the operations performed at the beneficiary, through the Telemonitoring Center, the communication between the supplier and the beneficiary being done through the Cyber Space (INTERNET and INTRANET).

The protocols to be transmitted between the beneficiary - respectively the cyber-mixmechatronic system and the provider - respectively the Telemonitoring Center, are cybernetically secured and moreover they are qualitatively assured, for the best cobot collaboration.

The collaborative processes of the technological platforms that make up the platform network, are ensured, at a distance, by the Integrative Center for telemonitoring, teleconfiguration, telecontrol and teleservices of COBOT type technology platforms.

Collaborative assistance [16] is provided by the active operator of the integrative Telemonitoring Center, by applying specialized collaborative software, for each COBOT type platform, in order to carry out specific activities for each one, for validation of activities, for service on request from the beneficiary, for collaboration beneficiary-supplier and to ensure the quality of all the activities carried out by each platform.

The productive and quality contribution of the entire Cobot Network is integrated with each SME-member of each Network, for the maturation of the digital transformation process of the SME and the Intelligent Industry (4.0).

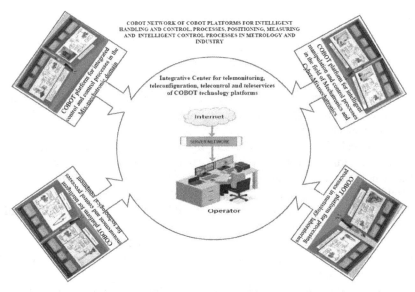

Fig. 2. COBOT network of COBOT platforms for intelligent handling and control, processes, positioning, measuring and intelligent control processes in metrology and Industry

4 COBOT Network of COBOT Technology Platforms for Control, Processes on Automotive Manufacturing Lines, Industrial Parts Assembly Processes, Manufacturing Processes (Welding and Handling) and Machine Tool Assistance Processes Web CNC, for the Machine Tools Industry

According to Fig. 3, where the Cobot Network of COBOT technology platforms is presented, its structure is based on four COBOT technology platforms specific to the operation of services for the fields of the machine building industry, as follows:

- COBOT technology platform for control processes and on automotive manufacturing lines;
- COBOT technological platform for assembly processes of industrial parts in the field
- of machine construction;
- COBOT technology platform for CNC machine tool assistance processes;
- COBOT technology platform for manufacturing processes (welding and handling);

Each COBOT technology platform, specific to the type of services served and the type of domain to which it is intended, has a typical organizational and functional structure, as follows:

- the intelligent mechatronic system/equipment serving the specialized service/services - measuring/controlling/assembling/assisting/manufacturing;
- cyber space with INTERNET and INTRANET based logistics;
- the center for remote monitoring, remote control and remote configuration.

Each intelligent mechatronic system/equipment is built and equipped with several domain-specific expression features and quality features, in accordance with European and international standards.

The output data flow from the system/equipment is transmitted/transferred, through the cyber space at the remote monitoring center to the beneficiary and supplier.

The remote monitoring center of each COBOT technology platform, collaborates with each of them, in order to be able to remote configure and to collaborate - system/equipment in the manufacturing process related to the operator of the center and the operator of the system/equipment.

Each COBOT technology platform, is placed in a specific and collaborative communication and enters into an active collaboration - platform - operator, based on the dedicated software used.

The collaboration between the four COBOT technology platforms is provided by the Integrative Center for t remote monitoring, remote configuration, remote control and remote service of COBOT technology platforms, through specialized software.

By integrating and implementing these cobot networks of COBOT technology platforms for intelligent control processes, intelligent manufacturing processes and all other ancillary processes, each SME will be transformed, and each SME will become a digitized SME, respectively all activities in SMEs, will be subjected to digital transformation,

Original Constructive Solutions for the Development of Industry 4.0 in Romania   275

from the supply, their transformation, the realization of products and product assemblies, to the specific activities of marketing, financial accounting, management, etc., which will become these, digital activities served by industrial robots, economic robots, and so on.

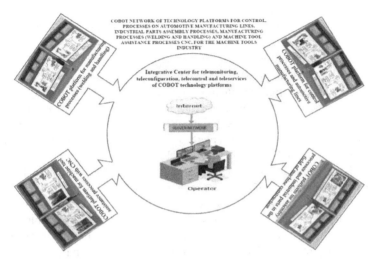

Fig. 3. COBOT network of technology platforms for control, processes on automotive manufacturing lines, industrial parts assembly processes, manufacturing processes (welding and handling) and machine tool assistance processes CNC, For the machine tools Industry

5   COBOT Network of COBOT Technology Platforms for Checking Leaks of Molded Parts of the Automotive Industry, Dimensional Control with Probes and 4D Intelligent Control

According to Fig. 4, where the Cobot Network of cobot technology platforms is presented, it consists of four cobot technology platforms/mechatronic robots (systems), as follows:

- technology platform/robot (system) mechatronic leak-proofing of molded parts of the automotive industry;
- technological platform/robot (system) mechatronic dimensional control with touch probes;
- technology platform/robot (system) mechatronic for dimensional control;
- technology platform/robot (system) 4D smart mechatronics.

Each cobot technological system (system) intelligent mechatronic robot has a typical structure based on intelligent mechatronic platform/robot (system), cyber space (INTERNET/INTRANET) and remote monitoring center.

The data flow, collected and transformed, is sent through the communication bus, to the respective beneficiary, the supplier for monitoring the processes carried out at the quality levels in accordance with European and/or international norms and standards, through the t remote monitoring and remote configuration center.

Cobot network of cobot technology platforms/mechatronic robots (systems), ensures collaboration with the network operator, respectively with the computer/server through specialized software and the Internet and/or Intranet.

Depending on the structure of the network, the collaboration is ensured for the performance of all the services related to the specialized fields.

The Cobot network of cobot platforms/systems (robots) is intended for the development of the digitized enterprise and the intelligent industry (4.0), while ensuring the modernization and development of intelligent specialized fields: the intelligent car industry; the mechatronics and cyber-mechatronics industry; advanced aeronautics industry; intelligent shipping industry; advanced railway industry; the smart medical industry; to.

The Cobot Network of cobot/robot platforms (systems) develops around it innovative groups of SMEs with intelligent specialized fields.

Each SME in Romania or in the European and international world, based on the integration and implementation of cobot technological platforms for intelligent electronic control and industrial technological platforms for the specialized intelligent fields, will pass in steps and development stages when digitizing all technical-technological activities - economic-logistics, etc., respectively as a digital enterprise and so on, as a smart industry (4.0) [17], respectively as a computerized and post-computerized society.

Each SME, thus transformed, will have a digitized structure, will have a staff trained in the digitization processes for all types of activities carried out on all the related levels and will have an adaptive management to the development level of the company.

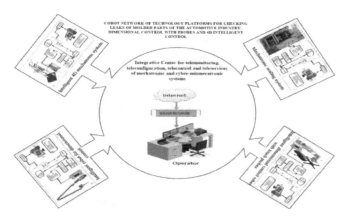

Fig. 4. COBOT network of technology platforms for checking leaks of molded parts of the automotive industry, dimensional control with probes and 4D intelligent control

6 COBOT Network of COBOT Technology Plat forms for the Analysis and Recovery of Walking in the Smart Medical Field and for Surgical Operations and Laboratory Analyzes in the Smart Medical Field

According to Fig. 5, where the Cobot Network is presented, it is structured on: COBOT technology platform for surgery and laboratory analysis in the smart medical field;

COBOT technological platform for the analysis and recovery of walking in the smart medical field.

Each COBOT technology platform, has a typical structure and functionality , based on:

- specialized equipment/system, with adaptive characteristics to the environment and domain, specialized;
- cyber space with logistics (INTERNET/INTRANET);
- remote control/remote configuration center.

The information flow collected and transformed, from the technological equipment is communicated to the communication bus through INTERNET and INTRANET, by using the Internet and the Intranet, to the beneficiary, respectively to the supplier.

The monitoring of the information is done in real time, by carrying out specific activities for the beneficiary and the supplier, respectively, according to the specialized intelligent fields served.

The Cobot network of technological platforms/robots (systems) develops groups of innovative SMEs, respectively value chains materialized by high quality levels.

The monitoring and collaboration of the Cobot Network is done through the Integrative Center for remote monitoring, remote configuration, remote control and remote services of the COBOT technology platforms and through the operator of all the activities carried out in real time, through the computer and the Network Server, with the operation of the Internet and Intranet as efficient logistics.

By integrating and implementing all the cobot networks of technological platforms cobot control and technological process, digital companies and intelligent industries (4.0) are obtained respectively, which together will build a digital and post-digital society.

The construction of the cobot networks of COBOT technological control platforms and of technological processes and/or of the mix networks, will contribute to the development of the digital SMEs and the intelligent industries respectively and together of the information and post-information society.

Then each SME and each smart industry (4.0) will jointly ensure the construction of intelligent products, advanced and advanced technologies, and intelligent services, and in the following periods, they will ensure the construction of the information and post-information society.

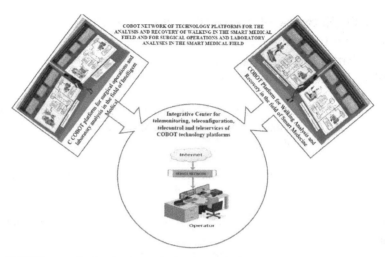

Fig. 5. COBOT network of technology platforms for the analysis and recovery of walking in the smart medical field and for surgical operations and laboratory analyzes in the smart medical field

7 COBOT Network of Technological Platforms/Intelligent Robots for Welding Services, Body Assembly, Engine Assembly and for Handling Parts in the Automotive Industry

The cobot network of COBOT type technology platforms/intelligent robots for welding services, body assembly, motor assembly and for handling parts in the automotive industry (Fig. 6), includes in its modular and typified structure, the following intelligent technological cobot platforms:

- platform/intelligent robots for welding operations in the automotive industry;
- intelligent platform/robots for bodywork assemblies in the automotive industry;
- platform/intelligent robots for motor assembly in the automotive industry;
- intelligent platform/robots for handling parts in the automotive industry;

Each of the intelligent platforms/robots for the automotive operations has a typical/modular structure, based on:

- the mechatronic/cyber-mixmechatronic equipment/system for performing the operations and services of the car, whose final data is transmitted through the communications bus to the other systems related to the platform, data that are dosed and which represent the Integration Database of the platform;
- cyber system/cyber space, which has subsystems consisting of antennas, modems, Internet, etc.
- the center for remote monitoring, remote control, remote configuration and remote maintenance of the related equipment/system of the platform.

Original Constructive Solutions for the Development of Industry 4.0 in Romania 279

The process of performing the operations/services in the field of technical/technological platforms, of services related to all the special and specific activities of the cobot network, comprises:

- the integration of the mechatronic/cyber-mixmechatronic system/equipment into the related technological platform;
- execution of the operation/service required in the technology platform;
- conducting the operation/service based on the specific software, in accordance with the technological guide;
- transmission/transfer of data to the Database of the integrative process and to the beneficiary of the equipment/system and/or to the supplier of the equipment/system;
- active monitoring of the activities carried out by the equipment/system for their validation, in accordance with the technical/technological guide of the platform or when errors/accidental occurrences occur, etc., for their resolution through the operator from the Telemonitoring Center, from the platform provider's space.

Within the Cobot Networks, monitoring/services are carried out, of all COBOT technical/technological platforms, through the Integrative Center for remote monitoring, remote configuration, remote control and remote service of the mechatronic and cyber-mixmechatronic systems of the Cobot Network.

Through this Integrative Center, each COBOT-type platform is remote monitored, remote configured and remotely controlled, in order to carry out specific activities in the best conditions, for each SME and for the Intelligent Industry.

Thus, each COBOTIC Network, is monitored, for all its functions and for its contribution, to the digital construction and transformation of the enterprise (SME) and the Intelligent Industry.

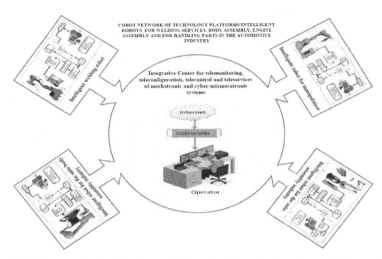

Fig. 6. COBOT network of technology platforms/intelligent robots for welding service, body assembly, engine assembly and for handling parts in the automotive industry

8 COBOT Network of COBOT Technology Platforms/Systems (Robots) for Positioning, Electronic Platelet Control, Ultra-Precise Positioning and Intelligent Handling

Figure 7 shows the Cobot Network of COBOT technological platforms/Mechatronic systems (Robots) for intelligent manipulation/ultra-precise positioning and electronic platelet control.

The structure of the Cobot Network comprises the following COBOT type Intelligent Technology Platforms (Robots):

- COBOT/Dodecapod intelligent ultra-precise positioning technology platform;
- COBOT technology platform/Intelligent Positioning System - Hexapod;
- COBOT technology platform/Intelligent mechatronic system for electronic platelet control control;
- COBOT/Robot intelligent technology platform for intelligent manipulation;

Each component of the COBOTICE Network, has a typical/modular structure consisting of three related buses, as follows:

Intelligent Dodecapod system of ultra-precise positioning, which performs ultra-precise measurements and positions, in metrology laboratories and industrial workshops;

The cybernetic space/cybernetic environment, having in its composition, elements and components that serve the data transfer from the intelligent pickling system to the communication bus to the other components of the COBOT platform, for example antennas, 4G and 5G modems, etc.;

The remote monitoring center of the platform, for its continuous dialogue with the supplier and the beneficiary of the related intelligent system.

The interrelation of COBOT platform systems, is carried out collaboratively between these platform components and the operator of the remote monitoring center on the one hand, and on the other hand, between the entire COBOT type platform and the Integrative Center of the Cobot Network, respectively through the operator of the platform center.

Depending on the applications of the Cobot Network and respectively the applications of the COBOT Platform, the special activities of the Cobot Network and of the COBOT Platform, respectively, will be carried out, all the activities of the Network, for their integration in each SME-component member of the network, respectively of the platform, in order to become in time, the digitalized SME and contributing to the development of Intelligent Industry 4.0 related to the SME domain.

The integration of Cobot Networks and COBOT Platforms, from the SME level, to the groups/clusters of intelligent specialized domains, this will contribute to the development of the digital enterprises and the Intelligent Industry (4.0), respectively of the Intelligent Industry (4.0).

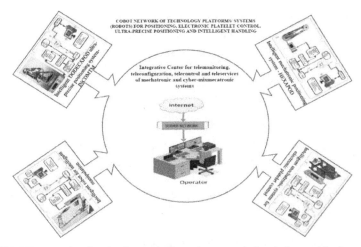

Fig. 7. COBOT network of technology platforms/systems (robots) for positioning, electronic platelet control, ultra-precise positioning and intelligent handling

9 COBOT Network of COBOT Technology Platforms/Mechatronics Systems (Intelligent Robots) for Organic Farming and Smart Agriculture

Figure 8 shows the COBOT Network of COBOT-type technological platforms/Mechatronic systems (Intelligent Robots) for modular/typical ecological agriculture consisting of:

- COBOT/intelligent robot platform for ecological agriculture (with agricultural robot);
- COBOT/intelligent robot platform for intelligent agriculture (with advanced robot); COBOT/intelligent robot platform for ecological agriculture (with intelligent robot);
- COBOT/intelligent robot platform for intelligent agriculture (with multi-application agricultural robot).

For intelligent and ecological agriculture, where cobot networks and COBOT technology platforms are applied, each cobot network and each COBOT technology platform makes a major contribution to the digitization or digital transformation of each enterprise or SME becoming a Digitized Enterprise or SME-the digitized and together participating in the development of the Intelligent Industry and respectively in the development of the Company in the computerized society or in the post-computerized Society.

Each COBOT type technology platform has a typical/modular structure and consists of:

- the intelligent equipment/mechatronic system/robot related to the cobot platform or network, which performs all the activities and services related to them, transmits within the technological process served, all the data related to it, through the communication bus between the component parts of the cobot network or network;

- the cybernetic space/system, which ensures the real/virtual connection of the cybernetic space with the equipment provided to the beneficiary so that the equipment/system can be monitored, through the remote monitoring center;
- the remote monitoring center, the COBOT platform, or the cobot network, permanently carries out remote monitoring, remote configuration, remote control, remote service activities, for the intelligent equipment/system/intelligent robot related to the platform or network.

The active and virtual dialogue of the component parts of the cobot platform or the cobot network is done through the remote monitoring center of the platform or through the remote monitoring center of the cobot network.

These telemonitoring centers of the cobot platform or network, carry out all operational activities on agricultural lands, in intelligent processes, to demonstrate the field of intelligent and ecological agriculture.

The support of intelligent and ecological agriculture through the conception, design and realization of intelligent agricultural technologies, will lead to the real construction of intelligent agriculture in Romania, and subsequently to the construction of smart agriculture in Europe and the World.

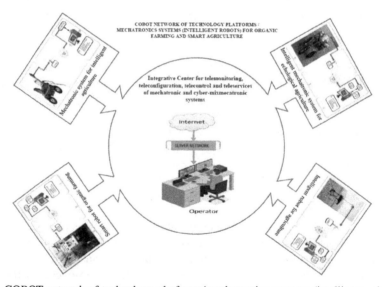

Fig. 8. COBOT network of technology platforms/mechatronics systems (intelligent robots) for organic farming and smart agriculture

10 Conclusions

In the next stage of Romania, the basic elements of the Industry 4.0 are initiated, through decisions at national level to elaborate the Strategy of Industry 4.0 for Romania and its integration/implementation in the National Development Strategies of Romania in the medium and base period, for a strategic start of Industry 4.0, of the National Action Plan of those foreseen in these strategies.

The Roadmap for Industry Strategy 4.0 will include all the elements necessary for the implementation of the Industry 4.0 strategy, the relative architecture for digitization and all other professional and technological components, human and social resources, for a sustainable support of Industry 4.0 in Romania.

Therefore, the solutions that are designed, realized and implemented in the industrial environments for the consolidation of the Intelligent Industry, become the basic pillars for the Smart Industry and for the Digital Enterprise.

References

1. Lepenioti, K., Bousdekis, A., Apostolou, D., Mentzas, G.: Prescriptive analytics: literature review and research challenges. Int. J. Inf. Manage. **50**, 57–70 (2020). https://doi.org/10.1016/j.ijinfomgt.2019.04.003
2. Jantunen, E., Junnola, J., Gorostegui, U.: Maintenance supported by cyber-physical systems and cloud technology. In: 2017 4th International Conference on Control, Decision and Information Technologies, CoDIT, pp. 708–713 (2017). https://doi.org/10.1109/CoDIT.2017.8102678
3. Cheng, B., Zhang, J., Hancke, G.P., Karnouskos, S., Colombo, A.W.: Industrial cyberphysical systems: realizing cloud-based big data infrastructures. IEEE Ind. Electron. 25–35 (2018). https://doi.org/10.1109/MIE.2017.2788850
4. Canadas, N., Machado, J., Soares, F., Barros, C., Varela, L.: Simulation of cyber physical systems behavior using timed plant models. Mechatronics **54**, 175–185 (2018). https://doi.org/10.1016/j.mechatronics.2017.10.009
5. Sanislav, T., Miclea, L.: Cyber-physical systems - concept, challenges and research areas. J. Control Eng. Appl. Inform. **14**(2), 28–33 (2012)
6. Zhu, W., Wang, Z., Zhang, Z.: Renovation of automation system based on industrial internet of things: a case study of a sewage treatment plant. Sensors **20**, 2175 (2020). https://doi.org/10.3390/s20082175
7. Iarovyi, S., Mohammed, W.M., Lobov, A., Ferrer, B.R., Lastra, J.L.M.: Cyber-physical systems for open-knowledge-driven manufacturing execution systems. Proc. IEEE **104**(5), 1142–1154 (2016)
8. Asghari, M., Yousefi, S., Niyato, D.: Pricing strategies of IoT wide area network service providers with complementary services included. J. Netw. Comput. Appl. **147**, art. no. 102426. (2019). https://doi.org/10.1016/j.jnca.2019.102426
9. Leão, C.P., et al.: Web-assisted laboratory for control education: remote and virtual environments. In: Uckelmann, D., Scholz-Reiter, B., Rügge, I., Hong, B., Rizzi, A. (eds.) ImViReLL 2012. CCIS, vol. 282, pp. 62–72. Springer, Heidelberg (2012). https://doi.org/10.1007/978-3-642-28816-6_7
10. Kunz, G., Machado, J., Perondi, E., Vyatkin, V.: A formal methodology for accomplishing IEC 61850 real-time communication requirements. IEEE Trans. Ind. Electron. **64**(8), 6582–6590 (2017). Art. no. 7878522. https://doi.org/10.1109/TIE.2017.2682042
11. Kunz, G., Perondi, E., Machado, J.: Modeling and simulating the controller behavior of an automated people mover using IEC 61850 communication requirements. In: IEEE International Conference on Industrial Informatics (INDIN), art. no. 6034947, pp. 603–608 (2011). https://doi.org/10.1109/INDIN.2011.6034947

12. Gheorghe, G., Bajeanru V., Ilie, I.: Ingineria Mecatronică și Cyber-MixMecatronică pentru Construcția Intreprinderii Digitale și Industriei Inteligente (4.0). CEFIN Pubishing House, Bucharest (2019)
13. Gheorghe, G.: Concepts and mechatronics and cyber-mixmechatronics constructions, integrated in COBOT type technology platform for intelligent industry (4.0). In: Gheorghe, G.I. (ed.) ICOMECYME 2019. LNNS, vol. 85, pp. 281–300. Springer, Cham (2020). https://doi.org/10.1007/978-3-030-26991-3_26
14. Costa, D., et al.: Performance evaluation of different mechanisms of production activity control in the context of industry 4.0. In: Gheorghe, Gheorghe Ion (ed.) ICOMECYME 2019. LNNS, vol. 85, pp. 82–103. Springer, Cham (2020). https://doi.org/10.1007/978-3-030-26991-3_9
15. Silva, M., Pereira, F., Soares, F., Leão, C.P., Machado, J., Carvalho, V.: An overview of industrial communication networks. In: Flores, P., Viadero, F. (eds.) New Trends in Mechanism and Machine Science. MMS, vol. 24, pp. 933–940. Springer, Cham (2015). https://doi.org/10.1007/978-3-319-09411-3_97
16. Constantin, A.: Telementenance and teleservice oriented design of dependable mechatronic system in automotive industry. In: Proceedings of the International Conference on Numerical Analysis and Applied Mathematics 2014 (ICNAAM-2014), AIP Conference Proceedings, vol. 1648, p. 620004 (2015). https://doi.org/10.1063/1.4912854
17. Gheorghe, G.: Challenges and research in the innovation of digital enterprise and smart industry (4.0). In: Proceedings of 2019 International Conference on Hydraulics and Pneumatics – HERVEX, 13–15 November 2019, Baile Govora, Romania, ISSN 1454–8003 (2019)

Metrology Information in Cyber-Physical Systems

João Sousa[✉], João Silva, and José Machado

MEtRICs Research Center, School of Engineering, University of Minho,
4800-058 Guimarães, Portugal
jsousa@dem.uminho.pt

Abstract. The implementation of Cyber-Physical Systems (CPS) applied across the manufacturing value chain helps to make processes more efficient since CPS enable the creation of more personalized, diversified and mass-produced products. This paper addresses the challenges of integrating Dimensional Measurement Equipments in a CPS manufacturing system, often addressed as Cyber-Physical Production Systems (CPPS). CPS and Industry 4.0 represent an opportunity for metrology. The challenge is to design a CPPS, considering the metrology information required to provide value-added services to the manufacturing system. The use of open information models and established standards will be addressed to enable data to flow seamlessly from the physical layer (measuring devices) to the cyber layer of a CPPS, mainly focusing on the Quality Information Framework (QIF) information model.

Keywords: Cyber-Physical Production Systems · Open information models · Metrology · Integration · Smart manufacturing

1 Introduction

The embracing of Cyber-Physical Systems, one of the fundamental approaches in the forth-industrial revolution, is stated with the term Cyber-Physical Production Systems (CPPS). The vision of manufacture as a CPPS is considered for standard production and also for the flexibility to adapt the production system to the customer requirements. This is an ongoing transformation of the field level in a factory, towards more intelligent, active (i.e. smart) devices, which provides benefits regarding process optimization. The digitization, interconnection, and the augmented flexibility of production assets, which lead to CPPS, represent an opportunity for the field of metrology. With an interconnected production, metrology is the foundation for a thorough valuation of production scenarios and is fundamental to the implementation of model-based manufacturing. Furthermore, the development emphasis of metrology will change from device engineering to the provision and linking of data. However, the vertical integration of information in a smart manufacturing environment, the linking, and the consequent harmonization of heterogeneous data persists as major challenges.

Industry 4.0 and specifically CPPS provide a new paradigm based on distributed manufacturing services [1, 2] where the integration aspects are considered some of the main features, namely [3]:

- the horizontal integration across business networks for inter-factory cooperation,
- the vertical integration of different hierarchical subsystems to provide flexibility and reconfigurability of manufacturing systems within a factory, and;
- the end-to-end integration of different engineering domains through the value chain for product customization.

The vertical integration inside a factory and the horizontal integration of factories enable the end-to-end integration of different engineering domains due to the product lifecycle embracing various phases that may be accomplished by diverse factories. In the fields of production and automation engineering and Information Technologies (IT), vertical integration deals with the integration of the diverse IT systems at different hierarchical levels, namely the machine, the actuator, the sensor, Programmable Logic Controllers (PLC), Supervisory Control and Data Acquisition systems (SCADA), Manufacturing Execution Systems (MES) and Enterprise Resource Planning (ERP) to deliver a top-down solution. Through vertical integration, manufacturing systems become flexible and reconfigurable within a factory, from the machine component to MES and ERP systems, thus allowing the network of manufacturing machines to become a self-organized system that can be dynamically adapt to different products, achieving an efficient custom production.

To cope with these requirements, new approaches for automation and production systems that exploit the developments in the fields of Industrial Internet of Things (IIoT) and CPPS need flexibility, agility, adaptability and interoperability. These solutions lean towards the integration and harmonization of data coming from a variety of heterogeneous components at distinct hierarchal levels. Some industrial standards have appeared over the last years, provide semantic definitions to some extent, for data modelling and data exchange for diverse areas of the manufacturing industry. The most notorious example being the IEC 62264 (ISA-95) standard, which provides a framework to assist the integration and interoperability of enterprise-control systems. The IEC 62264 is also one of the foundations of the RAMI 4.0 architecture [4].

Metrology or measurement information (e.g. the dimensions, measurements uncertainty data, process capability, measurement system capability, failed parts, etc.) need to be collected and fed back to the upper layers which isn't traditionally being performed. Research covering the use of open information models for metrology/inspection/quality domain is mainly concerned in closing the gap between design and the inspection process. A literature review shows a promising open information model addressing information requirements for the end-to-end chain of metrology activities in a manufacturing company – the Quality Information Framework (QIF). In [5, 6], authors propose a model-based inspection framework to integrate design, manufacturing, inspection and inspection process procedures. The framework uses the QIF information model to some extent and shows some results in terms of speeding up inspection related tasks. In [7] a semantic information model for manufacturing resources is built with QIF for metrology information, OPC UA[1] and MTConnect[2] for exchange data.

2 Metrology Information Modelling

The metrological domain in a manufacturing system involves different activities, each one with different information outputs and requirements from distinct sources, such as design, process planning, process execution, measurement and results reporting and analysis. In each of the activities, a multitude of commercial software systems exist usually referred to as Computer Aided Inspection systems (CAI) and Computer Aided Inspection Planning systems (CAIP). Still, the information exchange among these software systems is usually proprietary resulting in expensive data transformations for the software users, the suppliers, the vendors, and in the end to the customers. The existing metrology standards and initiatives that offer the definition and execution of the inspection activities and results analysis and improve the interchange of inspection-related information and measurement results are:

- DMIS - Dimensional Measuring Interface Standard;
- DML - Dimensional Markup Language;
- STEP-NC part 16, which is still only available as a working standard for GD&T and inspection procedures definition [8];
- STEP AP 219 Dimensional Inspection Information Exchange;
- QMD - Quality Measurement Data;
- I++ DME - Inspection Plus-Plus Dimensional Measurement Equipment;
- QIF - Quality Information Framework;

QIF is the most promising initiative to focus primarily on the activities associated with metrology and quality control and address metrology interoperability. The QIF information model is a neutral data exchange format that defines a set of XML schemas to address interoperability issues in the data flow through quality systems.

The QIF Information model enhances the exchange of metrology information along the product lifecycle from design, manufacturing, inspection, maintenance, and recycling/disposal phases. QIF is a feature and characteristic-based data format, that expresses four aspects of a characteristic: Definition, Nominal, Item, and the Actual allowing QIF to describe Product and Manufacturing Information (PMI) and report measurement results while STEP only states the PMI [9].

The QIF schemas (XSDs) are structured in six generic topics for the metrology domain: model-based definition (MBD), Rules, Resources, Plans, Results, and Statistics. And is generically classified into the following sub-systems: product definition, measurement process planning, measurement process execution and measurement results reporting [10]. A simplified QIF workflow is presented in Fig. 1.

Such as STEP, QIF supports the principles of GD&T as defined in ISO GPS standards and ASME GD&T standards (ASME Y14.5).

---
[1] https://opcfoundation.org/about/opc-technologies/opc-ua/.
[2] https://www.mtconnect.org/.

An example of a sheet metal thickness measurement using QIF is provided below (adapted from [10]). The thickness has a nominal representation and an associated tolerance that is required to be evaluated. The thickness specification is 5 ± 0,02 mm. The tolerance value is specified with the characteristic definition:

```
<ThicknessCharacteristicDefinition id="1">
  <Tolerance>
    <MaxValue>0.02</MaxValue>
    <MinValue>-0.02</MinValue>
    <DefinedAsLimit>false</DefinedAsLimit>
  </Tolerance>
</ ThicknessCharacteristicDefinition>
```

The nominal value is specified in the characteristic nominal:

```
<ThicknessCharacteristicNominal id="10">
  <CharacteristicDefinitionId>1</CharacteristicDefinitionId>
  <TargetValue>5</TargetValue>
</ThicknessCharacteristicNominal>
```

If the thickness is measured using a caliper, the tolerance condition can be assessed, along with the actual thickness value:

```
<ThicknessCharacteristicMeasurement id="20">
  <Status>
    <CharacteristicStatusEnum>PASS</CharacteristicStatusEnum>
  </Status> <CharacteristicItemId>1</CharacteristicItemId>
  <Value>5.01</Value>
</DiameterCharacteristicMeasurement>
```

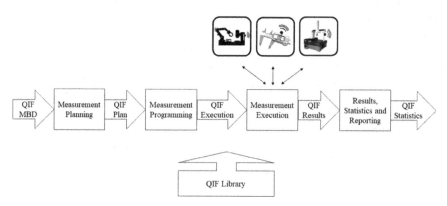

Fig. 1. Simplified QIF based workflow.

Several tests were conducted to assess how correct and complete is the specification that revealed that the QIF Results schema file is capable of representing diverse types of measurement features and characteristics [11]. As the format is independent of the data collection method, the QIF Results schema is capable of transferring measurement results from devices such as coordinate-measuring machines or vision scanning or optical devices demonstrating that QIF Results schema offers a solution for seamlessly data exchange and the collection of measurement results without costly subsequent data translation. Additionally, other information models define traceability data such as DMIS, DML, and ISO 10303 AP 238 though, the association between traceability and measurement data is unsatisfactory according to [11]. QIF has one important feature: the fact that all data generated throughout the process is linked to the authority model. This satisfies traceability requirements and gives room to further application with advanced analytical tools like data mining. This non-proprietary format can be used with an enterprise control system, specialized in quality and manufacturing management but also support the linking (at least) with the following domains [10]:

- FAIR - First Article Inspection Plan and Report;
- SPC - Statistical Process Control;
- MRP - Materials Resource Planning;
- MSA - Measurement Systems Analysis;
- MES - Manufacturing Execution Systems;
- CAM – Computer-Aided Manufacturing.

3 Towards the Central Role of Metrology Information in a CPPS

This paper addresses the pursuit of a CPPS architecture with measuring systems and the adjacent services playing the central role. The objective is the integration of Dimensional Measurement Equipments (DMEs) in a smart factory providing metrology/quality (web) services, by providing seamless metrology information exchange through the different enterprise control systems avoiding the current proprietary data block. These services can leverage the metrology information within a factory to optimize the manufacturing process since there is no standard method to act upon a manufacturing system using metrology/quality data analysis as support. Figure 2 shows the central role of metrology in a CPPS, linking the manufacturing and the inspection systems, and the metrology services. This approach will benefit from a common information model that links the aforementioned contexts.

Taking into account the number of information models and standards, addressing product and metrology data, one information model - or a combination of different information models - should be used, to enable information to flow seamlessly from the physical layer to the cyber layer of a CPPS. This paper addresses the pursuit of a CPPS architecture extensible to a high number of manufacturing SMEs.

Considering the different frameworks for Industry 4.0 implementation and CPS architectures, a methodology should be adopted, enabling the integration of DMEs in CPPS, providing added-value services to the company's stakeholders. An integrated solution considering the multiple dimensions of the manufacturing CPPS will provide a sustainable implementation methodology for the seamless integration of DMEs, further enhancing the industrial digitalization. The DMEs (either standalone systems or integrated in manufacturing machines) are connected to a network of distributed enterprise-control system modules and metrology related data is used through the different systems to provide optimization to the manufacturing devices.

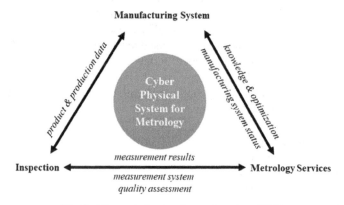

Fig. 2. The central role of metrology in a CPPS.

3.1 The Steel Tube Manufacturing Use-Case

The envisioned approach aims to measure the profile of the steel tube to evaluate if the shape of the steel tube conforms to the industrial steel tube standards (ISO 10305). This is performed by the input of the production order, either directly from the ERP or the MES system, or manually introduced by the operator. Then the evaluation of the product's conformance can start, the system can show in real-time the steel tube production conditions and can compare the product's specification to the measured steel tube profile dimensions. The dimensional measurement is performed employing an industrial camera as shown in Fig. 3.

Fig. 3. Measurement of a steel tube profile using optical methods

The required dimensional specifications can be provided by a STEP AP 242 MBD file translated to QIF MBD (XML file) that states that "the width the square tube is 20 ±

0,5 mm". The measured shape of the steel tube is then compared to the measured shape to identify if the dimensions are correct.

The measurement result can be provided by the industrial camera through a QIF Results file. A QIF Statistics file can then be generated. The file has the dimensional requirements for that specific product and the conformance result (Pass/Fail) based on the specified tolerance – that specific tube can be defined as "out of tolerance". The operator can be alerted and the information is stored.

A further level of development of the envisioned system requires the application of the collected information in the manufacturing system: the system must provide the information that "the velocity of the roller from the bending machine is responsible for the failed tolerance result". And ultimately, the application of specific knowledge towards the optimization of the manufacturing system, where the system can advise to "adjust the said velocity to 20 mm/s to correct the failed tolerance".

Another application is the use of the generated QIF Statistics file to feed the SPC system to monitor and control the steel tube production line.

For the envisioned system to work, the optical dimensional measurement system has to be able to measure the different dimensional specifications (e.g. width, height, diameter and corner radius) from different products (e.g. square, rectangular, round and special tubes). Traceability is also a big requirement since each result has to be traceable to the production order.

4 Conclusions and Outlook

CPPS and IIoT are considered major cornerstones of the new industrial revolution, with a significant impact in manufacturing systems, namely Smart Factories. To design a CPPS, several categories of real-time manufacturing and measuring data has to be acquired from the shop floor level, and passed to the cyber level. The main challenge is to define the information models that are independent of the manufacturer. Metrology related information models are being improved for interoperability mainly by the developments of QIF that show promising features for interoperability within the metrology domain.

This paper addresses the problem of integrating DMEs in a CPPS, with special attention to manufacturing SMEs, due to the lack of resources to use expensive (and extensive) commercial enterprise systems. The approach intends to contribute to the field, by exploiting current standards and open information models for the integration of metrological devices to exchange metrological information in a smart manufacturing environment.

A use case for the implementation metrology information is a CPPS is presented.

Acknowledgements. This work has been supported by FCT – Fundação para a Ciência e Tecnologia within the R&D Units Project Scope: UIDP/04077/2020 and UIDB/04077/2020.

References

1. Leão, C.P., et al.: Web-assisted laboratory for control education: remote and virtual environments. In: Uckelmann, D., Scholz-Reiter, B., Rügge, I., Hong, B., Rizzi, A. (eds.) ImViReLL 2012. CCIS, vol. 282, pp. 62–72. Springer, Heidelberg (2012). https://doi.org/10.1007/978-3-642-28816-6_7
2. Barros, C., Leão, C.P., Soares, F., Minas, G., Machado, J.: RePhyS: a multidisciplinary experience in remote physiological systems laboratory. Int. J. Online Eng. **9**(SPL.ISSUE5), 21–24 (2013). https://doi.org/10.3991/ijoe.v9iS5.2756
3. Kagermann, H., Wahlster, W., Helbig, J.: Recommendations for implementing the strategic initiative INDUSTRIE 4.0 (2013)
4. Hankel, M., Rexroth, B.: Industrie 4.0: the reference architectural model industrie 4.0 (RAMI 4.0). 2 (2015)
5. Liu, R., Duan, G.-J., Liu, J.: A framework for model-based integrated inspection. Int. J. Adv. Manuf. Technol. **103**(9–12), 3643–3665 (2019). https://doi.org/10.1007/s00170-019-03775-2
6. Rui, L., Guijiang, D.: An application of MBD based inspection in cloud manufacturing. In: IECON 2018 - 44th Annual Conference of the IEEE Industrial Electronics Society, pp. 4171–4175 (2018). https://doi.org/10.1109/IECON.2018.8591520
7. Hu, Y., Zheng, L., Wang, Y., Fan, W.: Semantic information model and mobile smart device enabled data acquisition system for manufacturing workshop. J. Phys.: Conf. Ser. **1074**, 012148 (2018). https://doi.org/10.1088/1742-6596/1074/1/012148
8. Kumar, S., Newman, S.T., Nassehi, A., Vichare, P., Tiwari, M.K.: An information model for process control on machine tools. In: Huang, G.Q., Mak, K.L., Maropoulos, P.G. (eds.) Digital Enterprise Technology, vol. 66, pp. 1565–1582. Springer, Heidelberg (2010). https://doi.org/10.1007/978-3-642-10430-5_118
9. Heysiattalab, S., Morse, E.P.: From STEP to QIF: product and manufacturing information. In: 2016 Annual Meeting on American Society for Precision Engineering, vol. 65, pp. 312–317 (2016). https://doi.org/10.13140/RG.2.2.34292.55682/1
10. DMSC: Quality Information Framework (QIF) 3.0 – A Data Model for Manufacturing Quality Information (2018)
11. Zhao, Y.F., Horst, J.A., Kramer, T.R., Rippey, W., Brown, R.J.: Quality information framework – integrating metrology processes. In: IFAC Proceedings Volumes, vol. 45, pp. 1301–1308 (2012). https://doi.org/10.3182/20120523-3-RO-2023.00113

Overview of Collaborative Robot YuMi in Education

Jiri Vojtesek and Lubos Spacek[✉]

Department of Process Control, Faculty of Applied Informatics, Tomas Bata University in Zlin,
Nad Stranemi 4511, 760 05 Zlin, Czech Republic
{vojtesek,lspacek}@utb.cz

Abstract. The collaborative robotics is together with digitalization a trending topic in the automation industry. It allows operators to safely work among robots and collaborate with them on various tasks. This solution combines the advantages of robots and humans and reduces space requirements, which results in boosted effectivity. Students of robotic fields should be aware of the usage conditions of collaborative robots to be fully prepared for their future work. They need to understand how different they are from classic industrial robots and what limitations they have. Collaborative robot YuMi from ABB is a perfect candidate for education as it is quite easy to program, with its low payload is harmless even when used inappropriately and has two arms that can be used for coordinated motion applications.

Keywords: Collaborative robot · YuMi · Rubik's Cube · Education

1 Introduction

This paper aims to incorporate the collaborative robot YuMi to education in a way that will draw students' attention, is easy enough to be covered in less time and can be extended on the software and hardware side by final projects or theses. There are many applications for robots, but most of them lack this extensibility because robots are meant to perform repetitive tasks. Once these tasks are programmed, there is not much room for improvement except path and process optimization.

A very good application in education for a two-arm collaborative robot could be Rubik's Cube solving. It is interesting enough to keep students busy, it is simple to understand and more importantly, it can be extended by optimization of algorithms and spin moves, implementation of new ones and usage of cameras or other sensors. Human-machine interaction can be also improved or developed. Everything mentioned can be at first simulated in a virtual environment (RobotStudio in ABB's YuMi case) as a digital twin and this paper covers the creation of the tool needed – a virtual 3D representation of Rubik's Cube in the RobotStudio. You can find a few solutions for YuMi solving the cube [1–3], but they are mostly for marketing purposes and lack a student-based approach and complete 1-to-1 virtualization.

The paper introduces industrial robot YuMi in Sect. 2, Rubik's Cube in Sect. 3 and describes the created component for RobotStudio in Sect. 4.

2 Collaborative Robot YuMi

2.1 Robot

YuMi is a collaborative industrial robot designed and produced by ABB (Fig. 1). YuMi has two 7-DoF manipulators (arms) with each having a payload of 0.5 kg which allows him to reach maximum TCP (tool center point) speed up to 1.5 m/s while maintaining the safety of human operators. It is programmed in the language called RAPID and offers full virtualization in the RobotStudio environment. Its two arms allow for generating coordinated movements of all 14 motors built in these two arms. Synchronization of both arms is built in the system, thus providing an excellent tool for the cooperation of both arms.

Fig. 1. Collaborative robot YuMi [4]

2.2 RobotStudio Environment

RobotStudio is a development environment for modeling, programming, and simulation of robotic applications (Fig. 2). It offers complete virtualization of ABB's robots and controllers and allows an online connection to real controllers [5]. It is more than suitable to be used for creating a digital twin with all its tools and features. Digital twin provides smart integration to the production by reducing design and optimization costs and increasing time effectiveness of the whole process [6, 7]. In education, this digital twin can play a big role during home assignments or larger final projects because students don't have usually access to industrial robots at their homes physically. They need virtual solutions and RobotStudio can be used anywhere. Most of the schools have probably a server license of RobotStudio installed for their robotics courses, thus students only need VPN access to the school's network. With the help of a digital twin that can be prepared in RobotStudio using its tools called Smart Components can students test their solutions and algorithms at home and when ready, just copy-paste the code to the real robot. Smart Components are RobotStudio objects (most of the time without a 3D graphical representation) that can simulate logic or behavior of physical objects or logic circuits used in the production. Several Smart Components can be aggregated together using a node-like structure with connections and bindings (Fig. 3).

Overview of Collaborative Robot YuMi in Education   295

Fig. 2. RobotStudio environment

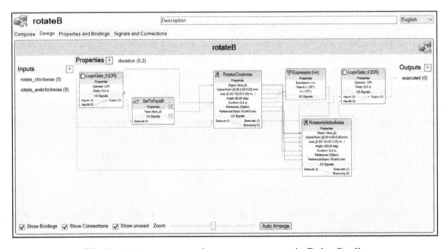

Fig. 3. Node structure of smart components in RobotStudio

3 Rubik's Cube

3.1 Skill Toy

Probably everybody knows or at least heard of a famous puzzle called Rubik's Cube (Fig. 4), originally invented in 1974 by a Hungarian professor of Architecture, Ernő Rubik [8]. It is famous even after 45 years, mostly as part of the skill toys group and has a large community of solvers with competitions in speed solving all around the world. It was also redesigned in numerous ways, adding different shapes and sizes (up to 33 × 33 × 33 cube) in the last decades.

Fig. 4. Rubik's Cube [9]

There are many solving techniques and algorithms, most of them used in a discipline called speedcubing [10]. These algorithms are devised for humans but can be easily rewritten to a computer algorithm, although they are not optimal in the number of moves (face twists). It was proven in 2013 that the minimal number of moves to solve every position of the cube is 20 [11]. The most optimal algorithm to solve the cube in the least number of moves is, of course, a logical way to design a Rubik's Cube solving robot, but for education purposes, general speedcubing algorithms are more interesting. It is not possible for a human to calculate the solution of the Rubik's Cube in the most optimal way possible and thus students will be forced to use algorithms developed for humans. In this way, they will learn more about the programming and be able to come up with own optimizations and implementations. To support their endeavor, a digital twin of the cube should be created in RobotStudio with the same behavior as the real cube.

3.2 Rubik's Cube Solving Algorithms

There are several algorithms used by professional Rubik's Cube solvers for speedcubing. All of them require the memorization of certain moves based on the specific layout of the cube which is advantageous when using a robot to solve it (note they are not optimized for time or number of moves). The most used and well-known methods for speedcubing are [12]:

- Fridrich method (also called CFOP),
- Roux method,
- ZZ method.

The comparison of these methods is needed to determine the best method for the robot. This comparison was done on multiple occasions by the community of cube solvers and this paper presents the comparison shown in Fig. 5. Although all these methods are suitable for the solution, it is advised to start with the CFOP method that is more comprehensive than the other two methods.

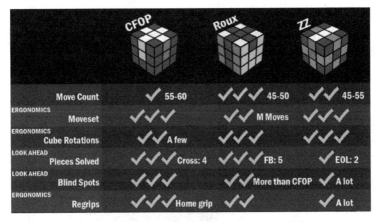

Fig. 5. Comparison of speedcubing methods [13]

4 Digital Twin Implementation

Following functions are needed to create a fully operational digital twin of Rubik's Cube in RobotStudio:

- Communication interface,
- input signals mapping,
- logical interface,
- graphical representation,
- output signals mapping,
- human-to-machine interface,
- simulation of physical properties (solved by RobotStudio),
- substitution of external signals (e.g. from a camera).

It may seem obvious that the graphical representation is the easiest part, but because RobotStudio has graphic controls similar to CAD programs, the implementation of rotations of faces is not so straight forward. Each sub-cube of the large cube should belong to a group representing one face, however, the rotation of one face distributes sub-cubes over other face groups, thus every time a face rotates, its group has to be filled with sub-cubes currently present in that face. The implementation of sub-cubes distribution to active faces is a challenging task in a CAD-oriented environment and would be otherwise not possible in RobotStudio without Smart Components.

The graphical representation of the Rubik's Cube in RobotStudio can be seen in Fig. 6, Fig. 7 and Fig. 8 and the controls for positive and negative face rotations are shown in Fig. 9, which also shows a reset button for the cube and duration property for the rotation of the face.

Fig. 6. Solved Rubik's Cube in RobotStudio

Fig. 7. Unsolved Rubik's cube in RobotStudio

Overview of Collaborative Robot YuMi in Education 299

Fig. 8. Rubik's Cube in RobotStudio in mid-rotation

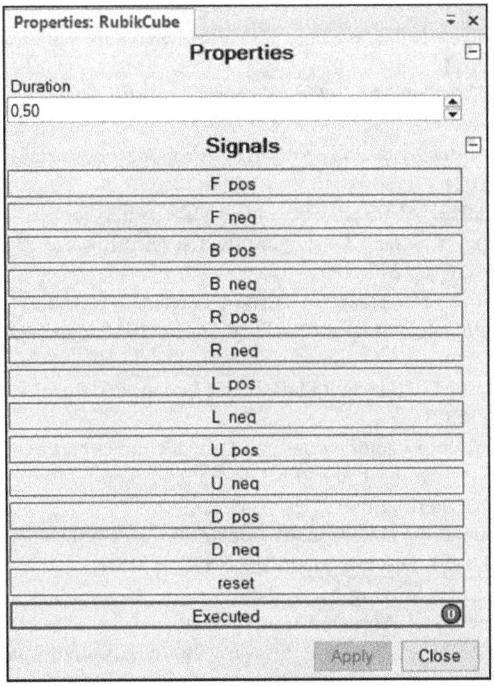

Fig. 9. Rubik's Cube face rotation controls in RobotStudio

5 Conclusion

This paper dealt with the preparation of graphical Smart Component in RobotStudio for students in robotics courses. This component simulates the behavior of 3 × 3 × 3 Rubik's Cube and provides an interface for a robot that helps with the manipulation of the object. This short overview of the whole concept covers a basic idea about the motives behind the creation of this type of application and its usage in classes. It is already prepared to be used in final projects and although it has not a practical purpose in the industry, it can teach students a lot more than few never changing applications for a demonstration that do not require extra research and studying from students' side. Students thus acquire deeper knowledge about processes behind and can develop their algorithms or optimizations.

The advantages of a collaborative robot in education and production were also briefly described. Dual-arm collaborative robot YuMi from ABB was also introduced, and it was shown that its potential for this kind of application in education is huge. Students and teachers could all benefit from exploiting its abilities during classes while maintaining safety standards.

References

1. YuMi vs Rubik's Cube. https://www.youtube.com/watch?v=GN7RKRFtZiQ. Accessed 15 Mar 2021
2. ABB's YuMi® solves Rubik's Cube. https://www.youtube.com/watch?v=_jIVCiEu2Wk. Accessed 15 Mar 2021
3. ABB YuMi vs Rubik's Cube. https://www.youtube.com/watch?v=0ghYbFUC9h0. Accessed 15 Mar 2021
4. ABB YuMi®. https://new.abb.com/products/robotics/industrial-robots/irb-14000-yumi. Accessed 15 Mar 2021
5. RobotStudio Manual, 3HAC032104-001. https://library.abb.com/en. Accessed 15 Mar 2021
6. Qi, Q., Tao, F., Zuo, Y., Zhao, D.: Digital twin service towards smart manufacturing. In: Procedia CIRP, pp. 237–242 (2018)
7. Kritzinger, W., Karner, M., Traar, G., Henjes, J., Sihn, W.: Digital twin in manufacturing: a categorical literature review and classification. In: IFAC-PapersOnLine, pp. 1016–1022 (2018)
8. Ewing, J., Kosniowski, C.: Puzzle It Out: Cubes, Groups and Puzzles. Cambridge University Press, Cambridge (1982)
9. Rubiks Cube Clipart. http://getdrawings.com/rubiks-cube-clipart. Accessed 15 Mar 2021
10. Harris, D.: Speedsolving the Cube: Easy-to-Follow, Step-by-Step Instructions for Many Popular 3-D Puzzles. Sterling Pub., New York (2008)
11. Rokicki, T., Kociemba, H., Davidson, M., Dethridge, J.: The Diameter of the Rubik's cube group is twenty. SIAM J. Discrete Math. 1082–1105 (2013)
12. Different Rubik's Cube Solving Methods. https://ruwix.com/the-rubiks-cube/different-rubiks-cube-solving-methods/. Accessed 15 Mar 2021
13. The BEST Speedcubing Method. https://www.youtube.com/watch?v=QKK8J3JKWi4. Accessed 15 Mar 2021

Reliability of Replicated Distributed Control Systems Applications Based on IEC 61499

Adriano A. Santos[1,2(✉)], António Ferreira da Silva[1,2], António Magalhães[3], and Mário de Sousa[3]

[1] CIDEM, School of Engineering of Porto, Polytechnic of Porto, 4249-015 Porto, Portugal
{ads,afs}@isep.ipp.pt
[2] INEGI - Instituto de Ciência e Inovação em Engenharia Mecânica e Engenharia Industrial, Rua Dr. Roberto Frias, 400, 4200-465 Porto, Portugal
[3] Faculty of Engineering, University of Porto, Rua Dr. Roberto Frias, 4200-465 Porto, Portugal
{apmag,msousa}@fe.up.pt

Abstract. The use of industrial and domestic equipment is increasingly dependent on computerized control systems. This evolution awakens in the users the feeling of reliability of the equipment, which is not always achieved. However, system designers implement fault-tolerance methodologies and attributes to eliminate faults or any error in the system.

Industrially, the increase in system reliability is achieved by the redundancy of control systems based on the replication of conventional and centralized programmable logic controllers. In distributed systems, reliability is achieved by replicating and distributing the most critical elements, leaving a single copy of the remaining components. On the other hand, given the nature of the distributed systems, it will also be necessary to ensure that the data set received by each of the replicas has the same order. Thus, any change in the order and data set received will result in different results, in each of the replicas, which may manifest in erroneous behavior.

In this paper, the interactions and the erroneous behavior of the replicas are explained, depending on the data set received, in a fault tolerant distributed system. Its tendency, behavior and possible influences on reliability are presented, considering the failure rate and availability based on the mean time to failure.

Keywords: Dependability · Distributed systems · Event-base control · Fault-tolerance · IEC 61499 · Industrial control · Real-time · Reliability · Replication

1 Introduction

Technological developments that have occurred in recent years have led to the proliferation of computerized systems both at the industrial level and for consumption and domestic use. At the industrial level, the technology used is based on a control system centralized in Programmable Logical Controllers (PLCs), sometimes redundant to ensure the reliability of the systems, running in a single machine. On the other hand, the use of computerized systems at the domestic and consumer level is more discreet. Embedded systems are present in our daily lives in the most diverse devices and with more visibility in traditional computers for daily use.

This proliferation makes us more and more dependent on computerized systems, becoming us more vulnerable to the occurrence of failures both in terms of the control of domestic systems and in terms of industrial systems. Failures in the control system of a reactor at a nuclear power plant will have more serious consequences than in domestic systems. The reactor, when subject to a failure, can put thousands of people in danger and seriously affect the ecosystem. On the other hand, a failure an embedded system in a washing machine or toaster, keeping the equipment energized, can also, at a lower level, put people and property at risk. Obviously, if a fire occurs due to system failure, the housing in question and the contiguous ones can be in great danger. From the point of view of the reliability of a computer system, the underlying idea is that the system behaves according to its specification, in view of the numerous problems that can occur (natural accidents, software and hardware errors, among others). In fact, what we expect from computer systems is that they will work correctly to guarantee integrity and availability.

Industrially, the process control is based on centralized PLCs, generally programmed according to the languages standardized in IEC 61131 [1]. On the other hand, with the proliferation of communication networks in the industrial domain, PLCs were interconnected with each other resulting in distributed control applications. However, according to the semantics of IEC 61131, programming languages and their execution are not presented as a good practice for the requirements used in flexible automation and distributed control applications. It was mainly for this reason that the International Electrotechnical Commission (IEC) developed the standard IEC 61499 [2] to facilitate the development of Distributed Industrial Process Measurement and Control Systems (DIPMCS). This standard proposes the use of Function Blocks (FBs) as a basis for the development of reusable software modules for the control system. Each function block is a functional software unit that encapsulates local data and algorithmic behavior within an event/data interface where operations, within a function block, are controlled by an event-driven state machine.

Given the nature of distributed control applications, many new problems must be considered. Therefore, when developing a distributed application, the designer must be aware of the possibility of partial failures, that is, the possibility of some device stopping its execution and the others continue with their normal processing. So, when developing IEC 61499 applications with high dependability requirements, the implementation of solutions to tolerate these partial failures should be considered. The increase in the reliability can be obtained by masking the failed devices, introducing fault-tolerance in the application architecture.

The purpose of this text focuses on the presentation and definition of some concepts related to the dependability attributes of computer systems. In this way, it is also intended to expose some of the classic mechanisms that allow to solve the dependability problems, namely the faut-tolerance techniques. The analysis presented in the next sections are an adaptation of traditional approaches to partial fault-tolerance. Some of the problems associated with the reliability of replicated systems will be explained considering the various possibilities of communication between the devices.

2 Dependability

Replication is widely used in distributed systems as a Fault-Tolerance (FT) mechanism to maintain the desired availability and reliability. Among other reasons for its use, the fact that replication fits naturally in the topology of distributed systems stands out. So, it is for these reasons that fault-tolerance techniques are generally used to satisfy dependability requirements. Dependability of systems engineering – computational, mechanical, physical, and human systems – must be expressed by a relatively small number of failures. On the other hand, it is necessary to consider that the acceptable levels of failure of the systems vary according to their measured specifications, in percentage, according to the *Mean Time Between Failures* (MTBF) and the *Availability* (A).

Dependability systems has been defined by [3] as "...*the trustworthiness of a computing system which allows reliance to be justifiably placed on the service it delivers* ...". Another similar definition is given by Avizienis *et al.* [4] "... *the ability to deliver service that can justifiably be trusted* ...". However, the main idea of dependable systems is that they must be able to ensure that the service provided meets the specifications and, as such, does not fail. This means that the use of dependable systems is always desirable, as they are "trustworthy" [5]. The perception of dependability can be very generic and understood in different ways by different people, however this must integrate the following attributes: Availability, Reliability, Safety, Integrity, Maintainability, and Confidentiality.

Availability is used to measure the probability that a system or component will become operational and perform its functionality after a failure. Mathematically, the "*availability is a measure of the fraction of time that the item is in operating condition in relation to total or calendar time.*" [6]. More formally, availability can be defined as the "... *proportion of time a system is in a functioning condition*" [7]. The availability of the system at time t is defined by $A(t)$ which represents the fraction of time that the system is available. It is also called inherent availability or steady-state availability and can be expressed at time t by:

$$A(t) = \frac{u}{u+d} \tag{1}$$

where u is the mean uptime and d is the mean downtime of the system.

The next attribute, reliability, is define as "... *the ability of an item (a product or a system) to operate under designated operating conditions for a designated period of time or number of cycles.*" [6]. More formally, reliability can be defined as "... *the ability of a system to function under stated time and conditions*" [7]. So, reliability is the probability that the system is operating normally in the range $[t_0, t]$, therefore, reliability at time t is denoted as $R(t)$. On the other hand, the abnormal operating condition of the system at $[t_0, t]$ is defined as unreliability degree at time t and it denoted as $F(t)$. The relationship between reliability and unreliability can be expressed at time t by:

$$R(t) = 1 - F(t) \tag{2}$$

Safety is defined as "*the nature of a system not to endanger personnel and equipment.*" [7], however Avizienis *et al.* [4] define safety considering the environment as the "...

absence of catastrophic consequences on the user(s) and the environment". Note that availability and reliability are related, however they are not synonymous with security. The availability or unavailability and the reliability or unreliability of a system does not translate into security. Therefore, high reliability results in high security, but high security does not necessarily result in high reliability [7]. Safety is denoted as $S(t)$.

Maintainability is defined as "*the ability of a system to recover its required function*" [7]. It is the ability of the system to be successfully repaired under certain conditions. The degree maintainability is denoted as $M(t)$. Integrity is the non-occurrence of undue changes to the information in the systems, they cannot be modified without authorization, while confidentiality is defined as the absence of unauthorized disclosure of information, that is, the information is not disclosed without authorization.

From the above, we can say that these attributes are interconnected, so, a product or system will be considered dependable if it has all the attributes. On the other hand, we must also consider that the degree of dependability for a system is not a binary phenomenon but is based on the degradation of these attributes and the limits that are considered acceptable [5]. It is a combination of availability, confidentiality, and integrity.

2.1 Relationship Between Fault, Error, and Failure

According to the above, a service will only be performed correctly when it meets the conditions mentioned in the specifications. The existence of an error (see Fig. 1), is due to the occurrence of a component failure caused by physical phenomena of mechanical or electrical origin, internal or external. This can spread, manifesting itself in the degradation of the service, such as, for example, limiting services, decreasing speed of service provision, etc., causing a failure. In software, faults resulting from wrong programming will translate into the existence of a latent fault that, when activated, called the wrong instructions or the use of the wrong data. This event will give rise to an error that in turn will propagate to produce a failure [8].

Fig. 1. Error propagation chain [4].

A *fault* is a static feature of a system and is the cause of an error. The *error* is an incorrect internal state of the system that can lead to failure. A *failure* is considered as the occurrence of an unexpected behavior of the system, that is, when the system does not fulfill its mission. This is caused by an error. So, if we look at a car tire, we can say that the failure in this system can be caused by an external event (a nail, a wire, or a glass) that causes a puncture or an internal cause like the tube's trail (*causation*). This fault will give rise to an error (*activation*), deflating the tire, compromising its use for the specified purposes (*propagation*) (see Fig. 1). The tire deflating causes the failure, preventing the vehicle moving forward. Therefore, we can say that the faults, however small, they can cause major errors, however the presence of an error will not necessarily be considered the cause of a failure [9].

Note, however, that the classification of the several internal states of a system for *fault*, *error* or *failure* situations will depend on the system and its specific characteristics. Therefore, based on the characteristics of the systems, for example, centralized, distributed and/or replicated, will be in the presence of organic software and/or hardware failures. In this perspective, we must consider them different since the software does not wear out. Thus, it can be said that there will be a software error or failure if consider the definition presented in [10] "... *work according to the original contract or according to the requirements documentation...*" whenever its MTBF, for example, exceeds the default value. In this assessment, will be in the presence of a fault, but the system will remain operational.

2.2 Means to Attain Dependability

Over the years, various means have been used to achieve the attributes of dependability. These attributes can be grouped into four groups [4]: *Fault Prevention* (use of design methodologies, techniques, and technologies to avoid faults), *Fault-Tolerance* (use of means to provide service even in the presence of faults), *Fault Removal* (use of review, analysis, and testing techniques to reduce the number and severity of faults) and *Fault Forecasting* (estimate of the number, incidence, and future consequences of faults).

The development of flawless computer systems is rarely achieved, so some of the attributes of dependability, such as fault-tolerance, are often used. Fault-tolerance is used to increase the probability that the final design of the application will show acceptable behaviors and thus produce correct outputs. On the other hand, correctness of the system is closely linked to the structure of the application, so, the greater or lower level of fault-tolerance will directly depend on the characteristics of itself and its design. Dependability based on fault-tolerance is generally achieved by redundancy. Redundancy is the duplication (triplication, quadruplication, etc.) of a specific component, considered critical, for which the failure of the component or subcomponent will not result in the failure of the entire system.

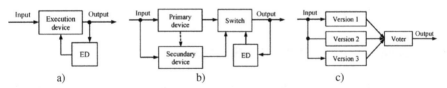

Fig. 2. Fault-tolerance architecture: a) Normal implementation; b) Standby implementation (Replica in standby, rum in parallel); c) Parallel implementation, voting validation.

From the perspective of the operational requirements of fault-tolerance systems, they can have classified into two classes of operability (Continuous and Non-continuous operations) and three main approaches. In the first case, the systems remain operational even in the presence of faults (provides a continuous service) while in the second case the systems interrupt their operation. On the other hand, the fault-tolerance implementation architectures will be based on the use of error detection (ED), with modules operating in standby or in parallel, and the use of voters that validate and consolidate the output values of the redundant elements (see Fig. 2).

The criteria for choosing one of these implementations will depend on the system itself. It will be obvious, and therefore logical, that the application designer only replicates devices with a high influence on the application, while maintaining a single copy of the remaining devices. However, regardless fault-tolerance approach used, the availability, reliability, and safety properties (see item 2) for the system, must be maintained.

The measurement of these properties involves knowledge of MTBF based on the Mean Times To Failure (MTTF) and Mean Time To Repair (MTTR). So, MTBF can be expressed as:

$$MTBF = MTTF + MTTR \tag{3}$$

then availability, according to (1), will translate into the following expression:

$$A = \frac{MTTF}{MTTF + MTTR}(\%) \tag{4}$$

On the other hand, from a reliable point of view, it can be said that the system's reliability represents the probability that there will be no fault during a given period, in which more than one failure is possible [11]. In this sense, it can be considered that the ratio of the number of faults (fault rate) will also be an indicator of reliability showing the portion of components or equipment that must survive in an instant t. The fault rate (λ) will be given by the following expressions:

$$\lambda = \frac{Number\ of\ faults}{Time\ of\ use} \quad or \quad \lambda = \frac{1}{MTBF} \tag{5}$$

Safety must be a propriety for the entire system, including hardware and software, and for its entire life cycle. Thus, when addressing security issues, a set of attributes must be considered that must be verified at the same time: 1) availability for authorized actions, 2) confidentiality and 3) integrity [4].

3 Overview of IEC 61499 Replications

In the industrial context, control processes are dominated by PLCs. With the increase in demand for new, more flexible productive markets, flexibility requirements become preponderant and lack real-time responses. On the other hand, it must be considered that these new needs for flexibility demand greater availability of productive systems and, consequently, greater reliability of them. Reliability or redundancy in centralized applications using PLCs, is usually achieved using a second or more identical PLCs. However, if the same criterion were used in a distributed system, the replication of all devices would become very complex and with exaggerated dimensions.

To address the problems associated with distribution and reliability based on device replication, the International Electrotechnical Commission (IEC) has developed the IEC

61499 standard. The distributed nature of this standard allows each device to perform a simple replica or a sub-application of other devices that are running on the same system.

Based on a several FB – Basic FB (BFB – which runs an Execute Control Chart (EEC) on the head and several algorithms on the body), Composit FB (CFB – several FBs running inside of the FB) and Special Interface FB (SIFB – communication FB) –, the application designer decides which component, as a software unit (FB), to be replicated. The designer replicates the FBs or sub-applications that have greatest influence on the dependability of the system (defining new input and output events and data port, (see Fig. 3a), keeping a single copy of the remaining (see Fig. 3b).

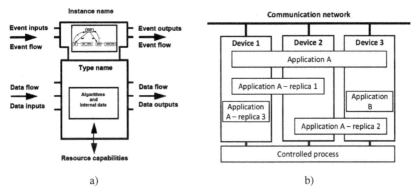

Fig. 3. a) Basic Function Block (BFB) with input and output ports; b) Distribution of replicated and non-replicated applications among divices.

A distributed IEC 61499 application can be distributed or replicated between divices. Thus, the replica of an application can be performed on a single device (application A – replica 3), while other replicas of the same application are distributed between two or more devices (application A – replicas 1 and 2), see Fig. 3b. Based on these distribution possibilities, several interaction scenarios can be identified.

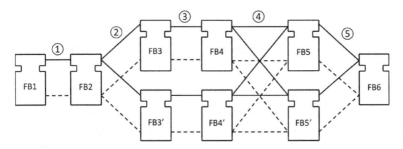

Fig. 4. Distributed system with the several interaction scenarios of the replicated FBs [12].

The interaction scenarios depend on the system distribution and the replication performed by the application designer, that is, on the components that send events and/or data ② *(one-to-many)*, the receivers ⑤ *(many-to-one)* or both ④ *(many-to-many)* are replicated components or not, as you can see in Fig. 4 and explained in [12] and [13].

4 Example for Redundancy Implementation

To implement redundancy, all replicas must be identical. Therefore, for this requirement to be met, all replications must be performed assuming a deterministic behavior, that is, the algorithm of each of the replicated FBs must be deterministic. In this sense, when executing the FBs, in IEC 61499 execution environment, if it is guaranteed that they will be executed in the same way, the determinism of the replicas will be guaranteed. This means that the order in which each event is run will be the same in all replicas and will produce the same results.

To illustrate the implementation of redundancy, based on IEC 61499, we will present a simple example. The fault tolerance approach will be based on simple active redundancy of software and hardware. This application is a small part of a more complex application for conveyor-based pieces distribution systems. In this sense, the case study will be limited to the analysis of the interconnection of two types of conveyors (linear, C1 and C2, and rotating, C3) that working, perpendicularly. C3 receives parts from both conveyors considering that all conveyors are relatively small and will only be able to transport one part at a time and its work according to the layout shown in Fig. 5.

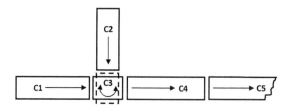

Fig. 5. Example conveyor layout.

Workpieces arrive from the left (C1) and the top (C2) and must be transferred to the rotating conveyor belt (C3) at the center. Each conveyor belt is controlled by a single FB based on the conveyor type (linear or rotating) that runs on an independent low-cost device (Raspberry Pi) that controls the entire mechatronic device of the conveyor.

The pieces to be transferred are simulated by a simple algorithm in each of the FBs. The linear conveyors feed (C1 and C2) provides two integers value corresponding to each of the conveyors, data 1 and 2, respectively. The rotating conveyor (C3) receives data from the feed conveyors. This conveyor processes the data according to the event's availability time (the oldest first) and transfers the part and data information to the following conveyor. Each of these timed messages will be ordered according to the time they were available. A conveyor will be free to receive a new piece as soon as the piece in transport is transferred to the next conveyor.

The design of the application and the replication of elements considered critical of the system are shown in Fig. 6. Note that the conveyors need to receive events and data from the previous ones and send events and data to the following ones. So, conveyors belts on the left, which are feed conveyors, are triggered by a start event sent simultaneously to C1 and C2 and the data produced are sent to the subsequent replicas. Device 3 and 4 are replicas of conveyor C3 that will receive information from the clients C1 and C2. The information received in each of the replicas is an indicator of availability for delivery a work piece. The data received in each of the replicas must belong to the same data set and in the same order.

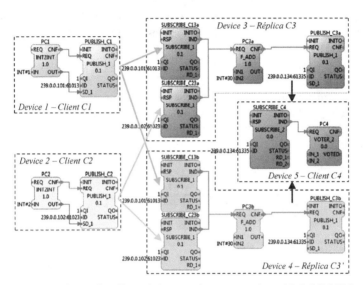

Fig. 6. Design application and replicated elements, interconnection with Publish/Subscribe pair.

4.1 Execution Semantics

In this item, a brief description of the design process and how the replication/distribution was performed in IEC 61499 application will be made. The application design is very simple, and it is focuses on the fault-tolerance approach. To develop the replicated distributed system, we used the open-source *Eclipse 4diac*™ and the FORTE IEC 61499 execution environment [14]. Application was replicated according to scenario ② *(one-to-many)*, shown in Fig. 4 and the final design of it is shown in Fig. 6.

As already explained, to maintain replicas determinism, it will be necessary to ensure that the execution of each replicated FBs occurs in a synchronized manner, that is, that each event received generates the same sequence of actions in all replicated FBs. So, to ensure communications between all replicated FBs and maintain design application, you must use Publish/Subscribe communication SIFBs pairs to implement timed-messages protocol (*one-to-many* and *many-to-one*) as well internal synchronization. On the other hand, all instances of the FORTE runtime environment and devices must be in the same multicast group and be synchronized [15], as well as internal clocks by Network Time Protocol (NTP) synchronization, for example. However, when we consider that "*FORTE* is relatively well suited for supporting the replication design ..." [13] the expected results will not be guaranteed, since "... its execution semantics almost guarantee a deterministic execution of event sequences." [13] as demonstrated in [16].

4.2 System Reliability

A system is usually composed of several components that will have different reliability and that, in principle, will be known. In this sense, knowing the reliability and trend of each component, determining the reliability of the system will be relatively simple, if the system can be represented by a reliability block diagram (RBD). However, when it is only possible to know the failure rate, without knowing the system's tendency (decreasing, constant or increasing), the adoption of any distribution to quantify the reliability will become an almost impossible task. The RBD representation of the implemented redundant system is shown in Fig. 7. This is a complex system composed of components in series and in parallel. Determining the value of system reliability is a succession of mathematical operations that combine redundancy in series and in parallel to obtain an overall value for it.

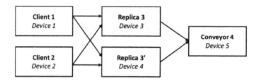

Fig. 7. Reliability for scenario 2, communication *one-to-many*.

However, the study carried out aims, essentially, to analyze the deterministic behavior of the Publish/Subscribe communication pairs, that is, quantify their guarantee of determinism. Therefore, all communication errors between clients 1 and 2 and the replicas C3 and C3' were recorded and based on this analysis, it was possible to build the graph shown in Fig. 8. The trend of the system can be determined through a process of statistical inference based on hypothesis tests, that is, to verify if it is possible to confirm or deny the formulated hypothesis. To verify H_0 (constant rate) we will have to confirm it or not using the statistical process called Laplace Test (ET). Following equation:

$$ET = \sqrt{12N}\left(\frac{\sum_{i=1}^{N} t_i}{N.t_0} - 0,5\right) = \sqrt{12 \times 567}\left(\frac{2711967}{567 \times 10000} - 0,5\right) = -1,789 \quad (6)$$

where N is the number of faults, t_i the time of fault i, and t_0 the total time. So, in this case the test is conclusive, as there is evidence of acceptance of H_0, since the ET value is outside the rejection region $[-ET(\alpha/2) < ET < +ET(\alpha/2)]$, that is, $-1,960 < ET < +1,960$ with $\alpha = 5\%$. When accepting H_0, rejecting H_1 (non-constant), we consider that the fault rate is constant, and the occurrences are Independent and Identically Distributed (IID).

The fault rate of the system is $\lambda = 0,0567\ fault/s$ (5) so, reliability components, considering only replicas, is given by $R(t) = e^{-\lambda t}$. According to Fig. 7, the reliability system (R_S) will be obtained, based on the received components, considering the rest with high reliability, by the expression: $R_s = 1 - \left(1 - e^{-\lambda t}\right)^n$. Data received in the active replicas, have an $R_S^{10000} = 0,951$, a MTTF = 17,637 s, and $F(t) = 0,049$ (2).

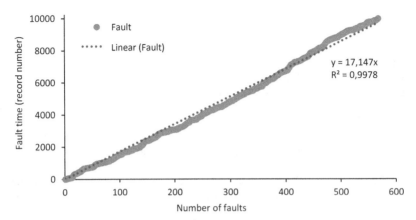

Fig. 8. Relation between fault number and time (F-T).

5 Conclusion

In a distributed system, there are usually partial failures that must be considered. Critical component redundancy is one of the measures that must be used for fault-tolerance. However, the adoption of redundancy measures will lead to replication with or without voters in which the results outgoing replicas must present the same values. This restriction will require that all replicas remain synchronized.

The use of timed messages protocol and NTP are an important contribution to ensure the synchronization of the replicas. However, it will be necessary to consider the different communication scenarios between replicated and non-replicated components and how to guarantee the determinism of IEC 61499 applications, considering their restrictions. Thus, to ensure the ordering of data in a FORTE execution environment, it will be necessary to ensure that the Publish/Subscribe communication pairs send and receive the same data set in the same order in which they are made available.

The analyses carried out shows that the design of replicated systems, using the FBs defined in the IEC 61499 standard is not able to guarantee its determinism. Thus, based on the lead time and the ordering of data received, it was possible to quantify the reliability of the system, considering only the data received in the replicated pairs, so the measured reliability was around 95%. These results define a 5% rate of unreliability (base in 10 K data received in each replica), which will indicate that the replication based on the IEC 61499 FBs and FORTE runtime do not guarantee reliability by itself. Reliability will be guaranteed by the development of additional FBs that implements it.

Acknowledgments. We acknowledge the financial support of CIDEM, R&D unit funded by the FCT – Portuguese Foundation for the Development of Science and Technology, Ministry of

Science, Technology and Higher Education, under the Project UID/EMS/0615/2019 and this work was supported by FCT, through INEGI and LAETA, project UIDB/50022/2020.

References

1. IEC 61131: Programmable Logic Controllers Part 3 (IEC 61131-3). International Electrotechnical Commission, 3rd edn. IEC (2013)
2. IEC 61499: International Standard IEC 61499-1, Function Block Architecture Part 1, 2nd edn. International Electrotechnical Commission. IEC (2012)
3. IFIP Working Group 10.4 on Dependable Computing and Fault Tolerance. http://wg10.4.dependability.org/. Accessed 07 Dec 2020
4. Avizienis, A., Laprie, J.-C., Randell, B., Landwehr, C.: Basic concepts and taxonomy of dependable and secure computing. IEEE Trans. Dependable Secure Comput. **1**(1), 11–33 (2004)
5. Farrukh Khan, M., Paul, R.A.: Chapter 4 - pragmatic directions in engineering secure dependable systems. In: Advances in Computers - Dependable and Secure Systems Engineering, vol. 84. Elsevier, USA (2012)
6. Modarres, M., Kaminskiy, M.P., Krivtsov, V.: Reliability Engineering and Risk Analysis, a Practical Guide. 3rd edn. CRC Press, Taylor & Francis Group (2017)
7. Yang, M., Hua, G., Feng, Y., Gong, J.: Chapter 1 - introduction. In: Fault-Tolerance Techniques for Spacecraft Control Computers. Wiley (2017)
8. Bloomfield, R., Lala, J.: Safety-critical systems: the next generation. IEEE Secur. Priv. **11**(4), 11–13 (2013)
9. Abdulhameed, O.A., Jumaa, N.K.: Designing of a real time software fault tolerance schema based on NVP and RB techniques. Int. J. Comput. Appl. **180**(26), 35–4 (2018)
10. ISO 9000-3:1997: Quality management and quality assurance standards - Part 3: Guidelines for the application of ISO 9001:1994 to the development, supply, installation and maintenance of computer software. ISO (1997)
11. O'Connor, P.D.T., Kleyner, A.: Practical Reliability Engineering, 5th edn. Wiley, UK (2012)
12. Santos, A.A., de Sousa, M.: Replication strategies for distributed IEC 61499 applications. In: IECON 2018 - 44th Annual Conference of the IEEE Industrial Electronics Society, pp. 2225–2230. IEEE. Washington, DC (2018)
13. de Sousa, M.: Chapter 9 – fault-tolerance IEC 61499 applications. In: Distributed Control Applications: Guidelines, Design Patterns, and Applications Examples with the IEC 61499, 2nd edn. CRC Press, Boca Raton (2016)
14. Eclipse 4diac. https://www.eclipse.org/4diac/. Accessed 29 Dec 2020
15. Pinho, L.M., Vasques, F., Wellings, A.: Replication management in reliable real-time systems. Real-Time Syst. **26**(3), 261–296 (2004)
16. Santos, A.A., da Silva, A.F., Magalhães, A.P., de Sousa, M.: Determinism of replicated distributed systems-a timing analysis of the data passing process. Adv. Sci. Technol. Eng. Syst. J. **5**(6), 531–537 (2020)

Inspection Robotic System: Design and Simulation for Indoor and Outdoor Surveys

Pierluigi Rea[3], Erika Ottaviano[1(✉)], Fernando J. Castillo-García[2], and Antonio Gonzalez-Rodríguez[2]

[1] University of Cassino and Southern Lazio, via Di Biasio 43, 03043 Cassino, FR, Italy
ottaviano@unicas.it
[2] University of Castilla-La Mancha, Avda. Carlos III. Real Fábrica de Armas, 45071 Toledo, Spain
Fernando.Castillo@uclm.es
[3] University of Cagliari, Via Marengo, 2, 09123 Cagliari, CA, Italy
pierluigi.rea@dimcm.unica.it

Abstract. In this paper, the design and simulation are presented for a wheeled robot designed for indoor and outdoor inspections. In particular, for a large number of cases, an automatic or tele-operated survey can be performed by wheeled mobile robots, which represent the most efficient solution in terms of power consumption, control, robustness and overall costs. Referring to the analysis of structures and infrastructure, such as bridges and pipelines, wheeled robots must be able to move on horizontal or sloped surfaces and overpass obstacles that in most of cases are steps, i.e. longitudinal internal stiffeners. In this paper, we present a mechatronic design and simulations of a wheeled robot being used in indoor and outdoor inspections, taking as illustrative example an infrastructure inspection. In particular, the wheeled robot is equipped with suitable sensors in order to take information on the main structural elements, avoiding the need of experienced personnel to get directly inside the site to be inspected.

Keywords: Mobile robots · Simulation · Robotic infrastructure inspection

1 Introduction

Mobile robots represent an interesting field of research due to their great potential features. They can be used for a wide range of applications including inspection [1], service [2], defense, manufacturing, cleaning, remote exploration [3], rescue [4] and entertainment. Wheeled and tracked systems can perform fast and robust motion on flat terrain, but they are less effective in overpass obstacles that are greater, or having a similar size, than the radius of the wheel. In a large number of applications, such as bridges or pipelines, when it is possible to move on horizontal surfaces, wheeled robotic solution can perform the inspection of structures and infrastructure. As reported in the literature, several solutions can be adopted for the inspection when moving on a horizontal surface, see for example [5–9]. Indeed, if a surface is relatively smooth, with suitable dimensions of the obstacles, wheeled solution is the most efficient, in terms of power consumption, costs, complexity of operation and control, robustness, travelling speed. A possible application of wheeled locomotion system for inspection is for box girder bridges and water distribution systems, where there is a relatively large area to inspect that is difficult to access. Box girders are used for bridges with large spans, having excellent high torsional stiffness and when the self-weight of the bridge needs to be minimized. The clean lines of box girders bridges, usually with no visible external stiffening, is generally considered to give an excellent appearance and durability, since there are no traps for dirt and moisture. Box girders can be used for highway bridges, railway bridges and footbridges. The selection of a box girder shape usually results in relatively thin plate panels (in terms of thickness to width ratio) for the webs and bottom flanges (and for top flanges, in all-steel boxes). Avoidance of local buckling in compression zones and in shear requires appropriate stiffening and longitudinal stiffeners are often required. Although box sections offer high torsional stiffness, internal cross frames are usually needed to prevent distortion when one web is subject to greater shear than the other, one diagonal dimension across the cell increases and the other decreases. The inspection of box girders can be taken into account for the simulation purposes, as a possible practical application of robotics inspection, which is an area of growing interest, as reported in [10–13]. In this paper, we propose the mechatronic design and simulation of a wheeled robot for infrastructure inspection; nevertheless, the robot can perform outdoor inspection carrying suitable instrumentation as well.

2 Mechatronic Design

2.1 Mechanical Design

Figure 1 shows the mechanical design of the robot with a sensorization described in Fig. 2. The mechanical design has been kept simple in order to have a robust and less expensive mechanical solution. Four independent direct-drive wheels operate the robot. The battery is located on-board. Tables 1 and 2 show robot and sensors specifications.

2.2 Sensorization

The mechatronic control scheme is built according to design principles reported in [14, 15]. The wheeled robot is commanded by remote-control and HMI interface is used for monitoring the sensors on board. Front and rear cameras are mounted; one is intended to be used for robot localization and motion into the environment, the second one is used for inspection purposes. They have both pan, tilt and zoom, which can be controlled by a virtual joystick shown in Fig. 2, as the red point on the screen of the HMI; the used cameras allow infrared vision. The RC controller is a Spektrum DXe 6-Channel 2.4 GHz coupled with a Spektrum Receiver Mk610 2.4 GHz 6 Channel. Tables 1 and 2 summarize the main characteristics of the robot and its sensorization.

Inspection Robotic System: Design and Simulation 315

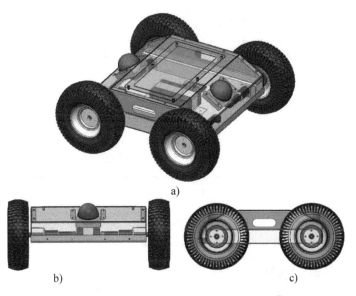

Fig. 1. Mechanical design: a) 3D view; b) front view; c) side view.

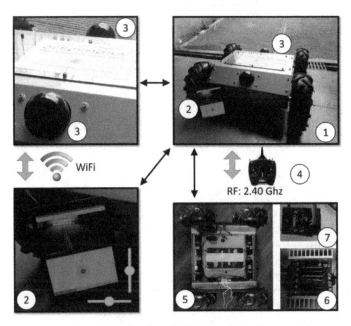

Fig. 2. Overall mechatronic system: 1) robot; 2) HMI interface; 3) rear and front cameras; 4) RC controller; 5) internal view of the robot-control boards; 6) electronic board for motor control; 7) Spektrum Receiver Mk610.

Table 1. Robot specifications.

Item description	Specification	
Robot system 4- wheeled robot	Size (LxHxW)	900x600x500mm
	Mass	45 kg
	Max speed	Up to 3mph
	Actuation	24VDC IG52
	DOFs	4 (wheels)
	Max step size	100 mm

Table 2. Sensor specifications.

Item description	Specification		
Internal sensors	Range	Resolution	Power
MTK Ver.1 Accelerometer	0...80	0.001	0.001mA
MTK Ver.1 Gyroscope sensor	0...34	0.001	0.001mA
MTK Ver.1 Gravity sensor	0...39	0.001	0.001mA
MTK Ver.1 Magn. field sensor	0...4000	0.150	0.001mA
MTK Ver.1 Lin. Accel. sensor	0...39	0.001	0.001mA
External sensors	Model		
Thermal camera (FLIR)	Thermal FLIR		
Front - Rear camera (Sony)	48/5 MP type		
Communication			
USB WiFi router	TP-LINK Model TL-WN821N		
Spektrum Receiver Mk610	AR6210 DSMX		
Control station			
Remote-controller	1		
HMI Interface	1		

3 Simulation

Simulations were carried out to test the engineering feasibility and application of the robot for infrastructure inspection. In particular, we consider an indoor evaluation inside a bridge deck. Steel box girders and steel and concrete composite box girders are used for long spans, where their excellent torsional stiffness is of particular benefit and the self-weight of the bridge needs to be minimized. A commonly adopted configuration uses the "open" trapezoidal girders. These girders have a steel bottom flange, inclined steel webs and a thin steel flange on the upper part of each web. Each of these elements need to be inspected, usually visually or with a thermal camera. The deck slab of the reinforced concrete bridge forms the closed cell. For any closed cell that requires internal access to construct it or to carry out inspection and maintenance, health and safety considerations require sufficiently large and well-placed openings for the access by trained personnel, alternatively, these sites can be inspected with mobile robots. The scenario that a robot encounters is a flat surface with steps to climb. Therefore, suitable simulations involve the use of flat surfaces with obstacles and steps to be surpassed.

A good standard evaluation method for certifying the locomotion characteristics of mobile robots is the step field pallets, as it is reported in [16]. They represent a suitable and repeatable topology of surfaces for testing the motion capabilities of a robot. In this context, we have considered a modified version than encounters steps simulating the internal stiffeners of the box girders, as in Fig. 3, simulations are shown in Figs. 4, 5, 6 and 7.

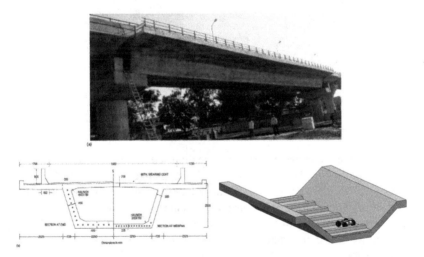

Fig. 3. Field designed for a test on the mobile robot representing a box girder bridge.

More specifically, Fig. 3 shows a boxed girder bridge deck, whose main dimensions are given. According to those dimensions, a realistic scenario is reproduced for testing the robot, as it is shown in the motion sequence in Fig. 4. A specific 3D contact between wheels and surface have been modeled and implemented.

In particular, Figs. 4, 5, 6 and 7 show the results of a dynamic simulation for the wheeled robot during a motion on the surface represented in Fig. 3. This present a worst case of obstacle with high = 150 mm and width = 150 mm.

The numerical results report a simulated motion inside a box girder bridge, where a robot should move for the survey. The simulations show the robot overpassing the

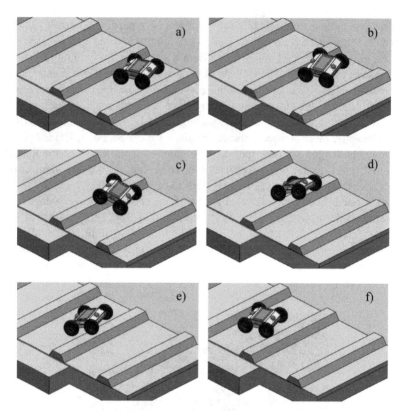

Fig. 4. Dynamic simulation snapshots for the robot to overpass the internal stiffeners in Fig. 3.

obstacles, which are the internal stiffeners, usually used in this context to reinforce the system. In particular, Fig. 5a) shows the required actuation power for the left and right front wheels, the rear ones are similar and then not repeated here for the sake of brevity. Figure 5b) shows the motion coordinates of the center of gravity of the wheeled robot. Figure 6 reports the velocity of the center of gravity during the simulation, while Fig. 7 shows the acceleration components a_x, a_y and a_z experienced during the simulation.

Simulation tools are very effective when dealing with large-scale system being designed and tested to operate in urban environment, as it is reported in [17–20]. Simulations are therefore of crucial importance to verify the robot capabilities, also in terms of motion and carriage of suitable sensors. In addition, they are useful tools for overthrowing verification. Future development of the work is the implementation of additional sensors onboard of the prototype and a trial on a real case of study.

Inspection Robotic System: Design and Simulation 319

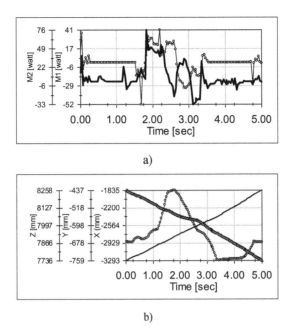

Fig. 5. Results for the simulation in Fig. 4: a) power for the left and right actuators; b) coordinates X, Y and Z of the center of gravity.

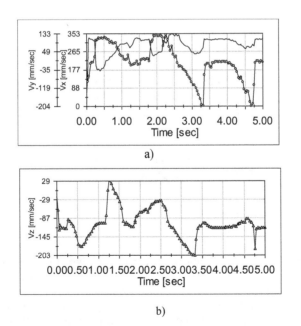

Fig. 6. Results of Fig. 4: velocity of the center of gravity a) V_x and V_y; and b) V_z.

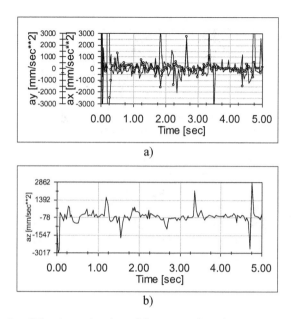

Fig. 7. Results of Fig. 4: acceleration of the center of gravity a) a_x and a_y; and b) a_z.

4 Conclusion

In this paper, we have proposed the simulation tests of a four-wheeled robot designed for inspection and survey of sites. In particular, a suitable 3D multibody model permits the development of the simulation of a realistic scenario and operation planning for a mechatronic survey. First simulations show a good behavior of the robot allowing the planning of operations that are more complex. The ongoing work with the proposed robot involves the integration of the sensor suit with thermal cameras and experimental tests on real site of interest.

Acknowledgments. This paper is a part of a project that has received funding from the Research Fund for Coal and Steel RFCS under grant agreement No. 800687.

References

1. Bekhit, A., Dehghani, A., Richardson, R.: Kinematic analysis and locomotion strategy of a pipe inspection robot concept for operation in active pipelines. Int. J. Mech. Eng. Mechatron. **1**(1), 15–27 (2012)
2. Morales, R., Feliu, V., Gonzalez, A., Pintado, P.: Coordinated motion of a new staircase climbing wheelchair with increased passenger comfort. In: IEEE International Conference on Robotics and Automation, Art. no. 1642315, pp. 3995–4001 (2006)
3. Wettergreen, D., Moreland, S., Skonieczny, K.: Design and field experimentation of a prototype lunar prospector. Int. Jnl. Robot. Res. **29**(12), 1550–1564 (2010)

4. Kececi, E.F.: Design and prototype of mobile robots for rescue operations. Robotica **27**(5), 729–737 (2009)
5. Ottaviano, E., Rea, P., Castelli, G.: THROO: a tracked hybrid rover to overpass obstacles. Adv. Robot. **28**(10) (2014). https://doi.org/10.1080/01691864.2014.891949
6. Figliolini, G., Rea, P., Conte, M.: Mechanical design of a novel biped climbing and walking robot. In: Parenti Castelli, V., Schiehlen, W. (eds.) ROMANSY 18 Robot Design, Dynamics and Control. CICMS, vol. 524, pp. 199–206. Springer, Vienna (2010). https://doi.org/10.1007/978-3-7091-0277-0_23
7. Rea, P., Ottaviano, E.: Design and development of an inspection robotic system for indoor applications. Robot. Comput. Integr. Manuf. **49**, 143–215 (2018)
8. Rea, P., Pelliccio, A., Ottaviano, E., Saccucci, M.: The heritage management and preservation using the mechatronic survey. Int. J. Archit. Heritage **11**(8), 1121–1132 (2017). https://doi.org/10.1080/15583058.2017.1338790
9. Ottaviano, E., Rea, P.: Design and operation of a 2-DOF leg-wheel hybrid robot. Robotica **31**(8), 1319–1325 (2013)
10. Granosik, G., Borenstein, J., Hansen, M.G.: Serpentine robots for industrial inspection and surveillance. In: Huat, L.K. (ed.) Industrial Robotics – Programming, Simulation and Applications, pp. 633–662. Pro-Literatur Verlag (2007)
11. Maurtua, I., et al.: MAINBOT - mobile robots for inspection and maintenance in extensive industrial plants. Energy Proc. **49**, 1810–1819 (2014)
12. PETROBOT (2021). http://petrobotproject.eu/
13. Biber, P., Andreasson, H., Duckett, T., Schilling, A.: 3D modeling of indoor environments by a mobile robot with a laser scanner and panoramic camera. In: Proceedings of IEEE/RSJ International Conference on Intelligent Robots and Systems, Sendai, pp. 3430–3435. IEEE Computer society (2004)
14. Sorli, M., Figliolini, G., Pastorelli, S., Rea, P.: Experimental identification and validation of a pneumatic positioning servo-system. Power Transm. Motion Control PTMC 365–378 (2005)
15. Thomas, F., Ottaviano, E., Ros, L., Ceccarelli, M.: Performance analysis of a 3-2-1 pose estimation device. IEEE Trans. Rob. **21**(3), 288–297 (2005)
16. Jacoff, A., Downs, A., Virts, A., Messina E.: Stepfield pallets: repeatable terrain for evaluating robot mobility. In: Performance Metrics for Intelligent Systems (PerMIS) Workshop, Gaithersburg, pp. 29–34 (2008)
17. Castelli, G., Ottaviano, E., Rea, P.: A cartesian cable-suspended robot for improving end-users' mobility in an urban environment. Robot. Comput.-Integr. Manuf. **30**(3), 335–343 (2014). https://doi.org/10.1016/j.rcim.2013.11.001
18. Ottaviano E., Ceccarelli M., De Ciantis M.: A 4-4 cable-based parallel manipulator for an application in hospital environment. In: 15th Mediterranean Conference on Control and Automation – MED07, Athens (2007)
19. Gonzalez-Rodriguez, A., Castillo-Garcia, F.J., Ottaviano, E., Rea, P., Gonzalez-Rodriguez, A.G.: On the effects of the design of cable-driven robots on kinematics and dynamics models accuracy. Mechatronics **43**, 18–27 (2017). https://doi.org/10.1016/j.mechatronics.2017.02.002
20. Ceccarelli, M., Ottaviano, E., Galvagno, M.: A 3-DOF parallel manipulator as earthquake motion simulator. In: Proceedings of the 7th International Conference on Control, Automation, Robotics and Vision, ICARCV 2002, pp. 944–949 (2002)

Dutch Auction Based Approach for Task/Resource Allocation

Eliseu Pereira[1,2](✉), João Reis[1,2], Gil Gonçalves[1,2], Luís Paulo Reis[1], and Ana Paula Rocha[1]

[1] FEUP, Faculdade de Engenharia, Universidade do Porto, Porto, Portugal
{eliseu,jpcreis,gil,lpreis,arocha}@fe.up.pt
[2] SYSTEC, Research Center for Systems and Technologies, Porto, Portugal

Abstract. The introduction of Cyber-Physical Systems (CPS) in the industry through the digitalization of equipment, also known as Digital Twins, allows for a more customized production. Due to high market fluctuation, the implementation of a CPS should guarantee a high flexibility in both hardware and software levels to achieve a high responsiveness of the system. The software reconfiguration, specifically, introduces a question: "With heterogeneous equipment with different capabilities - namely processing and memory capabilities - where a certain software module should execute?"; that question fits on the task/resource allocation area applied to CPS software reconfiguration. Although in task allocation issue several approaches address such a problem, only a few of them focus on CPS resources optimization. Given that, an approach based on the Dutch Auction algorithm is proposed, implemented at the CPS level enables the software reconfiguration of the CPS according to the existing equipment resources. This approach, besides the optimization of the CPS resources and the energy consumption, transforms the CPS in more reliable and fault-tolerant systems. As shown by the results, despite the demonstration of its suitability in task/resource allocation problems in decentralized architectures, the proposed approach also as a major advantage of quickly finding a near-optimal solution.

Keywords: Multi-agent systems · Resource allocation · Task allocation · Dutch auction · FIPA-ACL

1 Introduction

The fourth industrial revolution boosts the introduction of new mentalities in the manufacturing field. One of the key insights is the transformation of the mass production paradigm in a more flexible manufacture, allowing the customization of the products according to customer preferences. That level of customization requires a flexible shop floor layout, with constant monitoring of the processes. The usage of Cyber-Physical Systems (CPS) enables both the digitalization of the shop floor equipment and the reconfiguration of their functionalities. A self-organizing approach [1] allows the reconfiguration of the CPS according to the product characteristics. Regarding that, the self-organizing CPSs have to reconfigure at both hardware and software levels. The software reconfiguration, in particular situations, generates the overload of the equipment resources, due to the exceeding amount of processing modules to execute. That constraint requires an equilibrium of consumption resources across the machines [2]. Typically, that issue represents a task/resource allocation problem variant, where a set of tasks (functions) are placed in a collection of machines (agents), to optimize each machine resources and the system as a whole.

Regarding the CPS resources, the main beneficiary of the correct deployment of software functions according to the machine resources is the hardware itself, which can result in the perfect load for the available computational resources. This impacts in a reduction of the energy consumption, resulting from the avoidance of the machines over-processing allowing high energy demanding software application in a CPS to be more efficient. Additionally, regarding CPS maintenance, such an approach enables the automatic replacement of components/machines with minimal configuration of the new equipment, following the plug and play philosophy in the true sense.

The task/resource allocation issue fits in a multi-agent architecture, where each agent makes its decisions to accept or refuse the execution of a task. Traditionally, there are a multitude of different approaches to handle that problem, e.g., agent negotiation, voting or multi-agent reinforcement learning. The proposed approach uses a Dutch Auction-based algorithm, taking the advantages of the auction algorithms such as quick achievement of solutions. The present implementation differs from the original algorithm with the adoption of a decentralized architecture, where each agent can become the auctioneer.

The main contribution of that multi-agent system approach is the optimization of the processing consumption by the agents, allocating functions across the available agents uniformly and were the agent resources suite best. Also, the utilization of Python, as a general programming language, enables the usage of frameworks related to different fields, e.g., Machine Learning, Statistics, or Networking, that are increasingly used the Industry 4.0, for sensorization, maintenance or resources optimization. Regarding the Dutch Auction algorithm, typically the majority of action implementations adopt a centralized architecture, where the central agent (auctioneer) manages all the process, the actual approach uses a decentralized architecture which increases the fault tolerance in the system.

The paper is organized in six different sections, starting with the *Introduction*. The following section is the *Related Work*, which contains two subsections, 1) the *Background* concept, that describes the different types of auctions and introduces the Agent Communication Language (ACL) standard, proposed by the Foundation for Intelligent Physical Agents (FIPA), and 2) the *Related Work*, where other approaches used to address the task/resource allocation problem are compared. The next two sections are the *Problem Formalization* and the *Implementation*, where in the first one defines the problem and respective approach mathematically, the second one describes the implementation process of the whole multi-agent system. The last two sections are the *Results* and the *Conclusions & Future Work*, which validate the methods implemented and present their advantages and limitations, respectively.

2 Literature Review

The *Literature Review* focuses on two different concepts, the background context, and the related work. The background concept introduces the standard FIPA-ACL, which enables the agents' communications, and is crucial to understand the implemented protocols. The related work focuses on the existent algorithms that address the task/resource allocation issue, describing and comparing them in terms of architecture, environment vision, and scalability.

2.1 Background

The FIPA-ACL [3] standard defines the structure of the messages to exchange between agents, some protocols of interaction between agents (e.g., *request*, *query*, *contract-net* [4] or *English Auction*), and some agent types (e.g., *Agent Management System*). The normative FIPA-ACL specifies a standard name pattern to identify each agent (*agent_name@ip_address:port*), that agent ID contains the IP address and port where it is executing the agent and also a name to facilitate its recognition. To use of the Agent Communication Language (ACL) is crucial to understand the structure of the messages exchanged between the agents. The message structure contains: 1) the sender agent ID, 2) the receiver(s) agent ID(s), 3) the performative, which indicates for a particular protocol (e.g., *query*) what action to perform (e.g., *inform*, *query* or *failure*), 4) the protocol, and 5) the content of the message.

The protocols define the sequence of actions that the agent could perform. FIPA-ACL represents a broad set of interaction protocols; in this particular case, it's crucial to understand two of them, the *query* and the *Dutch Auction* protocols. The *query* protocol starts when one agent sends a message with the performative *query*. The receiver agents handle that message and respond with an *inform*, *failure*, *refuse*, or *not-understood* performative. Assuming that if everything okay with the *query* message, the receiver agent answers with an *inform* performative, with the requested data as content. The *Dutch Auction* protocol has a more complex sequence of performatives. The process starts when the auctioneer sends a *call for proposals* performative to the bidders, the interested ones answer with a *bid* message. The auctioneer selects the best one and sends an *accept* message to the winning bidder, and a *reject* message to the other ones, at the end of the auction, sends an *inform* message to notify all the agents about the results.

Regarding the FIPA-ACL advantages (e.g., standard communication, reliability), there are few implementations. The popular one is the Java Agent Development Framework (JADE) [5], which is an implementation using Java as a general programming language, which is a tool primarily adopted in the multi-agent system projects. However, due to the incremental employment of data analytics and machine learning in projects of several applications, the usage of Python as the main programming language is growing exponentially. The FIPA-ACL standard isn't an exception to the increasing usage of Python, and the Python Agent DEvelopment framework (PADE) [6] fills that gap, with an embryonic implementation of the protocols, mainly used for simulation purposes.

2.2 Related Work

Resource allocation and task allocation are relevant problems in multi-agent systems [7], due to its complexity (NP-Hard), and the numerous practical applications. Both issues have similar goals, which are to optimize the placement of the tasks through the available agents, somehow to maximize the utility function of each agent and the entire system—considering that each agent makes its own decisions and wants to increase its utility. The methods applied to address the resource allocation issue have different characteristics, namely their scalability, architecture, and information requirements. The architecture of the multi-agent system might be centralized or decentralized regarding the method properties, e.g., the auction protocol [8] is a typical example of a centralized architecture, where the auctioneer is the central entity. The method scalability evaluates how the multi-agent system performance progresses with the addition of new agents, and indicates its limitations concerning the number of agents. The agent information requirements indicate the level of knowledge essential for the agent to perform its decision, e.g., partial environment information, information about the entire environment, or also data about the other agents' decisions.

The conventional implementation [8] of resource allocation in auction-based approaches [9,10] has a centralized architecture composed by bidders and one auctioneer, where the bidders make offers, and the auctioneers coordinate the process and decide who receives the item. There are different types of auction, e.g., the English Auction, the Dutch Auction, where in the first one, the agents know the other offers instead of the second one, which is a blinded auction where the agents don't have information of the other bids. Typically that approaches are an incremental process (one item auctioned individually and in sequential order) where the agent only has information about the current and previous auctioned items, resulting in incomplete information about the environment.

Regarding decentralized architectures [11–14] applied to handle the task allocation issue, one of them is the distributed stochastic approach [11] that obtains quick results without the overload of messages in the network. However, the algorithm sometimes got stuck in a local minimum. Another method is the ant colony algorithm [12], which uses a global pheromone matrix to achieve the optimal solution. The approach based on the fast sum-max algorithm [13] returns an answer at any time, concerning that the more time the algorithm executes, more profitable the solutions will be. However, the algorithm consumes a substantial amount of memory. The usage of swarm intelligence [14] is another option; these implementation uses a cluster-based algorithm that reflects the behavior of forager bees in the agents, deciding jump to the next cluster or stay in the current one. That behavior indicates a well-balanced exploration/exploitation rat; nevertheless, it has low computational speed. Both fast sum-max and swarm intelligence based algorithms have good scalability, allowing the successive addition of new agents, where each one has a local vision about the environment.

Concerning the resource allocation issue, different approaches [15–18] handle to solve that problem. One them uses reinforcement learning [15] as the main method, where each agent has several predictors to estimate the consumption of each server, and based on that estimations, the agent makes its allocations. Then, based on the real consumption of each server, the agent updates its predictor parameters. As a result, that implementation has a reduced exchange of messages; however, it uses a substantial value of computational resources. The negotiation [16] is another approach where the agents deal with different agreements and concessions based on the agents' vicinity environment. The agents have the option to decommit agreements receiving a paying penalty, using that the algorithm can obtain the optimal solution, regardless of the high exchange of messages between agents. As an alternative, the voting approach [17] models a distributive justice method, where the agents vote on the weight associated with the scoring function. A similar technique is the fair division algorithm [18], which expands the concepts from the fair division theory. Both the voting [17] and the fair division [18] algorithms obtain an exact equilibrium between the agent resources; however, they require a high number of messages exchanged. Regarding the agent information requirements, three of the implementations [15–17] use a local vision of the environment, besides the fair division algorithm [18], which needs a global image of the environment. Those methods have good scalability, allowing the addition of new agents.

In general, the described implementations address the task/resource allocation issue, using different architectures, algorithms, and archiving different levels of scalability. Considering that, a new approach based on the auction algorithm, with modification to fit in a decentralized architecture, would be an excellent alternative to the previous algorithms.

3 Problem Formalization

Regarding the variants of the resource allocation and task allocation issues, the addressed one is the function assignment problem, where the environment is a set of different machines, and the algorithm finds a more suitable machine to execute each function. The functions and their dependencies fit in a pipeline structure (e.g., $fa \rightarrow fb \rightarrow fc \rightarrow fd$). The notation used traduces the work pipeline in a graph data structure (G), where each vertex corresponds to one function and each edge to the dependency/precedence between functions. The machines that execute the functions correspond to agents, where each agent has different capabilities. Summarizing, each function from the graph (G) matches an agent from the agent set (A).

An approach to solve the function assignment problem is the Dutch Auction algorithm, where the functions are the bid items, and the agents are the bidders that make offers for each item. In the Dutch Auction, the auctioneer establishes a base price (threshold) for each item, and waits for proposals, if there are not offers the auctioneer decreases the amount. The adaptation of that algorithm to the function assignment problem forces each bid to be the agent cost to execute the corresponding function, so the agent that offers a lower cost will win the function bid (Eq. 1), which corresponds to a minimization algorithm. The threshold allows the agent to decide if it bids or not, depending on if the offer is lower than his own. If there are no offers for that threshold, the auctioneer increases the threshold. In the end, the auctioneer stores the accepted proposals and the corresponding bidders in the bid winners set (BW).

$$BW = \{min(bid_{00}...bid_{i0})...min(bid_{0j}...bid_{ij})\}$$
$$i \in A, j \in G$$
$$bid \leq threshold$$
(1)

As already stated, the agent profile depends on several attributes, more concretely their RAM (a_ram), CPU (a_cpu), and processing time (a_time) capabilities. Also, the functions have their properties, more precisely the RAM (f_ram), CPU (f_cpu), and processing time (f_time) requirements. According to the agent specifications and the function requirements, each agent calculates its bid for a particular function; if the bid value is lower than the current threshold, the agent sends the offer to the auctioneer. The bid value (Eq. 2) depends on several factors; the most determinant is the function cost, which is the quotient between each function requirement (f_ram, f_cpu and f_time) and the agent specifications (a_ram, a_cpu and a_time). Additionally, the bid function considers the previously accepted bids from the agent as a way to balance the functions over the agents. The formula also attends to the neighbours' functions associated with the bidder agent, considering the network delay associated with executing the two linked functions in different machines. This way, if the agent has neighbour functions mapped in itself, a neighbour cost (nb) is decreased in the bid formula, reducing the communications between agents, when they execute the work pipeline.

$$bid = \frac{f\_ram}{a\_ram} + \frac{f\_time}{a\_time} + \frac{f\_cpu}{a\_cpu} + a\_cost - nb$$
(2)

The theoretical analysis of the problem demonstrates that the usage of the Dutch Auction algorithm solves the problem of function assignment in a multi-agent system. However, a more practical description is useful to understand the algorithm and implementation and also some validation tests to prove the concept demonstrated.

4 Implementation

The problem of function assignment is a combinatorial optimization problem, which has several approaches to solve it. In this case, the adopted method was the Dutch Auction, which fits in a multi-agent system, where each agent is a bidder, and one of them is also the auctioneer. tas The functions negotiated between agents use Python as a general programming language, which enables the usage of an extensive number of frameworks applied to different fields, e.g., Machine Learning, Computer Vision, or Networking. Given that, the implementation uses as bases the Python Agent DEvelopment framework (PADE) [6], which implements the message structure and some FIPA-ACL interaction protocols. Due to the peer-to-peer network scenario, which requires bidirectional communication between agents, each agent uses the Twisted network framework [19] for communication purposes.

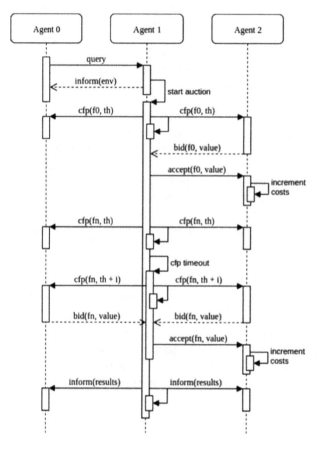

Fig. 1. Dutch auction protocol (agents interaction)

4.1 Agents Interaction

The agents' interaction, Fig. 1, uses TCP/IP as the transport layer, executing each agent in a different process to guarantee the isolation between agents. In the auction architecture, all the agents are bidders, and one of them is also the auctioneer. The auction starts when a new agent (*Agent 0*) joins the network and queries anyone about the environment information. The asked agent (*Agent 1*) answers with the environment data, and starts an auction; this agent becomes the auctioneer and sends the *call for proposals (CFP)* message to all the agents (including him), asking about bids for the first function. Each CFP message contains the bidding function ID and the current threshold, allowing to identify the current bidding function and the limit cost threshold to perform a bid (the agents only send proposals smaller than the threshold). The auctioneer waits, a fixed interval, for submissions; then, if there are bids, it chooses the lowest one as the winner; if there are no bids, the auctioneer increments the threshold allowing agents with higher costs to bid. The auction is a sequential process, offering each function individually; this way, the agents with accepted bids will increase their cost (current bid plus previously accepted bids), limiting their future bids. When the auctioneer sold all the functions, it communicates the results, and all the auction winners can start executing the respective functions.

The entire implementation follows the FIPA-ACL standards, e.g., the protocols used or the messages exchanged. The agents interact using two protocols, the query protocol, and the dutch auction protocol. Where the query protocol allows new agents to join the network and receive the actual pipeline used as environment, using a message to query the actual state of the system/environment and other performative to inform the result (first part of Fig. 1). The dutch auction protocol includes all the interaction between agents to find the best bidder for each function: 1) the *call for proposals (CFP)* performative requests to each agent its bid for a particular function; 2) the *bid* performative sends the information about an agent bid (function ID and bid value); 3) the *accept* performative notifies the bidder winner to update its internal costs; 4) the *inform* performative to share the final results of the auction.

4.2 Agent Architecture

The agent architecture that allows this interaction contains several components, described in Fig. 2. In detail, the agent architecture splits into five different parts, 1) the agent itself, 2) the behaviors, that determine the agent actions, 3) the peer, which allows the communication between agents, 4) the environment, more concretely the graph representation of the pipeline of work, and 5) the auction, which is only used by the auctioneer to manage the auction process.

In detail, the communication component (peer) uses a TCP/IP server to receive the requests/messages from the other agents. Due that each agent needs to use a different URL (IP address and port), e.g., in the same machine, every agent must use a different port. Also, to send requests/messages, each agent uses a list of clients, one for each agent in the network. That approach allows bidirectional communication between agents, using messages with a FIPA-ACL structure.

The environment component allows the extraction of the cost value to perform the bids by the agent. Each agent has an updated copy of the graph data structure that represents the environment. When a new agent enter in the network, it receives the current version of the environment in the content of a FIPA-ACL message encoded as a JSON structure. The graph contains vertices associated with the pipeline functions and edges to the precedence/return between functions. Also, the graph stores the agent that executes each function as node attributes.

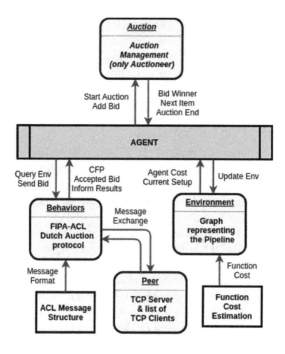

Fig. 2. Agent architecture

The estimation of the bid value for the current bid item provides from the agent specifications and the function requirements. The agent specification is the hardware description, which the user can edit in a configuration JSON file of each agent. The function requirements provide from a simulation component that selects one value from the normal distribution (mean and standard deviation) of each function requirement. Those distributions result from the experiment of executing several times a function in certain hardware; the variables collected are the processing time, RAM usage, and CPU usage. The bid value also depends on the previous accept bids, and the neighbour accepted functions. This way, during the auction, each agent increases its costs if their proposals are accepted.

Hence, the bid formula contains the values of the bids accepted to balance the auctioned functions through the agents equally. Also, the previous auctioned functions allow the agent to understand if the flow has many transitions between agents. If the pipeline is $fa \rightarrow fb \rightarrow fc \rightarrow fd$ and there are two agents in the network ($a1$ and $a2$) its profitable to map the agent $a1$ to the functions fa and fb and the agent $a2$ to the functions fc and fd, because it decreases the communication costs between agents (only need to communicate to pass data between the function fb and fc).

As mentioned before, one of the agents its both a bidder and an auctioneer. So, the auctioneer needs to manage all the auction process, due that each agent has a component named *Auction*, which is only used by the auctioneer to handle the process. This component receives the bids from the other agents and selects the profitable (lower one); if there are no bids, increases the threshold allowing the agents with the highest cost to bid. This approach has the goal to dispatch all the function through the agents.

5 Results

The multi-agent system implemented has several properties that require a specific validation. Given that, the experiments to perform are a proof of concept, where for a particular pipeline, the method collects the agent costs during the auction process. In detail, the experiment validates two different work pipelines, where the first has two independent flows that join at the end of the graph (left topologies in Fig. 3), with two available agents. The second has a more complex topology, with three available agents, which allows a more refined analysis of the system decisions. In each topology, two different situations were evaluated one using the neighbours' cost (described in Eq. 2) and the other without using that policy. Regarding those topologies, the expected results for the first pipeline, considering the neighbour cost, are the placement of each independent flow in a unique agent, ensuring the minimal communications between machines, reflected in a faster processing time of the whole pipeline. When the bid function does not contain the neighbour cost, the expected result is that the algorithm balances the costs in all the agents. In the second scenario, the results must prove the same conclusions, balancing the costs between agents and minimizing network communications.

Given the initial assumptions about the behaviors of the system, the results demonstrate that they are aligned with the real system decisions. The analysis of the first topology in Fig. 3 denotes that when the bid function considers the neighbour cost, it obtains the optimal solution (second column), instead of the version without neighbour cost, which only concerns about balancing the resources through all the agents. Regarding the second topology the results provide similar conclusions, where the functions are equally balanced across the agents (agent aid_0 was 4 functions associated, the aid_1 was 4 and the aid_2 was 5). Concerning the network delays, that topology presents a higher number of links between separated agents (14 edges) in the experiment without neighbour costs (column 3), when compared to the experiment with costs (13 edges).

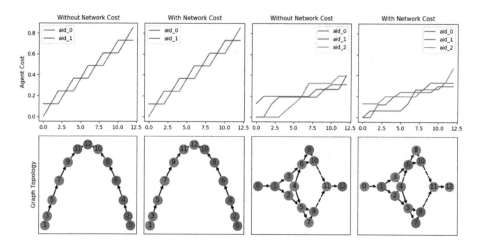

Fig. 3. Agents cost vs graph topology

6 Conclusions and Future Work

Overall the Dutch Auction allows the quick auctioning of items, which gives a smaller response time by the multi-agent system. Also, this algorithm enables the dispatching of all the functions, which is a crucial requirement of this problem (every function needs to have one associated agent). That version of the Dutch Auction algorithm enables the usage of a decentralized architecture, which transforms the CPS in a more fault reliable system.

Nevertheless, this approach has limitations, mainly due to the incremental essence of the algorithm; some solutions are ignored, which could result in suboptimal solutions. To address that issue, the usage of combinatorial auctions, with a global vision of the environment, would minimize that problem.

References

1. Zhang, J., Yao, X., Zhou, J., Jiang, J., Chen, X.: Self-organizing manufacturing: current status and prospect for industry 4.0. In: 2017 5th International Conference on Enterprise Systems (ES). IEEE (2017). https://doi.org/10.1109/es.2017.59
2. Zhang, Y., Guo, Z., Lv, J., Liu, Y.: A framework for smart production-logistics systems based on CPS and industrial IoT. IEEE Trans. Industr. Inf. **14**(9), 4019–4032 (2018). https://doi.org/10.1109/tii.2018.2845683
3. O'Brien, P.D., Nicol, R.C.: FIPA - towards a standard for software agents. BT Technol. J. **16**(3), 51–59 (1998). https://doi.org/10.1023/A:1009621729979
4. Smith, R.G.: the contract net protocol: high-level communication and control in a distributed problem solver. IEEE Trans. Comput. **C–29**(12), 1104–1113 (1980). https://doi.org/10.1109/tc.1980.1675516
5. Bellifemine, F., Poggi, A., Rimassa, G.: JADE – A FIPA-compliant agent framework. In: Proceedings of PAAM, London, vol. 99, p. 33 (1999)

6. de Freitas, B.K., Venturini, L.F., Domingues, M.A., da Rosa, M.A., Issicaba, D.: Exploiting PADE to the simulation of multiagent restoration actions. In: 2019 11th International Symposium on Advanced Topics in Electrical Engineering (ATEE). IEEE (2019). https://doi.org/10.1109/atee.2019.8724852
7. Ye, D., Zhang, M., Vasilakos, A.V.: A survey of self-organization mechanisms in multiagent systems. IEEE Trans. Syst. Man Cybern.: Syst. **47**(3), 441–461 (2017). https://doi.org/10.1109/tsmc.2015.2504350
8. Krishna, V.: Auction Theory. Academic Press, Cambridge (2009)
9. Fatima, S.S., Wooldridge, M.: Adaptive task resources allocation in multi-agent systems. In: Proceedings of the Fifth International Conference on Autonomous Agents, AGENTS 2001, pp. 537–544. Association for Computing Machinery, New York (2001). https://doi.org/10.1145/375735.376439
10. Wurman, P.R., Wellman, M.P., Walsh, W.E.: A parametrization of the auction design space. Games Econ. Behav. **35**(1–2), 304–338 (2001). https://doi.org/10.1006/game.2000.0828
11. Chapman, A.C., Micillo, R.A., Kota, R., Jennings, N.R.: Decentralised dynamic task allocation: a practical game: theoretic approach. In: Proceedings of The 8th International Conference on Autonomous Agents and Multiagent Systems - Volume 2, AAMAS 2009, pp. 915–922. International Foundation for Autonomous Agents and Multiagent Systems, Richland (2009)
12. Wang, L., Wang, Z., Hu, S., Liu, L.: Ant colony optimization for task allocation in multi-agent systems. China Commun. **10**(3), 125–132 (2013). https://doi.org/10.1109/cc.2013.6488841
13. Macarthur, K.S., Str, R., Ramchurn, S.D., Jennings, N.R.: A distributed anytime algorithm for dynamic task allocation in multi-agent systems. In: In Proceedings of AAAI, pp. 356–362 (2011)
14. dos Santos, D.S., Bazzan, A.L.C.: Distributed clustering for group formation and task allocation in multiagent systems: a swarm intelligence approach. Appl. Soft Comput. **12**(8), 2123–2131 (2012). https://doi.org/10.1016/j.asoc.2012.03.016
15. Schlegel, T., Kowalczyk, R.: Towards self-organising agent-based resource allocation in a multi-server environment. In: Proceedings of the 6th International Joint Conference on Autonomous Agents and Multiagent Systems, AAMAS 2007. Association for Computing Machinery, New York (2007). https://doi.org/10.1145/1329125.1329147
16. An, B., Lesser, V., Sim, K.M.: Strategic agents for multi-resource negotiation. Auton. Agent. Multi-Agent Syst. **23**(1), 114–153 (2010). https://doi.org/10.1007/s10458-010-9137-2
17. Pitt, J., Schaumeier, J., Busquets, D., Macbeth, S.: Self-organising common-pool resource allocation and canons of distributive justice. In: 2012 IEEE Sixth International Conference on Self-adaptive and Self-organizing Systems. IEEE (2012). https://doi.org/10.1109/saso.2012.31
18. Kash, I., Procaccia, A.D., Shah, N.: No agent left behind: dynamic fair division of multiple resources. J. Artif. Intell. Res. **51**, 579–603 (2014)
19. Fettig, A., Lefkowitz, G.: Twisted Network Programming Essentials. O'Reilly Media, Inc., Newton (2005)

MOBEYBOU

Tangible Interfaces for Cognitive Development

Hugo Baptista Lopes[1], Vítor Carvalho[1,2(✉)], and Cristina Sylla[3]

[1] 2Ai - School of Technology, IPCA, Barcelos, Portugal
a15999@alunos.ipca.pt, vcarvalho@ipca.pt

[2] Algoritmi Research Centre, School of Engineering, University of Minho, Campus of Azurém, Guimarães, Portugal

[3] Research Centre on Child Studies/ITI/LARSyS, University of Minho/ITI/LARSyS, Campus de Gualtar, Braga, Portugal
cristina.sylla@ie.uminho.pt

Abstract. This paper presents the development of a low-cost tangible interface to enhance children's learning through collaborative storytelling, using a set of movable blocks, with a high degree of autonomy and the ability to exchange information which other. Taking into account the context of the existing game model and on the basis of which this study was developed, the function of these removable blocks is to allow children who use them to create and recreate their own narratives in permanent interaction between the representations associated with these blocks ("characters" = figures of animals, objects, locations, or others) and those of the existing virtual platform, that is, children, in the way they best understand/succeed in the construction of their spontaneous narratives, will be able to project these representations on a screen through the activation and manipulation of the blocks, where they add, remove, move and interact with each other and with those in it already introduced at every moment.

To that end and using freely accessible hardware and Bluetooth® technology, we have attempted to apply a method that would somehow revolutionise the educational games' industry, through RSSI (Received Signal Strength Indicator) based triangulation to determine the relative position of various removable device/blocks with each other and with respect to the virtual platform which is supposed to interact, but were unable to secure results that would have made it possible to obtain a precise location. We thus instead developed a simpler format that achieves an interaction within a graphic environment as a result of the proximity of the physical objects in question, but not of any other factors, thereby essentially fulfilling the study's objective.

Keywords: Tangible interfaces · Children · Cognitive development · Game · Narrative

1 Introduction

Make-believe play in early childhood, which involves the use of fantasy and symbolism, is important for children's overall cognitive development. According to Fein [1] play is a symbolic behaviour in which one thing is treated as if it were another. He also underlined the importance of the sensations related to play and considered that affection linked to the game itself is a natural form of creativity [1]. Russ postulated that play is important in the development of creativity as many of the cognitive and affective processes involved in creativity occur when playing [2, 3]. Based on literature review, Russ identifies, in the model of affection and creativity, the main cognitive and affective processes involved in creativity and their relationship [3].

Playing should facilitate divergent thinking because, in play, children exercise the capacities of divergent thinking using objects that represent different things and imagining various scenarios throughout the process [4]. Some studies point to a relationship between affective processes during play and creativity. Leiberman found a relationship between play, containing affective components of joy and spontaneity, and divergent thinking in preschool children [5]. Christie and Johnson also concluded that there is a relationship between play and creativity [6]. J. L. Singer and Singer uncovered that children with a high degree of imagination recognize, in the context of playing games, more themes involving danger and power than children with less imagination [7].

Creative narratives may be the most appropriate context to stimulate language development in young children, as well as narrative competences [8]. Storytelling is a social activity, based on shared experiences that can introduce new vocabulary and may also include elements that help to develop phonological awareness, as are often found in repeated patterns, rhymes or funny stories [9]. The narratives that occur naturally in children's play is particularly important to stimulate their imagination, to acquire and practice different linguistic styles and to explore their social roles [10]. All kinds of narratives are the basis for understanding and creating our experiences and the world around us. At the individual level, the narratives that we listen to throughout life allow us to build our role in the world. At the family level, narratives are often involuntarily used to socialize children, to teach them values and principles. When exposed to these stories, children develop their own narrative skills. At the cultural level, the narrative gives cohesion to the transmission of principles and values [11]. It has been argued that children can use more elaborate language in their narratives than in their everyday conversations [12].

Collaboration is an important competence for young children's learning, and working in pairs or small groups can have very beneficial effects on children's learning and development, particularly in the early years and in primary education [13–15].

Technology offers opportunities to support and facilitate collaborative learning in many ways [16, 17]. The computer can provide a common frame of reference and be used to support the development of ideas among children. However, neither learning nor collaboration will take place simply because two children share the same computer. Learning and collaboration will only occur if the technology is designed according to the contexts of use and the purpose for which it is intended. Otherwise, the interface can become a barrier to learning [17].

The difference between a graphical user interface (GUI) and a tangible user interface (TUI) comes from the fact that the latter combines the technology of a virtual interface with the physical world. Traditional GUIs are characterized by virtual elements such as windows, icons, menus, and the user's interaction with the computer is done using devices such as mouse and keyboard. By contrast, the manipulation of a physical (tangible) interface is done through objects that combine the physical with the virtual world.

TUIs were initially explored by Fitzmaurice [18] with the "Bricks" prototype. Marshall [19] later stated that TUIs have great potential to support learning due to their "hands-on" nature. However, it would be necessary to prove its usefulness concretely. According to Piaget, cognitive structuring through the absorption and assimilation of schemes requires physical and mental actions [20].

TUI, combining the physical world with the virtual, can play a very important role in literacy and other areas in education. The development of technology and the widespread access to increasingly small electronic components makes it possible to lead to innovative teaching methodologies, which could translate into an improvement in cognitive development in children or even adults. The creation of tangible interfaces endowed with autonomy, capable of recognizing each other and interacting with a graphic environment can prove to be an important step in the context of games to support and facilitate collaborative learning.

Knowing how to read and write remains the basis of education and a prerequisite in other scientific fields, such as science, mathematics and technology. Through participation in language games, narratives, children's interaction with peers contributes to the development of these skills. In this context, of strong interaction, in order to be sure that their messages are clearly perceived [21] children surpass themselves and stimulate each other, thereby developing their capacity to build and tell stories [22]. The same process occurs through storytelling between tutors and children. Children who showed greater interest and involvement in activities related to narratives demonstrated greater activation in the association areas on the right side of the cerebellum during the task and greater functional connectivity between this activation cluster and the areas of language and executive functions. Recent studies suggest a potential cerebellar response mechanism at the level of child involvement that can contribute to the development of emerging literacy during childhood and synergy between guardian and child during story sharing [23].

For children, the theory suggests that supporting physical actions on computational objects, therefore virtual, makes it difficult to perform difficult mental tasks. However, experiences have recently emerged in which traditional teaching methods are compared to the manipulation of physical objects in which any aspect of their shape represents abstract concepts. Tangible interfaces can play a very important role in literacy and other fields in education. With the development of increasingly smaller technology and electronic components, the application of these in everyday objects facilitates the interaction between the user and the virtual world. Examples such as mobile phones, Tablets, Nintendo Wii, demonstrate the ability to apply technology to objects as a way of interacting with the virtual world in a non-traditional way. This bring forth a reduction in the cost of developing new devices and devices and their use is becoming more and more common. Transposing these types of characteristics to teaching methodologies may translate into an improvement in cognitive development in children or even adults. The widespread access of these components allows and increasingly facilitates the creation of new prototypes of TUI interfaces [18].

The main goal of this study is to design a tangible, technologically advanced object, framed in the Mobeybou project, which currently uses the TOK platform hardware [9]. In order to optimize the already developed hardware in the scope of this project, this study describes the development of a hardware solution that meets the existing system, acting as a complement that can be integrated into the existing graphic environment. To this end, it was taken into account not only what has already been developed in terms of hardware as well as consider the main features of the system.

TOK is a tangible interface, which includes a graphical component and two main tangible objects, the pieces/blocks, and a board. Each block has a capacitive sensor at its base that contains information about its identity (ID). Each block is associated with a graphic element. When placing a block on the board, it is recognized by the sensors of the electronic board and its identification is communicated to the graphical environment which in turn triggers a virtual graphical representation of the respective element that is displayed on the computer/tablet screen. The application allows users to mix all the narrative elements, and this depends on the choice of blocks placed on the platform, and this potential way to engage children in exploring different narratives.

What we propose is to make the game scalable, mouldable, customizable, modellable and flexible, making use of the most of the elements of the work done up to that point.

The use of currently available technology, namely computers, graphics and interfaces environments offer an opportunity to support and facilitate collaborative learning.

In order to reinforce the importance that tangible objects bring to children's cognitive development, it is proposed to elaborate an object that can be integrated in a graphic environment already created, but with an advantage over the solutions tested so far.

2 Background

TOK has enormous potential in terms of the ease of creating different blocks. The total number of different blocks that can be created is quite high, but limiting the number of blocks that can be read by the board is a disadvantage in the case where the pieces identify objects and not characters.

Taking into account that the aim of this work is to continue the Mobeybou project, it proved essential to look for an innovative alternative in relation to the existing solutions, through a TUI prototype with a significant degree of autonomy - which the state of the art does not reveal to have been achieved until now - verifying, on the other hand, that the identification of this object requires a study of the different types of approach using technology currently on the market.

Autonomy, as both energetic as functional, is the key issue in any technological object and this kind of object is no exception, insofar as an object that is restricted to a physical connection (by cables, for example) and powered by the power grid has "infinite" autonomy, while an object that it is movable has a limited energetic autonomy, although it can be recharged/powered by renewable energy resources, for example. On the other hand, the way that an object identification is generated requires a study of the different types of possible approach using technology currently on the market. It is also important to use cross-cutting and unified technology.

3 Development

3.1 Bluetooth® 5 Technology

Bluetooth® 5 technology has significant advantages over other solutions, when, as is the case, the aim is to locate and exchange information between TUI objects, namely in indoor environments [24]. Bluetooth®, in addition to low power consumption, omnipresence, low cost and easy use, is immune to electromagnetic noise and is easily scalable for scenarios with multiple agents. Other solutions, such as Wi-Fi, GPS (Global Positioning System), RFID etc., are not suitable for locating devices for the following reasons: GPS requires a line of sight with at least four satellites, so it is not suitable for indoor environments; RFID is not capable of communication; although Wi-Fi has a higher data rate, high power consumption is inadequate; other technologies, such as Zigbee, are not widespread and are not available on laptops, smartphones, desktops, among others [25]. Bluetooth® 5 brings vast advances to this technology and as the potential to become an alternative [26–33]. Using an approach based on the Received Signal Strength Indicator (RSSI) [34], one of the simplest and most widely used for indoor location, we tried to use Machine Learning and Deep Learning algorithms, most used in Artificial Intelligence, analysis of large collections of data and image processing, that can make RSSI-based localization more accurate. In addition, Kalman filters were applied to the RSSI values, aiming to improve the accuracy of the readings. All the technology was applied to the TUIs, the blocks, to make them "smart" in some way.

Bluetooth® devices can play different profiles as part of a communication, differently from a communication between a server and a TCP (Transmission Control Protocol) client. The most usual connection can be described by a master and a slave, in which the slave periodically sends his identity and some characteristics about himself allowing the master to discover him and proceed with the connection. This connection can be originated by any device and ends when the master disconnects. This condition is called pairing between Bluetooth® devices. However, this requires greater energy expenditure as it relies on a constant exchange of information. Instead, a much simpler method can be resolved by exchanging information through observer and advertiser, since a device advertises a packet with information and other collects this packet, thus reducing greatly energy consumption.

3.2 Received Signal Strength Indicator

RSSI (Received Signal Strength Indicator) is a unit signal power measurement. Unlike dBm (decibel-milliwatts), which also represents a signal strength measurement unit but in absolute value, the RSSI is used to measure the relative quality of the signal strength between devices. The scale is set by the manufacturer and varies between 0 and −255, but in all cases, it can be inferred that the closer to 0 the better the signal. Therefore, not being a scale that allows to quantify exactly the distance between two devices, it has been studied to obtain an approximation of an indoor location in substitution of other technologies that do not allow it, as the case of GPS.

The approach based on RSSI is one of the simplest and widely used for indoor location [35, 36]. Using the RSSI and a simple model of signal propagation loss over space [37], the distance d between the transmitter (TX) and receiver (RX) can be estimated by

$$RSSI = -10n \log 10(d) + A \qquad (1)$$

where n is the exponent of propagation loss (varies between 2 in open line and 4 with many obstacles) and A is the reference transmission value. Finding a device using RSSI can be done in different ways. In the case of trilateration or trilateration by N-points, used for example in the GPS signal, RSSI is used to estimate the absolute distance between the device to be located and at least three reference points, usually referred to as anchors. This method applies basic geometry and trigonometry. Although this approach is simple and easy to implement, it suffers from little precision due to the characteristics of the RSSI signal, especially in conditions where there is no line of sight, in which the signal attenuation results from obstacles and walls causing large fluctuations in the RSSI [35, 38]. However, different types of filters can be applied to mitigate these effects. Nevertheless, it is unlikely to obtain location with high precision without the use of complex algorithms.

One approach that has been studied is the use of Machine Learning and Deep Learning algorithms, most used in Artificial Intelligence, analysis of large data collections and image processing, but which can make the location based on RSSI more accurate.

In the case of neural networks, ANN (Artificial Neural Networks), the location is obtained through classifiers and forecasting scenarios. The neuronal network is trained with the different RSSI values corresponding to a given location in an offline process [39]. Once trained, it can be used to obtain the location of any other device that obviously has the same characteristics as RSSI. The MLP (Multi-Layer Perceptron) network with a hidden layer is one of the most used and common networks in this type of approach [40]. As an input, a vector with RSSI values is multiplied by weights and added to an input layer bias, as long as the bias is selected. The result obtained is then placed in the transfer function of the hidden layer. The output product of the transfer function and the trained weights of the hidden layer are added to the hidden layer bias. The network's output is the device's intended estimated location.

On the other hand, the application of clustering algorithms can also help in obtaining more accurate results for the location of devices. This is the case with the kNN (k-Nearest Neighbours) algorithm. The algorithm depends on RSSI online to obtain the closest matches, based on the measured RSSI values, previously stored in a database in the offline process, from known locations using the Root Mean Square Error - RMSE [40, 41].

3.3 Extended Kalman Filter

Kalman filters have been applied in navigation situations. This method, developed by Rudolf Emil Kalman in 1960, was originally designed and developed to solve the navigation problem in the Apollo project [42].

Although it can be described by a single equation, the Kalman filter is usually divided into two distinct phases: Prediction and Update and its purpose is to use measurements of quantities performed over time, with a high degree of noise and other associated uncertainties to obtain results that are close to the actual values. It performs efficiently, accurate inferences about a dynamic system. It is computationally lightweight and relatively easy to implement on systems with few input variables.

Prediction uses the previous state estimate to obtain an estimate of the current state in time, which is called an *a priori* estimate. The Update is called *a posteriori*, as it combines the current observation to improve the estimate. Prediction is given by:

$$x_k = x_{k-1} \tag{2}$$

$$P_k = P_{k-1} + Q \tag{3}$$

where x_k and x_{k-1} are the estimate of the state and the *a priori* state, P_k and P_{k-1} are the estimate of the covariance error and the *a priori* state and Q covariance of the system noise. And the update by:

$$K_k = P_k(P_k + R)^{-1} \tag{4}$$

$$x_k = x_k + K_k(z_k - x_k) \tag{5}$$

$$P_k = (1 - K_k)P_k \tag{6}$$

where R is the covariance of *a posteriori* noise, therefore the observation, K_k is the Kalman gain and z_k the measured value. To use the Kalman filter model, it is only necessary to provide the state matrices, the initial parameters and the covariance matrices Q and R. In order to form a well-structured opinion on the characteristics of the RSSIs in these devices, some samples were collected between two devices that showed a rapid change of positions. In fact, RSSI values change quite sharply when they are in a fixed position and often overlap. Since the values are integers, it is difficult to obtain a resolution that allows accurate distances to be inferred. After applying the Kalman filter and adjusting the parameters: Q - Covariance of the system noise; R - Covariance of the observation noise; P - Estimate of the covariance error; we can see an improvement of the signal values.

3.4 Requirements

The objective of this work is to design and develop a tangible, low-cost interface, with the ability to be integrated into a virtual platform. In order to make the game competitive in terms of the market, it is important to choose the technology that allows to obtain validation of the proposal and that has the minimum associated cost.

In order to make the game scalable and flexible, we thought of a solution that keeps up with the existing model. To this end, the idea would be to transform each block into a device that is autonomous and that collects information from other blocks in the vicinity.

Due to the RSSI characteristics, different from manufacturer to manufacturer, we chose to use only the RSSI values between devices and not considered the RSSI between computer - devices.

Every physical object, block, must meet the following parameters: transmitting information of their identification; recognize adjacent parts for interaction between them; switching status (on/off); be autonomous and have reduced price; easy integration in a graphic environment; meet the physical size appropriate to the game.

3.5 Approach

In order to design physical objects, a prototype was developed at an early stage using nRF52840 Chip/SoC, not only due to its reduced price, but also due to the characteristics in terms of development.

Different tests were carried out to measure the RSSI values in order to be able to draw conclusions about the methodology to be followed. After registering the RSSI values of these components, based on the Chip/SoC nRF52840, it was developed a functional prototype of the blocks, Fig. 1.

Fig. 1. Block case

The case consists of a printed circuit board, a 2700 mAh battery, a TP4056 based Chip/SoC board, with charge protection diode, to charge the battery and a button.

Each BLE Chip/SoC was programmed so that it would perform the role of observer and advertiser at the same time with appropriate intervals. For the sake of autonomy, it is intended that both the observer and advertiser profiles consume as little energy as possible without changing the game dynamics.

A block that finds another one in the vicinity is able to determine the RSSI value, filter it and transmit the value in the advertisement package to the API.

Due to the size of the advertisement package, 31 bytes were attempted to find a solution that would allow to collect and transmit as much information about the respective object and others in the vicinity. The advertisement package was configured to transmit the information of the game name, "MBBou", the object identification and the identification of the objects in the vicinity, together with their RSSI value. Each block manages the IDs between the blocks and the RSSI values that pass through a Kalman filter before they are included in the package. All information happens dynamically, that is, if a block meets another block, it adds the identification of the second block and the filtered RSSI to the package, when the second disappears, the package undergoes the corresponding change.

Thus, the processing shifts, Fig. 2, are to be done in the graphical environment itself and not in each block. In this case, the UWP application must be able to collect all the information of the blocks and all that is around them and make conclusions about these data and be able to transmit them to the game engine.

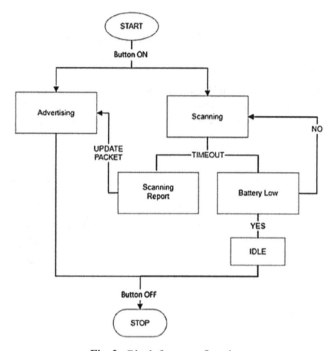

Fig. 2. Block firmware flowchart

When the button is pressed, the block switches to both the Advertising and Scanning profiles. At the end of the Scan time, the Chip/SoC reports the data to a variable that contains the information of the other blocks found, if applicable. The instant the data reported by the scan is received, the information is organized to allow it to be included in the advertisement package. The block is turned off when the button is pressed again.

Thus, the operation of the blocks is almost independent of the rest of the application so it can create different ways of reading the information transmitted by them.

3.6 RSSI Values at Different Distances

In order to automate the system to the maximum and achieve the best possible performance, a test was carried out in which devices separated by distance intervals were placed at different signal strengths. The BLE Chip/SoC allows you to change the TX transmission power, which in turn can result in a reduction in energy consumption.

It is therefore important to establish a comparison between the power of the TX signal and the different positions. Initially, the test was carried out with nRF52840 - dongle. 100 samples of the RSSI were extracted at different transmission powers for eight positions approximately 10 cm apart. The readings include the Kalman filter. Between 40 and 60 cm there is a signal distortion, and it can be seen that RSSI values overlap. On the other hand, if the block undergoes a rotation, the RSSI's behaviour also changes. We can see that the values at different emission intensities have almost the same behaviour, so choosing a lower intensity allows a lower energy consumption taking into account the variation of the RSSI values at given distances. One of the characteristics that is also important to mention is that along an axis and at different distances, if the block undergoes rotation there is a discrepancy in the values read. This becomes relevant when the block is positioned in different ways. Since it will display a sticker with the object's representation, it is important to know if a rotation in the block position influences the RSSI values.

Tests were carried out on the RSSIs values from some blocks to others. The values differ between blocks and, therefore, the readings present some differences. The values between the blocks with identification 1, 2 and 3 were measured. The Kalman filter was adjusted to obtain a better filtration. For each position, 100 samples of the measured RSSI values were taken. The values measured between blocks 1 and 2 are different from those measured between 2 and 1.

In general, the Kalman filter greatly attenuates the error associated with the RSSI values, but depending on the blocks (in this case different Chip/SoCs), even after filtering, there are quite relevant differences between the values. It will be necessary to infer about the value after filtering and the theoretical value, being able to affirm that at the beginning a triangulation by geometric shape will have errors too large to be considered in any way an accurate measurement.

Therefore, inference of the positions of blocks will have to be determined by an application which will allow to obtain satisfactory results and thus will reduce to maximum propagated errors of real positions.

The relationship between the different RSSI values shows that it is possible to infer at least two things: whether a block is near or far from another. In this sense, the approach contemplates a solution whereby, after reading the values by any device that is able to read the information of the blocks and worked on these data, it reports the state of the blocks to the game engine so that it subsequently reacts accordingly. For this, the API was developed in UWP, not only to allow access to Bluetooth® in the Windows operating system, but also to allow visualization of the data in a graphical environment to be later integrated in the Unity platform.

Following a Graph Theory model, in which the relationships between objects of a given set are described, each block represents a vertex (node) and an RSSI value between nodes represents an edge with a weight that is the RSSI's own value. The data can be visualized in a simple way by representing an adjacency matrix.

When the application is started, Fig. 3, it checks whether the Bluetooth® module on the device it is running, whether it is compatible with Bluetooth® Low Energy or if any other error has occurred. After successful confirmation from the BLE module, the application goes into scanning mode and, in parallel, starts a timer to communicate the status of the devices to the game platform. The scanner reads the information transmitted by the devices, these readings being almost random in the sense that it does not have a predetermined time (a block can be read 4 times in one second and 9 times in the other).

This allows the scanner of the devices to work in parallel with the gaming platform so any changes to the way the data is organized and the information read does not influence the way the scanner works. On the other hand, you can configure the time when the platform accesses the device information and the time for a block to be considered off. In the game platform it is only necessary to create a function to register the event.

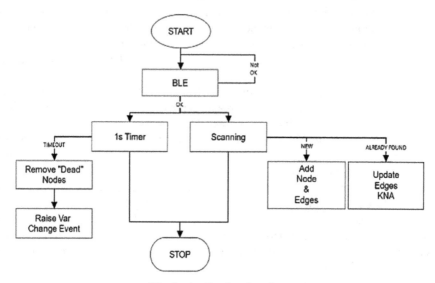

Fig. 3. Application flowchart

After performing tests of the hardware components, it was realized that it would be extremely difficult to build a neuronal network that contemplated all possible variations to determine the distances between blocks with the smallest possible error because, due to the characteristics of the devices, all classifications would depend on the number and the position of the blocks. For example, for a set of 10 blocks with a 10-position matrix, the neuronal network would have to be classified with several different values, since the RSSI value between block 1 and block 2 differs from block 2 and block 1, Fig. 4.

Although some tests were carried out with a multilabel classifier, the results proved to be inconclusive for the reasons described. From all the previously studied articles on indoor location with neural networks using RSSI values, it was concluded that, although it is possible to obtain better results than other approaches, training with a Bluetooth® Low Energy Chip/SoC gives different results that of from training with another Bluetooth® Low Energy Chip/SoC. In all cases, there is no specific description of this type of detail, so it is not shown to work on any device without having to undergo prior training.

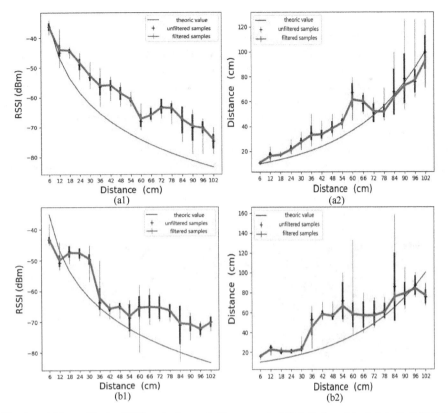

Fig. 4. $RSSI = -10n \, log10(d) + A$; $A = -4$ dBm; $n = 4$; (a1) (a2) from block 1 to 2; (b1) (b2) from block 2 to 1;

4 Field Test

The game consists of blocks with different representations in the form of stickers with correspondence in the graphic environment. To play, simply turn the different blocks on or off, which will then appear or disappear in the graphical environment, allowing the player to observe this behaviour and interactions between characters and consequently develop a story from these elements.

Tests were carried out with children from two different age groups, 10 years and 5 years, from two classes with different educators, Fig. 5. Although the graphic environment was not completed in order to achieve the expected, since the interactions between the characters occurs in a random way when the blocks are connected and not when there is an approximation between them, it was possible to demonstrate that there is a good reaction on the part of the children. Each child took a block at will and, almost as if taking possession of it, turned them on and off in order to understand what the game provided. Supposedly (this was the intention behind the experimentation) they should have had to turn on the block, place it on the table and move it closer or further away from the others.

To the 10-year-old children was asked, after handling the game, to write a group story with the theme "Christmas in India" in which the game characters would be included and the 5-year-old children to draw individually, with the same theme, a representation of what they had played. The 5-year-olds understood the intention better, perhaps because the 10-year-olds were already familiar with TOK, and were able, despite not seeing the characters in order to copy them, to draw all the elements included in the game, one of them represented even in its design, the flute notes. One of the problems they mentioned, in both groups, was the time it took the character to disappear when they disconnected the block. Among the 5-year-olds was a child with special needs, who had a hard time figuring out how to connect the block so he needed help at first; however, although he had this kind of difficulties, he realized that the representation of the character in the block was the same that appeared in the graphic environment, which brought him visible satisfaction.

Fig. 5. Field experiment

In general, the game was very well accepted by both children and educators. The report in both groups on the way they played and how they observed/interpreted the interactions between the characters was revealing that bonds were created between the children, which, in the long run, may even improve their social behaviours. Note that family and school environments determine children's narrative discourse. One of the children reported family experiences around the characters in the game. According to this, the snake represented a member of the family and, the interaction of the snake with the elephant, led her to say, in a group, that … she was going to prison for having made the elephant pass out.

5 Conclusion

Creating interactive solutions using technologies for the development of tangible interfaces is an asset for cognitive development in children of different age groups. The development of the narrative supported by these supports allows children to enhance communication, discourse and the emotional and affective relationship between them. For this project, an attempt was made to create a physical interaction that could be represented graphically in order to engage social and cognitive development. The use of features and details of some existing technologies allows, with some creativity, to develop applications for which they were not foreseen or which for that purpose have not been developed until now.

In this project, an attempt was made to create an alternative to a method previously developed, in order to make it flexible and scalable, and simultaneously compatible with the reduced cost for which it is intended. For this, a widely disseminated technology, Bluetooth®, was used, so that it could be applied to a game engine without the need to introduce "adapters" and other types of readers. An attempt was made to adopt a method that would somehow revolutionize the educational games industry, to create a triangulation to determine the position of Bluetooth® devices, but no results were obtained that would allow to obtain a precise location. This gave way to a more simplified form, in which interaction in a graphic environment is achieved only by the proximity of physical objects, fulfilling the purpose of the study.

Due to the characteristics of the new Bluetooth® devices, some difficulties were found in the integration between the operating systems, namely Windows 10, and the Unity game engine. The use of a system such as the Universal Windows Platform is very reductive in the sense that one does not understand why one has to use a single platform and different from other systems to access the peripherals of a computer, such as Bluetooth®. Since Unity allows us to create games for different operating systems, we would expect that there would be no difficulties when you want to use a game engine for Windows 10. On the other hand, the characteristics of RSSI of Bluetooth® Low Energy do not allow the location of devices in a small space, although it is possible to infer about the relative proximity between two devices. This raises some doubts in studies on indoor location even when other types of distances are involved and other types of sensors may be associated, such as accelerometers and gyroscopes, which for this case were not considered due to the associated cost. Some other methods, beside Kalman filtering, should be tested, namely MARS, CMARS and/or RCMARS, to improve the readings of Bluetooth® RSSI values.

The prototype met some limitations; therefore, some limitations were found in the way the block casing was built, namely in the way the on and off button was developed. The children experienced some difficulties in the connection phase of the blocks. This was evident by the child with special needs. In developing the prototype on a 3D printer is denoted a lack of rigor and many imperfections associated with the actual printing, although in this way it was possible to quickly create a working prototype. Due to the size of the battery, the prototype was slightly larger than what was expected, compared to the blocks of the previous TOK project.

There are some improvements and modifications that can be made to the blocks, namely in the way they connect and disconnect. One of the proposals would be the integration of a capacitive sensor that would allow to connect the block through touch. Regarding the size of the block, it could be improved by choosing a battery of different capacity, which in turn changes to a different size.

From these first tests it is possible to prove the functioning of this prototype, even if the graphic environment is not prepared for these blocks. In this sense, it would be necessary to adapt the game for the use of these blocks and to elaborate the exploratory tests again to draw conclusions. The same is true with the number of characters, that is, the number of blocks should be greater because one of the children's remarks was exactly the reduced number of characters in relation to the other game, TOK.

Although conclusive results on the use of neural networks for the location of the blocks were not obtained, the prototype allows to compare a group of a certain age to another group of players with another age. By recording the proximity between the blocks and the relationships in which children use them, one can, in some way, create relationships between groups of children with and without special needs, for example. This method can help to understand the level of narrative development between classes with identical or diversified characteristics and even among children of different ages.

Acknowledgments. This work was funded by projects "NORTE-01-0145-FEDER-000042", "NORTE-01-0145-FEDER-000045" and "POCI-01-02B7-FEDER-053284", supported by Northern Portugal Regional Operational Programme (Norte2020), under the Portugal 2020 Partnership Agreement, through the European Regional Development Fund (FEDER). It was also funded by national funds, through the FCT – *Fundação para a Ciência e Tecnologia* and FCT/MCTES in the scope of the project "UIDB/05549/2020".

References

1. Fein, G.: Pretend play: creativity and consciousness. In: Gorlitz, P., Wohlwill, J. (eds.) Curiosity, Imagination, and Play, pp. 281–304. Lawrence Erlbaum Associates Inc., Hillsdale (1987)
2. Russ, S.: Development of creative processes in children. In: Runco, M. (ed.) Creativity from Childhood Through Adulthood: The Developmental Issues, vol. 72, pp. 31–42. Jossey-Bass, San Francisco (1996)
3. Russ, S.: Affect and Creativity: The Role of Affect and Play in the Creative Process. Lawrence Erlbaum Associates Inc., Hillsdale (1993)
4. Singer, D.L., Singer, J.: The House of Make-Believe. Harvard University Press, Cambridge (1990)
5. Liebennan, J.N.: Playfulness: Its Relationship to Imagination and Creativity. Academic, New York (1977)
6. Christie, J., Johnson, E.: The role of play in social-intellectual development. Rev. Educ. Res. **53**(93–1), 15 (1983)
7. Singer, J.L., Singer, D.L.: Television, Imagination, and Aggression. Lawrence Erlbaum Associates Inc., Hillsdale (1981)
8. Morais, J.: L'Art de Lire. Odile Jacob, Paris (1994)
9. Sylla, C., Pereira, I.S.P., Coutinho, C.P., Branco, P.: Digital manipulatives as Scaffolds for Preschoolers 2019; Language development. IEEE Trans. Emerg. Top. Comput. **4**(3), 439–449 (2016). https://doi.org/10.1109/TETC.2015.2500099
10. Meltz, B.F.: Pretend play enriches development. Boston Globe, Boston; Section C: 1 (1999)
11. Cassell, J., Ryokai, K.: Making space for voice: technologies to support children's fantasy and storytelling. Pers. Ubiquit. Comput. **5**(3), 169–190 (2001). https://doi.org/10.1007/PL0 0000018
12. Peterson, C., McCabe, A.: Developmental Psycholinguistics: Three Ways of Looking at a Child's Narrative. Plenum, New York (1983)
13. Rogoff, T.: Apprenticeship in Thinking: Cognitive Development in Social Context. Oxford University Press, Oxford (1990)

14. Topping, K.: Cooperative learning and peer tutoring: an overview. Psychologist **5**(4), 151–157 (1992)
15. Wood, D., O'Malley, C.: Collaborative learning between peers: an overview. Educ. Psychol. Pract. **11**(4), 4–9 (1996)
16. Barfurth, M.A.: Understanding the collaborative learning process in a technology rich environment: the case of children's disagreements. In: Proceedings of CSCL 1995 (1995)
17. O'Malley, C.: Designing computer systems to support peer learning. Eur. J. Psychol. Educ. **VII**(4), 339–352 (1992)
18. Cottam, M., Wray, K.: Sketching tangible interfaces: creating an electronic palette for the design community. IEEE Comput. Graph. Appl. **29**(3), 90–95 (2009)
19. Marshall, P.: Do tangible interfaces enhance learning? In: Proceedings of TEI 2007, pp. 163–170 (2007)
20. Piaget, J.: The Origins of Intelligence in Children. University Press, New York (1952)
21. Goncu, A.: Development of intersubjectivity in the dyadic play of preschoolers. Early Childhood Res. Q. **8**, 99–116 (1993)
22. Cassell, J.: Towards a model of technology and literacy development: story listening systems. J. Appl. Dev. Psychol. **25**(1), 75–105 (2004)
23. Hutton, J.S., et al.: Story time turbocharger? Child engagement during shared reading and cerebellar activation and connectivity in preschool-age children listening to stories. PLoS One **12**(5), 1–20 (2017)
24. Bluetooth® SIG, Bluetooth® 5 Core Specification See. https://www.bluetooth.com/specifications/bluetooth-core-specification.
25. Raghavan, A.N., Ananthapadmanaban, H., Sivamurugan, M.S., Ravindran, B.: Accurate mobile robot localization in indoor environments using Bluetooth®. In: Proceedings - IEEE International Conference on Robotics and Automation, pp. 4391–4396 (2010). https://doi.org/10.1109/ROBOT.2010.5509232
26. Simon, H., Robert, H.: Bluetooth tracking without discoverability. In: LoCA 2009: The 4th International Symposium on Location and Context Awareness, May 2009
27. Anastasi, G., Bandelloni, R., Conti, M., Delmastro, F., Gregori, E., Mainetto, G.: Experimenting an indoor Bluetooth-based positioning service. In: Proceedings of the 23rd International Conference on Distributed Computing Systems Workshops, April 2003, pp. 480–483 (2003)
28. Bargh, M., Groote, R.: Indoor localization based on response rate of bluetooth inquiries. In: Proceedings of the First ACM International Workshop on Mobile Entity Localization and Tracking in GPS-Less Environments, September 2008
29. Jevring, M., Groote, R., Hesselman, C.: Dynamic optimization of Bluetooth networks for indoor localization. In: First International Workshop on Automated and Autonomous Sensor Networks (2008)
30. Huang, A.: The use of Bluetooth in Linux and location aware computing. Master of Science dissertation (2005)
31. Bruno, R., Delmastro, F.: Design and analysis of a Bluetooth-based indoor localization system. In: Conti, M., Giordano, S., Gregori, E., Olariu, S. (eds.) PWC 2003. LNCS, vol. 2775, pp. 711–725. Springer, Heidelberg (2003). https://doi.org/10.1007/978-3-540-39867-7_66
32. Hallberg, J., Nilsson, M., Synnes, K.: Positioning with Bluetooth. In: Proceedings of the 10th International Conference on Telecommunications, vol. 2, no. 23, pp. 954–958 (2003)
33. Pandya, D., Jain, R., Lupu, E.: Indoor location estimation using multiple wireless technologies. In: The 14th IEEE Proceedings on Personal, Indoor and Mobile Radio Communications, vol. 3, pp. 2208–2212, August 2003
34. Bahl, P., Padmanabhan, V.N.: RADAR: an in-building RF-based user location and tracking system. In: Proceedings of IEEE Infocom 2000, pp. 775–784, March 2000
35. Yang, Z., Zhou, Z., Liu, Y.: From RSSI to CSI: indoor localization via channel response. ACM Comput. Surv. (CSUR) **46**(2), 25 (2013)

36. Ladd, A.M., Bekris, K.E., Rudys, A., Kavraki, L.E., Wallach, D.S.: Robotics-based location sensing using wireless ethernet. Wirel. Netw. **11**(1–2), 189–204 (2005)
37. Kumar, P., Reddy, L., Varma, S.: Distance measurement and error estimation scheme for RSSI based localization in wireless sensor networks. In: 2009 Fifth IEEE Conference on Wireless Communication and Sensor Networks (WCSN), pp. 1–4. IEEE (2009)
38. Xiao, J., Wu, K., Yi, Y., Wang, L., Ni, L.M.: Pilot: passive device-free indoor localization using channel state information. In: 2013 IEEE 33rd international conference on Distributed computing systems (ICDCS), pp. 236–245. IEEE (2013)
39. Altini, M., Brunelli, D., Farella, E., Benini, L.: Bluetooth indoor localization with multiple neural networks. In: 2010 5th IEEE International Symposium on Wireless Pervasive Computing (ISWPC), pp. 295–300. IEEE (2010)
40. Liu, H., Darabi, H., Banerjee, P., Liu, J.: Survey of wireless indoor positioning techniques and systems. IEEE Trans. Syst. Man Cybern. Part C (Appl. Rev.) **37**(6), 1067–1080 (2007)
41. Zafari, F., Gkelias, A., Leung, K.K.: A survey of indoor localization systems and technologies. IEEE Commun. Surv. Tutor. **21**(3), 2568–2599 (2019). https://doi.org/10.1109/COMST.2019.2911558
42. Grewal, M.S., Andrews, A.P.: Applications of Kalman filtering in aerospace 1960 to the present [historical perspectives]. IEEE Control Syst. Mag. **30**(3), 69–78 (2010). https://doi.org/10.1109/MCS.2010.936465

Model of Acquiring Transversal Competences Among Students on the Example of the Analysis of Communication Competences

Marek Goliński, Małgorzata Spychała(✉), and Marek Miądowicz

Poznan University of Technology, Poznan, Poland
malgorzata.spychala@put.poznan.pl

Abstract. The article presents a model of a method of accelerating the acquisition of transversal competences by students, based on the example of communication competences. The analysis uses the MARS method to create the best model describing the impact of various variables on accelerating the acquisition of communication competences. The following variables were taken into account: the rank of the training method in the developed matrix, the number of students and the weighted average acceleration rate of acquiring the remaining transversal competences, i.e. teamwork, entrepreneurship and creativity of a given student. The results confirm that our new method accelerates the acquisition of transversal competences by students. This method can be used by various higher education institutions in different countries. The developed results can be used in the curriculum of already planned vocational courses and to develop the skills required by employers in various positions and in different professions.

Keywords: Transversal competences · Communication skills · Modelling data mining MARS · Competence management optimization

1 Competences in Modern Economy

Dynamically developing economy imposes constant changes in the functioning of enterprises. Maintaining a competitive position on the market, which is dominated by automation, digitalization, Big Data and Industry 4.0, forces business owners to implement changes in personnel management. Narrow specialization and technical competences, the scope of which is very diverse, currently play a less significant role. Much more required and valued are skills related to the speed of adaptation to the requirements of the enterprise and market environment. The growth of both headcount and wages in the last few decades has been particularly intensive in case of jobs requiring high mathematical and social skills (Deming 2017; Borghans et al. 2014). Teamwork ability, which is indispensable nowadays, requires the involvement of social skills. This translates into reduced coordination costs, enabling employees to specialize and cooperate more effectively (Karabarbounis and Neiman 2014; Sadłowska-Wrzesińska et al. 2017).

Among competences of future employees on the researched markets, certain competency groups dominate, including language and communication competences (Pater et al.

2019). The need to use communication competences occurs in all working conditions (Babbar-Sebens et al. 2019; Majchak et al. 2019), at all levels of management (Bjekić et al. 2015) and using all available techniques and conditions (Westerman et al. 2018). Communication competences and their development, e.g. by simulating real working conditions, is both an effective pedagogical strategy for the development of communication competences in preparation for internships or support in adaptation for recently recruited employees (Barker et al. 2018).

Improving and developing communication competences is also an important psychological aspect in motivating employees and increasing their commitment to work (Chen 2017). Research confirms the relationship between character traits (self-sufficiency, resilience, optimism) and factors required in the workplace, e.g. keeping safe (safety compliance, participation in safety), while improving the role of communication competences (He et al. 2019, Gabryelewicz et al. 2017, Lyon et al. 2018).

Both in the labor market and at the education stage, particular attention is paid to ensuring that employees meet the enterprise requirements related to currently desired and forecasted competences (Graczyk-Kucharska et al. 2018; McCoshan et al. 2008; Valente et al. 2014). In order to guarantee effective human resource management, it is important to correlate company goals, efficient management of human teams and mutual motivation, trust and development of communication competences (Luong et al. 2019).

Issues related to effective communication within the team enable the implementation of the most important functions of the organization, such as knowledge sharing in teams (Del Giudice et al. 2013) and efficient management of innovative projects (Sousa and Rocha 2018).

All the above examples indicate very high practical significance of communication competences and justify the principal objective and subject of the article, which is: to characterize communication as a very important transversal competence and to examine, using statistical analysis, the relationship between selected transversal competences such as: communication skills, entrepreneurship, creativity and teamwork.

These analyzes preceded research focused on determination of educational methods that positively affect the development of transversal competences and factors which characterize the surveyed students and influence the development of competences (Szafrański 2017; Graczyk-Kucharska et al. 2019; Szafrański et al. 2019).

2 Characteristics of Selected Transversal Competences

Engineers no longer manage their daily tasks with plain substance expertise; instead they must be adept at communication, collaboration, networking, feedback provision and reception, teamwork, lifelong learning, and cultural understanding (Lappalainen 2009). Therefore, technical competences are no longer sufficient to meet the requirements of the modern labor market (Platonoff 2011; Mumford et al. 2000; Christensen et al. 2006; Whetten and Cameron 2011). Individuals must also develop transversal competences, i.e. general or interdisciplinary competences, required for different positions and occupations, hence their transversal character (European Parliament and Council 18 December 2006/962/EC). Key competences were expressed in the European reference framework, which was inherited from the development of OECD, UNESCO and the Member States themselves (Pepper 2011). The European reference framework has defined key competences as knowledge, skills and attitudes relevant in the context of: communication in the mother tongue, communication in foreign languages, mathematical skills and basic competences in science and technology, digital competences, learning skills, social competences, the capacity for initiative and entrepreneurship, as well as cultural awareness and expression. All of the above emphasize critical thinking, initiative, problem solving, risk assessment, decision making and constructive management of emotions (Pepper 2011).

As part of the project *The acceleration method of development of transversal competences in the students' practical training process*, four transversal competences were selected from those proposed by the European Parliament: entrepreneurship, creativity, communication skills and teamwork. These competences are required for various jobs, also used to identify key competences for these jobs, called reference models (Spychała et al. 2017; Graczyk-Kucharska et al. 2018). These are the most frequently mentioned competences by employers in various countries as key requirements on the labor market (Szafrański et al. 2017a, b).

3 Characteristics of Communication Competences

Canale and Swain (1981) and Canale (1983) understood communicative competence as a synthesis of an underlying system of knowledge and skill needed for communication. Communication competence is the ability to convey thoughts, information, and intentions. It is the ability to use verbal and non-verbal competences to achieve desired goals in a manner adequate to the context. An indispensable element of communication competence is the ability to establish and maintain relationships with people, empathy and assertiveness (Morreale et al. 2007).

Communication competence involves the quality of communication, which is commonly composed of the elements of appropriateness and effectiveness (Spitzberg and Cupach 1984; Wiemann and Kelly 1981). An interpersonally effective person is the one who accomplishes his goals in contact with others, while doing so in an adequate and appropriate manner in a given social context. It depends on many factors, such as: attitude toward the interlocutor, personal and cultural preferences, his interests, attitudes and goals, clearly expressed intention through the choice of verbal and non-verbal signals. Effective communication is often cited as one of the most important attributes of a successful manager or supervisor (DiMeglio 2007), the most central qualifications in the workforce (Curtis et al. 1999; Hawkins and Fillion 1999) and linked with occupational success (Spitzberg and Cupach 1989; Griffith 2002).

It is necessary to develop communication competences which help shape relationships between company employees and the environment, build an assertive attitude, better solve problems and take on new challenges in life and in cooperation with other people in the team. The development of communication competences offers significant value added when hiring new employees and allows for maintaining social cohesion, which justifies the necessity of lifelong learning, also in relation to adapting to changes and participating in social life. By expanding communication competences, the individual becomes a conscious and responsible entity, choosing the means and predicting the effects of his own actions (Dickson and Hargie 2004).

4 Methodology

4.1 ACT Erasmus Method

Figure 1 shows the main stages of the ATC ERASMUS method, which essence is the development of the key competences of students, supervised by the team, including the authors of this article. This process consists of 7 steps. The first stage was to analyze the teaching methods of transversal competences and practical teaching methods. 88 methods of improving transversal competences were selected. Next, the requirements of employers regarding key competences were identified. After developing a matrix of dependencies between practical teaching methods and teaching transversal competences, models of processes improving transversal competences were designed and tested. These processes were tested at universities in Poland, Finland, Slovenia and Slovakia.

After testing, all comments from testers were collected, results of testing process models were developed, and conclusions were drawn. The next stage was the selection of the most effective teaching processes and their application. The final result of the project was the development of the ACT Erasmus guide, which can be used at every university, improving the transversal competences of students as well as in enterprises, shaping social competences of employees. The presented model is a proof that cooperation between Business and Education began. Educational Institutions have increasingly begun to address the needs of employers in developing the learning process (Goliński et al. 2016).

In the case of the ATC project, the main focus was on the study of accelerating the acquisition of selected transversal competences: entrepreneurship, group work, communication skills and creativity, which are important for enterprises and constantly sought after on the labor market in various positions (Szafrański et al. 2017b). In this study, the analysis of acceleration in acquiring transversal competences (communication skills) using the MARS method was performed. Applying the MARS method involves scanning the space of all input and output values and relationships between these values. Further base functions are added to the model of reality described by variables, selected from the set of all allowable functions, in such a way as to maximize the overall level of adjusting the model to the described reality. Such behavior leads to finding the most important independent variables and the most important interactions between variables (Hastie et al. 2009).

4.2 Multivariate Adaptive Regression (MARS)

Multivariate Adaptive Regression Splines (MARS) is a special regression method used to solve regression and classification problems in order to find the values of dependent variables on the basis of independent variables. The basic advantages of the MARS method compared to other methods is the fact that its use does not require any specific and restrictive special assumptions about the functional connections between dependent and independent variables. An important feature of MARS is the nature of the variables it operates on - they can have a nonparametric form and a multidimensional range. When developing a database for calculations at the input to the analysis, no list of assumptions about the type of dependencies between independent and dependent variables is required.

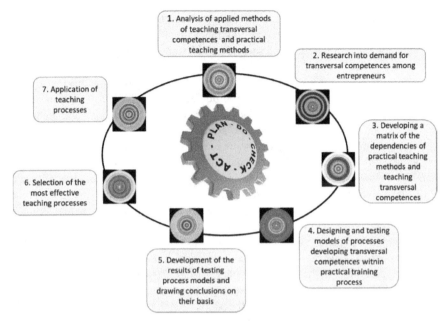

Fig. 1. Method model of processes developing transversal competences within practical training process http://www.awt.org.pl/wp-content/uploads/2018/07/Guide-for-implementation-O8-FINAL-ENG.pdf.

Thanks to these MARS features, the method is very useful in applying to inferences using large amounts of various data.

The MARS method is often interchangeably used with methods using an artificial neural network imitating a biological neural network (Merolla et al. 2014). Both methods are successfully used in social research and technical sciences because as accurate tools they allow the identification of many, even very small dependencies, and ultimately the development of an over-fit model. The MARS method meets the requirements of multi-criterion analysis thanks to operating on a large number of relationships between data and based on it drawing conclusions which independent variables have the greatest impact on the dependent variable. The studies cited in the article compared which of the analyzed factors have the greatest impact and connection with communication skills. On this basis, a statistical model was obtained showing the best reflection of reality.

The research also referred to the conclusions of similar studies on transversal competences regarding entrepreneurship and creativity (Graczyk-Kucharska et al. 2019).

4.3 Collection of Data

For the purpose of the analysis, data have been collected from testing five processes at six universities (Poznan University of Technology, Centria Ostrobotnia University of Applied Sciences, University of Maribor, University in Banska Bystrica, Wrocław University of Economics, Czestochowa University of Technology) from four European Union countries (Poland, Finland, Slovakia and Slovenia). Research conducted under the ATC project assumes that there are at least three methods of practical education within the process. The test on accelerating the acquisition of transversal competences was conducted on a relatively small sample of 113 students using a total of 10 different methods of practical education of students tested from February to October 2017. The development and collection of data on the pace of acceleration of acquiring transversal students' competences, guidelines were defined in the ATC project as basis for a harmonized way of data collection in the different countries. This also applies to a situation where the same methods were used at the same time. The students, on the basis of their self-assessment, determined the extent to which the level of their skills and, consequently, of the researched competences had increased after the application of each method. A sixstage scale from 0 to 5 was addressed in the study. It is assumed that in the case of transversal competences, the level of possessed skills, shaping the competence, cannot be reduced. The degree of change expresses the rate of change perceived by the students. After testing the process, in addition to the above data, information was also collected on the number of students per group, the year of study or the cultural factors that may influence the acceleration of the acquisition of the transversal competences in different EU countries. In this case, secondary data were used based on Hofstede factors (Hofstede 1984). For modelling, in a first step, the analyses selected basic data for the development of a model for accelerating the acquisition of transversal competences. One of the transversal competences, i.e., communicativeness, was selected as the researched independent variable applied by MARS algorithm on 100 data considering 26 inputs. The matrix of educational methods takes into account the research conducted in the 4 partner countries on the importance of transversal competences among employers (N 135). The expert method (Szafrański et al. 2017a, b) was used to evaluate each of the 85 educational methods by assessing the impact of a particular method on the growth of each of the analyzed transversal competences (Graczyk-Kucharska et al. 2019).

4.4 A Subsection Sample

Communication is a very important transversal competence. In the first part of the article, based on numerous scientific papers, its significance and impact on other competences sought on the labor market were demonstrated. The MARS statistical method used in the study allows estimating values from a certain range of data, enabling both additive and interactive effects of predictors that determine the variable response (Weber et al. 2012; Akyıldız et al. 2015). An important issue when determining the form of a nonparametric regression model while using MARS is the submodel portfolio assessment to select the best submodel with the appropriate number of nodes relative to a subset of predictors. In typical regression modeling, when the model has a large number of independent (predictive) variables, and there is no accurate information on the exact functional relationships between variables, many criteria of choosing the optimal model are difficult to determine. In the analyzed study, in order to find the simplest model which most accurately estimates the adjustment to reality, the number of parameters was sought based on the data compiled in the table (see Table 1).

Two models were used for data analysis: the first based on multiple (linear) regression and the second based on MARS multivariate adaptive regression.

Table 1. Sample variables used in research.

Type of variable	Symbol	Variable
Dependent variable	Y_k	Avarage acceleration of Communicativeness
Independent variable	X_2	No. of student
Independent variable	X_3	Rank of the method in matrix
Independent variable	X_4	No. of the method in the process
Independent variable	X_5	No. of the process
Independent variable	X_6	Size of tested group
Independent variable	X_7	Number of meetings
Independent variable	X_8	Number of test groups
Independent variable	X_9	Duration of testing (min.)
Independent variable	X_{10}	Average acceleration of Creativity
Independent variable	X_{11}	Average acceleration of Entrepreneurship
Independent variable	X_{12}	Average acceleration of Teamwork
Independent variable	X_{13}	Power Distance (Hoffstede)
Independent variable	X_{14}	Individualism (Hoffstede)
Independent variable	X_{15}	Masculinity (Hoffstede)
Independent variable	X_{16}	Uncertainty Avoidance (Hoffstede)
Independent variable	X_{17}	Long Term Orientation (Hoffstede)
Independent variable	X_{18}	Indulgence (Hoffstede)

Table 2 describes the results of the linear dependence model. This model shows that the increase in communication acceleration is most closely related to teamwork (X12) with a factor of 0.69, followed by X10, X11, and X5. Very weak relationship was demonstrated for variables X6 and X7.

Based on descriptive statistics for multiple regression, the relationship between the predicted raw data and the dependent variable was also determined - the results are shown in Fig. 2.

One of the values characterizing the flow of linear regression is the R squared coefficient, which is a measure of the model fit, i.e. the given model explains 88%. Variability of the phenomenon. Summarizing the results of data analysis and assessing the usefulness of the model, it should be noted that multiple regression results confirm the fit of the model. The r square factor is a measure of the quality of the fit of the model. It defines what percentage of one variable explains the variability of the other variable.

The second data analysis was based on Multivariate adaptive regression MARS. Table 3 contains the values characterizing the numerical measures of the regression model, while Table 4 shows independent variables based on which the model can be built. The most important factor that measures the model fit, the R squared factor, which

Table 2. Summary of dependent variable regression.

N=339	b*	Standard error b*	b	Standard error b	t(332)	p
Free indicator			-0,504898	0,354091	-1,42590	0,154837
X12	0,690822	0,036091	0,703683	0,036736	19,14114	0,00000
X11	0,129498	0,041804	0,132943	0,042916	3,09774	0,002116
X10	0,134745	0,041641	0,139706	0,043174	3,23587	0,001335
X5	0,122700	0,052806	0,095001	0,040885	2,32360	0,020750
X6	0,074444	0,052066	0,004452	0,003114	1,42980	0,153714
X7	-0,024969	0,019060	-0,053617	0,040930	-1,30998	0,191110

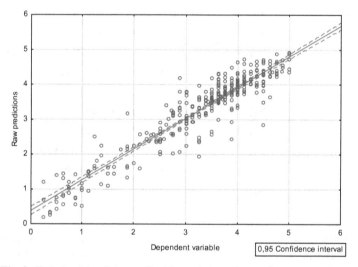

Fig. 2. Relationship of the predicted raw data to the dependent variable Y_k.

is the measure of the model fit, equals 0.88, which explains in a highly satisfying way the 88% variability of the phenomenon.

Based on the obtained data and taking into account descriptive (independent) variables that had rational measures of partial correlation, the following model was obtained:

$$Y_k = 3{,}16027669629377e{+}000 + 6{,}34980317952070e{-}001 * \max(0; X_{12} - 3{,}14285714285714e{+}000) - 7{,}99184901257433e{-}001 * \max(0; 3{,}14285714285714e{+}000 - X_{12}) + 2{,}96467637047732e{-}001 * \max(0; X_{11} - 3{,}13163349859628e{+}000) - 2{,}37216855731993e{-}001 * \max(0; 3{,}25515368953512e{+}000 - X_{10})$$

Table 3. Numeric quality measures Multivariate adaptive regression MARS.

Variable type	Value
Average (dependent variable)	3,29
Standard deviation (dependent variable)	1,15
Average (independent variable)	3,29
Standard deviation (independent variable)	1,08
Average (the rest)	0,00
Standard deviation (the rest)	0,40
Coefficient of determination R-squared	0,88
Coefficient of determination R-squared Adjusted	0,87

Table 4. Coefficients underlying the model's construction (cell highlighting indicates max base function)

Coefficients, nodes and base functions	Nodes Yk	Nodes X10	Nodes X11	Nodes X11
Free factor	3,160277			
Factor No. 1	0,634980			
Factor No. 2	-0,799185			-3,142857
Factor No. 3	0,296468		-3,131633	-3,142857
Factor No. 4	-0,237217	-3,255154		

The obtained model and the results indicate that only the variables X10, X11 and X12 are statistically significant, while the remaining variables should be removed from the regression model.

The resulting model indicates the key role played by the strengthening effect in the teaching process between the mutual impact of accelerating the acquisition of specific key competences. Accelerating the acquisition of communicative competence simultaneously influences the acquisition of other competences – in the described study mainly the competences in the area of entrepreneurship, creativity and teamwork.

Taking into account the results obtained in Figs. 3, 4 and 5, the correlation between the communicativeness competency and the other competences studied is presented.

The obtained results (Tables 2, 3 and 4) indicate that both based on multiple (linear) regression and multivariate MARS adaptive regression, models can be developed that with statistical reliability (identical value of determination coefficient $R2 = 0.88$) reflect the flow of the studied phenomenon, confirming the assumptions of the article.

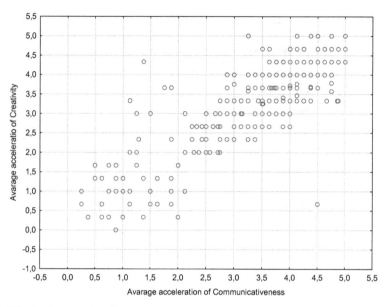

Fig. 3. Scatter plot of the dependent variable Y_k against the independent X_{10}.

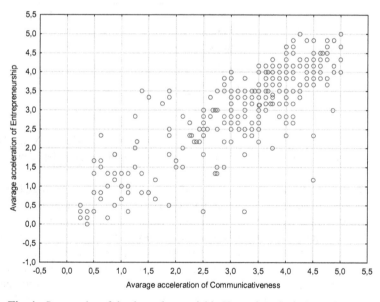

Fig. 4. Scatter plot of the dependent variable Y_k against the independent X_{11}.

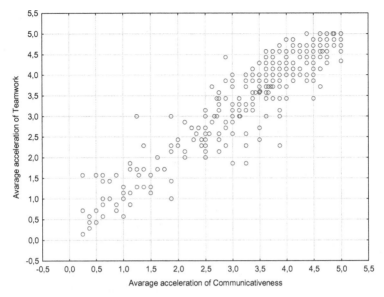

Fig. 5. Scatter plot of the dependent variable Y_k against the independent X_{12}.

5 Conclusions and Outlook

The article indicates only a few results of research on the issue of transversal competences, relationship between communication skills and other competences and their importance for the functioning of the enterprise.

The choice of the presented analyzes and sample results was to indicate the possibilities of managing employee competences in resource management of an enterprise. The research results presented in the article confirmed the assumptions made in the article that it is possible to develop a multiple regression linear model indicating statistically significant relationships between transversal competences:

$$Y_k = 3,16027669629377e+000 + 6,34980317952070e-001*max(0; X11- 3,14285714285714e+000) - 7,99184901257433e-001*max(0; 3,14285714285714e+000-X12) + 2,96467637047732e-001*max(0; X11- 3,13163349859628e+000) - 2,37216855731993e-001*max(0; 3,25515368953512e+000-X10)$$

The obtained model has a high value of the determination coefficient R2 (0.88), which confirms its correlation with the analyzed data. The model does not take into account many variables, so it does not show a complex description. From the analyzed 17 independent variables, only three were included in the model, also related to transversal competences (X10, X11, X12). In case of the obtained model, significant measures of partial correlations were also shown.

The stronger the correlation among the individual data analyzed, the more difficult it is to change one variable without changing the other. By introducing modeling, it is difficult to estimate the relationship between each independent variable and the dependent variable independently, because when the independent variables are correlated they change in accordance. High correlation of independent data can negatively impact the correct inference for modeling.

The practical aspects of using the results obtained are associated with both statistical confirmation of the assumptions and representativeness of the study group. The obtained results were entirely based on data from studies on students entering the labor market.

The ability to use the solutions described in the article related to accelerating the acquisition of transversal competences has a very significant impact on the systemic and innovative approach in resource management of an enterprise. This applies in particular to the transversal competence analyzed in the article, because a strong emphasis is placed on indicating the importance of the problem of communication between labor market entities and the vocational education system (Don-levy et al. 2016; Miranda et al. 2017). The importance of improving transversal competences is recognized and taken into account in the projects implemented since 2010 by the Poznan University of Technology.

To sum up, certain attention should be paid to the need of simultaneous analysis of data on education methods and tools, and the needs of the economy, which to the greatest extent builds its development on human resources. When implementing activities in the field of analysis and teaching of transversal competences, and as a result prevention in competency management, it is necessary to take into account a time horizon of several years, and to rely on the verified needs of managers in particular enterprises. The effectiveness of teaching transversal competences depends not only on the used methods but also on the content being taught, which must be created in cooperation with the university's economic environment (Spychała et al. 2019; Szafrański 2017).

The results described in the article have impact on both the scientific and the practical area. The research method, using the MARS tool, confirmed the theoretical assumptions related to the interdependence of the analyzed transversal competences. Regardless of the theoretical approach, the results obtained gave grounds for modification of educational programs implemented at the university, which included the method of accelerating the acquisition of transversal competences. In further research, it would be important to examine additional factors that could have impact on accelerating the acquisition of transversal competences, and the practical use of the developed model would be to apply it in relation with raising employees' competences in enterprises as a part of professional development.

References

Akyıldız, E., Gebizlioğlu, Ö.L., Karasözen, B., Uğur, Ö., Weber, G.W.: Recent advances in applied and computational mathematics: ICACM-IAM-METU. J. Comput. Appl. Math. **259**(15), 327–328 (2015)

Babbar-Sebens, M., et al.: Training water resources systems engineers to communicate: acting on observations from on-the-job practitioners. J. Prof. Issues Eng. Educ. Pract. **45**(4) (2019)

Barker, M., Fejzic, J., Mak, A.S.: Simulated learning for generic communication competency development: a case study of Australian postgraduate pharmacy students. High. Educ. Res. Dev. **37**(6), 1109–1123 (2018)

Bjekić, M., Bjekić, D., Zlatić, L.: Communication competence of practicing engineers and engineering students: education and evaluation. Int. J. Eng. Educ. Part B **31**(1), 368–376 (2015)

Borghans, L., Ter Weel, B., Weinberg, B.A.: People skills and the labormarket outcomes of underrepresented groups. Ind. Labor Relat. Rev. **67**(2), 287–334 (2014)

Canale, M., Swain, M.A.: Theoretical framework for communicative competence. In: Palmer, A., Groot, P., Trosper, G. (eds.) The Construct Validation of Test of Communicative Competence, pp. 31–36 (1981)

Canale, M.: From communicative competence to communicative language pedagogy. In: Richards, J.C., Schmidt, R.W. (eds.) Language and Communication, pp. 2–27. Longman, London (1983)

Chen, S.-L.: Cross-level effects of high-commitment work systems on work engagement: the mediating role of psychological capital. Asia Pac. J. Hum. Resour. **56**(3), 384–401 (2017)

Christensen, J., Henriksen, L.B., Kolmos, A.: Engineering Science, Skills, and Bildung. Aalborg University, Denmark (2006)

Croucher, S.M., et al.: A multi-national validity analysis of the self-perceived communication competence scale. J. Int. Intercultural Commun. 1–12 (2019)

Curtis, D.B., Winsor, J.L., Stephens, R.D.: National preferences in business and communication education. Commun. Educ. **38**, 6–14 (1999)

Del Giudice, M., Della Peruta, M.R., Maggioni, V.: Collective knowledge and organizational routines within academic communities of practice: an empirical research on science–entrepreneurs. J. Knowl. Econ. **4**(3), 260–278 (2013)

Deming, D.J.: The growing importance of social skills in the labor market. Q. J. Econ. **132**(4), 1593–1640 (2017)

DiMeglio, F.: New role for business school research. Business Week Online (2007). http://www.businessweek.com/stories/2007-08-14/new-role-for-businessschool-researchbusinessweek-business-news-stock-market-and-financial-advice

Donlevy, V., Curtarelli, M., McCoshan, A., Meierkord, A.: Study on Obstacles to Recognition of Skills and Qualifications, Final Report, Directorate-General for Employment, Social Affairs and Inclusion, Luxembourg: Publications Office of the European Union (2016)

Gabryelewicz, I., Krupa, P., Sadłowska-Wrzesińska, J.: Online measurement of work safety culture - statement of research. In: MATEC Web of Conferences, vol. 94, p. 06008 (2017)

Goliński, M., Miądowicz, M.: Management of employee competencies in resource management of an enterprise. In: Proceedings of the European Conference on Knowledge Management, ECKM, vol. 1, pp. 405–414 (2019)

Graczyk-Kucharska, M., Szafrański, M., Goliński, M., Spychała, M., Borsekova, K.: Model of competency management in the network of production enterprises in industry 4.0-assumptions. In: Hamrol, A., Grabowska, M., Maletic, D., Woll, R. (eds.) Advances in Manufacturing, pp. 195–204. Springer, Cham (2018). https://doi.org/10.1007/978-3-319-68619-6_19

Graczyk-Kucharska, M., Özmen, A., Szafrański, M., Weber, G.W., Goliński, M., Spychała, M.: Knowledge accelerator by transversal competences and multivariate adaptive regression splines. CEJOR **28**(2), 645–669 (2019). https://doi.org/10.1007/s10100-019-00636-x

Griffith, D.A.: The role of communication competencies in international business relationship development. J. World Bus. **37**(4), 256–265 (2002)

Guo, B.H.W., Yiu, T.W.: Developing leading indicators to monitor the safety conditions of construction projects. J. Manag. Eng. **32**(1) (2016)

Hastie, T., Tibshirani, R., Friedman, J.: The Elements of Statistical Learning. Springer, New York (2009)

He, C., Jia, G., McCabe, B., Chen, Y., Sun, J.: Impact of psychological capital on construction worker safety behavior: communication competence as a mediator. J. Saf. Res. **71**(12), 231–241 (2019)

Hofstede, G.: Cultural dimensions in management and planning. Asia Pac. J. Manag. **1**(2), 81–99 (1984)

http://www.awt.org.pl/wp-content/uploads/2018/07/Guide-for-implementation-O8-FINAL-ENG.pdf

Karabarbounis, L., Neiman, B.: The global decline of the labor share. Q. J. Econ. **129**(1), 61–103 (2014)

Lappalainen, P.: Communication as part of the engineering skills set. Eur. J. Eng. Educ. **34**(2), 123–129 (2009)

Luong, T.T., Sivarajah, U., Weerakkody, V.: Do agile managed information systems projects fail due to a lack of emotional intelligence? Inf. Syst. Front. **23**(2), 415–433 (2019). https://doi.org/10.1007/s10796-019-09962-6

Lyon, S., Hon, C., Chan, A., Wong, F., Javed, A.: Relationships among safety climate, safety behavior, and safety outcomes for ethnic minority construction workers. Int. J. Environ. Res. Public Health **15**(3), 484 (2018)

Majchrzak, J., Goliński, M., Mantura, W.: The concept of the qualitology and grey system theory application in marketing information quality cognition and assessment. Central Eur. J. Oper. Res. 1–24 (2019). https://link.springer.com/content/pdf/10.1007%2Fs10100-019-00635-y.pdf. Accessed 12 Dec 2019

McCoshan, A., Drozd, A., Nelissen, E., Nevala, A.M.: Beyond the Maastricht Communique: developments in the opening up of VET pathways and the role of VET in labour market integration: consolidated final report, ECOTEC Research and Consulting, Brussels, Belgium, European Commission, Directorate General for Education and Culture (2008)

Mikkelson, A.C., York, J.A., Arritola, J.: Communication competence, leadership behaviors, and employee outcomes in supervisor-employee, relationships business and professional. Commun. Q. **78**(3), 336–354 (2015)

Miranda, S., Orciuoli, F., Loia, V., Sampson, D.: An ontology-based model for competence management. Data Knowl. Eng. **107**, 51–66 (2017)

Morreale, S.P., Spitzberg, B.H., Barge, J.K.: Communication between people, PWN (2007)

Multivariate Adaptive Regression Splines (MARSplines). https://www.statsoft.pl/textbook/stathome_stat.html?https%3A%2F%2Fwww.statsoft.pl%2Ftextbook%2Fstmars.html. StatSoft Electronic Statostic TextBook. Accessed 22 Dec 2019

Mumford, M.D., Marks, M.A., Connelly, S., Zaccaro, S., Reiter-Palmon, R.: Development of leadership skills: Experience and timing. Leadersh. Q. **11**(1), 87–114 (2000)

Özmen, A., Yılmaz, Y., Weber, G.W.: Natural gas consumption forecast with MARS and CMARS models for residential users. Energy Econ. **70**, 357–381 (2018)

Parlament Europejski i Rada 18 grudnia 2006/962/WE

Pater, R., Szkola, J., Kozak, M.: A method for measuring detailed demand for workers' competences. Econ. Open-Access Open-Assessment E-J. **13**(27) 1–29 (2019)

Pepper, D.: Assessing key competences across the curriculum-and Europe. Eur. J. Educ. **46**(3), 335–353 (2011)

Platonoff, A.L.: Czas na rozwój. Człowiek, społeczność, organizacja. Podstawy zrównoważonego rozwoju w odniesieniu do jednostki, społeczności i organizacji, Master of Business Administration, Akademia Leona Koźmińskiego 1/2011 (116), 2–9 (2011)

Sadłowska-Wrzesińska, J., Gabryelewicz, I., Krupa, P.: The use of IT tools for the analysis and evaluation of psychomotor efficiency of employees. In: MATEC Web of Conferences, vol. 94, p. 06017 (2017)

Sousa, M.J., Rocha, Á.: Skills for disruptive digital business. J. Bus. Res. **94**, 257–263 (2018)

Spitzberg, B.H., Cupach, W.R.: Interpersonal Communication Competence. Sage, Beverly Hills (1984)

Spitzberg, B.H., Cupach, W.R.: Handbook of Interpersonal Competence Research. Springer, New York (1989)

Spychała, M., Szafrański, M., Graczyk-Kucharska, M., Goliński, M.: The method of designing reference models of workstations. In: Proceedings of the 18th European Conference on Knowledge Management, ECKM 2017, pp. 930–939. Academic Conferences and Publishing International Limited, Barcelona (2017)

Spychała, M., Goliński, M., Szafrański, M., Graczyk-Kucharska, M.: Competency models as modern tools in the recruitment process of employees. In: Sargiacomo, M. (ed.) Proceedings of the 10th European Conference on Intangibles and Intellectual Capital ECIIC 2019, Chieti-Pescara, Italy, pp. 282–291. Academic Conferences and Publishing International Limited (2019)

Szafrański, M., Goliński, M., Simi, H.: The acceleration of development of transversal competences. Centria University of Applied Sciences, Kokkola (2017a)

Szafrański, M.: Problem of language used to describe competences in the management of acceleration in the creation of knowledge resources in businesses. Procedia Eng. **182**, 679–686 (2017)

Szafrański, M., Bogurska-Matys, K., Goliński, M.: Problems in communication between businesses and technical education system. Manag. Prod. Eng. Rev. **8**(2), 9–18 (2017b)

Szafrański, M., Goliński, M., Graczyk-Kucharska, M., Spychała, M.: Cooperation of education and enterprises in improving professional competences - analysis of needs. In: Hamrol, A., Grabowska, M., Maletic, D., Woll, R. (eds.) MANUFACTURING 2019. LNME, pp. 155–168. Springer, Cham (2019). https://doi.org/10.1007/978-3-030-17269-5_11

Szafrański, M.: Threefold nature of competences in enterprise management: a qualitative model. In: Proceedings of the 20th European Conference on Knowledge Management, Universidade Europela de Lisboa, Lisbon, Portugal, vol 2, pp. 1006–1015 (2019)

Valente, A.C., Salavisa, I., Lagoa, S.: Education quality and economic performance in Europe. In: Proceedings of the European Conference on Knowledge Management, ECKM 2014, vol. 3, pp. 1028–1036 (2014)

Weber, G.-W., Batmaz, I., Köksal, G., Taylan, P., Yerlikaya-Özkurt, F.: CMARS: a new contribution to nonparametric regression with multivariate adaptive regression splines supported by continuous optimization. Inverse Prob. Sci. Eng. **20**(3), 371–400 (2012)

Westerman, C.Y.K., Reno, K.M., Heuett, K.B.: Delivering feedback supervisors' source credibility and communication competence. Int. J. Bus. Commun. **55**(4), 526–556 (2018)

Whetten, D.A., Cameron, K.S.: Developing Management Skills. Prentice Hall/Pearson, Upper Saddle River (2011)

Wiemann, J.M., Kelly, C.W.: Pragmatics of interpersonal competence. In: Wilder-Mott, C., Weakland, J.H. (eds.) Rigor and Imagination: Essays from the Legacy of Gregory Bateson, pp. 283–298. Praeger, New York (1981)

Mobile Applications in Engineering Based on the Technology of Augmented Reality

Tetiana Zhylenko, Vitalii Ivanov(✉), Ivan Pavlenko, Nataliia Martynova, Yurii Zuban, and Dmytro Samokhvalov

Sumy State University, 2, Rymskogo-Korsakova Street, Sumy 40007, Ukraine
ivanov@tmvi.sumdu.edu.ua

Abstract. The modern engineering educational environment ensures an extensive use of augmented reality for the successful and deep memorization of graphic material. There are many platforms for implementing augmented reality technology, the most popular of which were analyzed. As a result, the research work is devoted to the visualization of second-order surfaces and engineering objects. The article describes the scenario of augmented reality both with QR codes and authors' markers. Moreover, the components of the markers and their significance in reproducing the image were described in detail. The proposed methodology includes a step-by-step approach for using the applications based on augmented reality for downloading through the public domain. As a result, the examples of tasks that can be performed using AR were realized. Finally, many markers are proposed in the article, which can be used in the engineering field's educational process, especially with the mathematical peculiarities.

Keywords: Web application · Engineering graphics · Second-order Surfaces

1 Introduction

During the second decade, augmented reality (AR) was used in an educational process in many countries of the world. This technology came into use in business, medicine, industry, banking, education, and everyday life. Now every person who has a smartphone has at least one AR-application. Modern children have many applications that make it possible to combine reality and cartoons, games, stories, and emotions. AR entered the educational environment first in the library business, which now seems natural and not surprising. Then they began to use it for various games as hints when searching for a key.

Today, many students and teachers are engaged in using existing projects, but they also create their applications with and without QR-codes. Each teacher is looking for more and more assignments for an exciting lesson. Today students cannot be surprised simply by the encrypted text in the QR-code. Therefore, we decided to create 3D-models of second-order surfaces and machine-building demands using AR [1]. It diversifies the lesson by working with code and mastering a new application and representing surfaces in three-dimensional space where you can view an object from different angles, at different scales, from all sides. With all the problematic features depict on a piece of paper. The term "Augmented Reality" (AR) was first used in 1990 by Tom Codel, who worked for Boeing Corporation. The essence of the concept of "augmented reality" is in the name, which means a technology that complements reality with virtual elements.

Displaying virtual objects on the screen of devices such as a computer, tablet, or smartphone, visualization of real objects, and viewing virtual objects with special glasses and helmets becomes possible. AR is used to combine virtual objects with natural objects to create a unique combination of them. It is the main difference between AR and virtual reality.

You can specify the following features of AR: the ability to influence virtual objects, create photos of a real object with a virtual object, the ability to navigate the sites of companies, medical institutions, libraries, etc.

This paper aims to study recent approaches and AR formation methods to develop new AR-applications for the educational/retraining process of engineers in analytical geometry, mathematical analysis, descriptive geometry, engineering, and computer graphics.

2 Literature Review

The main problem is creating a mobile application that would be interesting for students and engineers in the educational process and make learning more fun and exciting. Firstly, a comparative analysis of the platforms for the creation of AR was carried out. The most popular AR-libraries ARToolKit, EasyAR, Kudan, Maxst, Vuforia, Xzimg, Wikitude, and ThingWorx Studio were analyzed. ARToolKit is open-source software. There is no watermark (a translucent image superimposed on top of the camera image is represented by the company logo and is present in free AR-library licenses). Only 2D-objects can be recognized. Access is free, but developer documentation is limited. It is not possible to recognize volumetric objects and labels from the cloud. AR-technology can be useful for manufacturing, assembling, quality control, maintenance [2, 3].

EasyAR can recognize only 2D-objects. There is no watermark. It supports up to 1,000 tags for recognition. Research is possible only following the documentation, which is not convenient and not productive. It is free and easy to use the alternative to Vuforia from Chinese developers [4]. The corresponding supporting platforms are Android, iOS, UWP, Windows, OS X, and Unity [5, 6]. The latest version of EasyAR has the following features: image recognition, 3D-object recognition, perception of the environment with a simultaneous orientation to the terrain, cloud recognition, working on smart glasses, cloud application deployment. The library is entirely free, and to get started with EasyAR. It is needed to register an account and generate a plug-in key for a particular application under development. EasyAR integrates easily since it does not work with tags but recognizes objects based on the 3D-model inside the project and generates target images as the program runs. Using EasyAR, it is needed to write more code and create a project in the IDE in a slightly different way. The developer manages the corresponding behavior, determining his behavior for the object recognition/loss event. The library recognizes objects in proper lights, but the percentage of successful tag detection is lower than Vuforia.

Kudan [7] has positive features, including the absence of restrictions on the number of recognized images and the small amount of memory required to store files on the device. It is possible to recognize 3D-objects and markerless tracking of objects. Developers can use the necessary documentation when accessing the library. However, the user manual is not precisely detailed and requires additional information. The mapping of add-ons is implemented through a separate wrapper component over OpenGL using a watermark. However, the free version is available for application testing only. To publish the application, purchasing a license is needed.

The platform Maxst [8] allows optimizing the AR-technologies specifically for mobile platforms for recognizing and tracking the image but at the same time no more than three goals. However, it is possible to track a flat surface and place the desired object on it. Additionally, the free version allows developing the solution only for Android and iOS. Moreover, there is support for the simultaneous creation and saving of 3D-space maps. The platform mentioned above describes only works with the 32-bit version of the Unity editor. However, it does not have a highly precise level of mark recognition quality.

Xzimg [9] represents the following three products for working with applications based on augmented reality: Augmented Face, Augmented Vision, and Magic Face. Augmented Face recognizes and tracks the faces, Augmented Vision identifies and tracks flat images, and Magic Face is designed to replace facial features and apply the corresponding makeup. It should be noted that it is possible to develop a project for HTML5. However, a watermark with a free trial is available for demonstration only (including the inverting of color and changing an image).

Wikitude [10] includes image recognition and tracking, 3D-model rendering, video overlay, and AR and SLAM technology [5], enabling object recognition, tracking, and instant position tracking devices without a marker. Additionally, the cross-platform SDK is available for Android and iOS operating systems. Moreover, it is also optimized for several points. Unlike Vuforia 3D-objects, one of the main features is available to download videos to the database. The target marker is generated from the video. However, to successfully create markers, a solid background is needed to be considered.

Wikitude recently released a relatively new powerful SLAM solution for the AR-application of the Wikitude SDK 7. The corresponding Wikitude SDK has the following features: excellent image recognition and tracking, 3D SLAM-based tracking, GEO Data (enhanced geo-referenced data), and cloud recognition. The main features include the execution of native code and JavaScript, examples of image recognition, 3D-models, many demonstrations, and the simplicity of launching augmented reality from a URL. Moreover, there is a free version of the developer library with watermarks on the screen.

ThingWorx Studio [8] uses a clear graphic editor and Vuforia computer vision technology. Recognition tags and 3D-objects are stored and used through the cloud database. However, this toolkit has a free trial only.

Vuforia software is widely used as the industry-leading AR-solution with the best-in-class machine vision technologies, powerful tracking features, and a wide range of platform support. It is one of the most popular AR-platforms [11], which includes an SDK for creating AR-applications. It uses computer vision algorithms to detect and track flat images (tags) and simple 3D-objects in real-time. Vuforia has its API that allows one to generate a code in C++, Java, Objective-C, and *.Net. Vuforia SDK can be installed as an extension for IDEs, i.e., Android Studio, XCode, Visual Studio, and develop software for a specific mobile platform (e.g., Windows Phone, Android, and iOS). The last ones can be used as a Vuforia extension for the Unity games with the corresponding engine and cross-platform mobile applications. The Vuforia Engine interacts with the real world. The AR interfaces mean the elements through which the interaction of the real (physical) world with the Vuforia Engine and the user is carried out, which brings the level of interactivity of the human-machine to a new level [6].

Additionally, it should be noted that the scientific novelty and practical significance of the proposed approach are highlighted by recent research works in the field of AR- and VR-applications. Remarkably, in the paper [12], the methodology of automated training of convolutional networks by virtual 3D-models for parts recognition in the assembly process is proposed. Importance and challenges of implementing the Industry 4.0 strategy using augmented reality, big data, and the internet of things, particularly in building a smart city, as well as recent trends for realizing intercultural management in a globalized world with the possibilities of intelligence implementation in manufacturing the role of the simulation were substantiated in the research works [13–15]. A systematic review and reflections in collaborative manufacturing based on a cloud platform with the Industry 4.0 oriented principles and technologies were presented in the paper [16]. In the research papers [17, 18], the fundamental approach for ensuring the reliability of technological equipment in the engineering field using artificial intelligence systems was developed.

The ways of industrial implementation of the AR-technology in computer-aided fixture design systems were proposed in the research [19]. The software for the automated training of deep learning networks by virtual 3D-models for object recognition was presented in the article [20]. The mixed reality implementation for managing the product lifecycle was proposed in the research [21].

According to the assessment strategy's quality-oriented rethinking, providing engineering disciplines at universities was realized in the paper [22]. The methods of industrial engineering used in network organizations were proposed in the research work [23]. The development of computer graphics management systems and the approach for increasing image recognition quality were presented in the articles [24, 25]. The recent research [26], based on the deep learning method, focused on recognizing objects and elements during the visual product inspection. Finally, general approaches for ensuring technological equipment reliability using the numerical simulation environment were developed in the research papers [27, 28].

3 Research Methodology

Vuforia has several methods of recognizing and inputting visual information based on reading from the camera of a device and the consequent processing by the Vuforia Engine. Image Targets dials with the real flat images printed on any suitable plain surface. It should be noted that the markers do not need exceptional monochrome (black and white) zones to be tracked in Vuforia Engine. The markers are convenient to use anywhere, but they can be potentially used for tips, informative help, create interactive books and manuals, and visualize concepts.

The VuMark Tag is a more advanced version of regular markers [29]. It has an outline, border, free area, code elements and may include a logo. The Vuforia software module searches for the tag's outline in the image obtained from the device's camera and, if detected, reads the code elements. The code elements are divided into "light" and "dark". The aggregate represents a binary code, which decryption can obtain a unique product ID, links to the website with helpful information about it, and more.

Multi-Targets software is similar to conventional markers, but it uses more than one marker on one real object, such as a cube to use. This is especially useful for sizeable augmented reality objects that need to be covered from all sides, such as houses, monuments, museums, etc.

Cylinder Targets are real flat images on cylindrical surfaces. They allow one to track markers in the form of twisted cones and cylinders. Such markers are used on cups, bottles, wide legs, etc.

Object Scanner recognizes 3D-objects and makes them simplified primitives of the geometric skeleton (tag objects). This allows one to identify a specific physical object in the physical space in the future. To create a new object marker, the first scan can be realized using an existing Object Scanner with the object data file's consequent automatic generation. This file is uploaded to Vuforia Target Manager (up to 20 object data files can be downloaded). The virtual objects' position supplementing the real space should be related to a specific fixed real object concerning which they will be placed. These objects are so-called "labels".

The "Surface math AR" application was created using the following components: a combination of the Android operating system, the Unity cross-platform development environment, and the VuMark technology. The Vuforia software was chosen as the development environment and toolkit.

Marker mockups were created in Adobe Illustrator using the VuMark Designer tool and then uploaded to Target Manager. Design Elements of the VuMarks has five key design components that contribute to its uniqueness, detectability, and data encoding capabilities (Fig. 1).

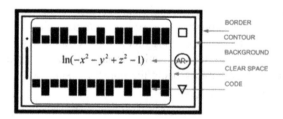

Fig. 1. Key design components of the VuMarks Designer tool.

Vuforia initially detects the Contour circuit. After finding the contour, the algorithm searches for the code and "reads" it to determine the value or "identifier" encoded in VuMark. The outline is not displayed explicitly or is not visible in the VuMark design. The Contour is the line that appears where the other two parts of VuMark meet (Border and Clear Space). The outline is determined by the contrast between two different colors of a border and a transparent space. Remarkably, the border is usually the most recognizable and defining in comparison with VuMark. The border is an external shape consisting of six straight lines forming a hexagon. The clean space is a mandatory empty area that appears next to the border along its entire length. It can be located both inside and outside the border and is required to guarantee the sufficient contrast of the algorithm to detect the contour.

Each individual VuMark contains a unique code as a visual representation of the encoded identifier. The code consists of the element, the data type, and the encoded value/identifier's length, determining the number of elements. The greater the length of the value, the more elements are required.

Each element has the following two states: "dark" and "light". A unique code is generated by the setting elements in a dark or light state (approximately 50% of the elements in each state).

The background or design area is the layer into which you can place any parts of VuMark that are not used for detection.

After VuMark was developed in Adobe Illustrator, they were exported as a Scalable Vector Graphics (SVG) file and then uploaded to Target Manager. VuMarks were added to the corresponding databases that were configured online and then downloaded from Target Manager. They were added to the authors' Vuforia project using Unity, Android Studio, Xcode, and Visual Studio 2015. They were packaged with an executable file of the previously developed Surface Math AR application.

Students download the Surface Math AR application to their smartphone or tablet from the Play Market, and the educator gives the assignment. It is necessary to find the area of existence of the function and draw it. Students calculate this area and draw an image or drawing on paper. The educator gives out markers, and students turn on the application and point the camera at the marker. A 3D-image of the figure appears. This figure explores from the outside, as well as viewed from the inside, close and far. Students appreciate the beauty and accuracy of the graphic design of the model.

4 Results

Modern engineering education has opened inexhaustible possibilities for using various visualization tools, such as VR and AR, based on generating accurate 3D-models [30, 31]. In contrast to traditional educational and methodological materials, the main advantages of AR-technology are simplicity, interactivity, and high performance [32, 33]. AR can be fully implemented in the complete cycle of students' engineering training, starting with the creation of models of products [34–37]. The ability of the future engineer to create and recognize the graphical representation of the products is essential. Moreover, with the lack of classes, traditional methods of graphic training are unacceptable.

Given the foregoing, it becomes apparent that the AR is another inevitable step towards improving the quality of engineering education. The introduction of a new methodological approach to improve students' spatial representation or listeners of training programs, understanding the relationship between the 3D-model, its drawings, and personal views is relevant.

Mobile application "AR in Engineering Graphics" has been developed. It allows demonstrating the visualization of 2D-drawings in 3D-objects using AR-technology. This technology is implemented by installing the mobile application from Google Play (https://play.google.com/store/apps/details?id=arieg.unity3d.sumdu.edu.ua) a smartphone or tablet on the OS Android. When pointing the camera on a 2D-drawing, the corresponding 3D-model will be visualized. The rotation of the drawing on the desktop allows inspecting the 3D-model from all sides, analyze all design and technological features [38, 39].

This application allows you to work with drawings and appropriate static 3D-models of different configurations: flat parts (Fig. 2 a); prismatic parts (Fig. 2 b); rotational parts (Fig. 2 c). With the proposed solution, complex geometric objects become understandable and straightforward, even when working with complex parts (Fig. 2 d).

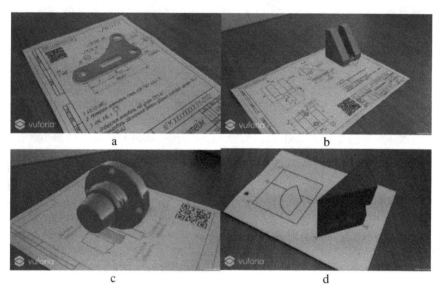

Fig. 2. Visualization of flat (a), prismatic (b), rotational (c), and complex (d) parts.

Using the developed application when working with assemblies allows showing the look and showing all the structural elements belonging to this product. Thus, the process of developing the technological scheme of assembly becomes simple and straightforward. The application's functionality allows us to present/hide items with the same design attributes, for example, show only standardized elements, or hide certain parts (Fig. 3).

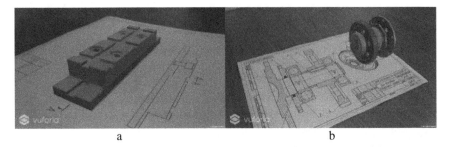

Fig. 3. Visualization of assemblies.

AR allows carrying out the structural elements of the assembly unit, and therefore, to demonstrate the process of assembly-disassembly (Fig. 4 a). If necessary, it's possible to make the unit's cross-section, which will be realized as a dynamic mode (Fig. 4 b).

Fig. 4. Visualization of the functions "assembly/disassembly" (a) and "cross-section" (b).

5 Conclusions

The mobile applications extend the functionality of engineering education. It makes the learning process more informative and exciting. The proposed and implemented approach is more productive than the use of traditional, static axonometric images. The appropriate form of education/training/retraining on the transfer of information in a virtual environment is useful in avoiding the difficulties of perception of graphical information by visualizing complex objects, contributing to a deeper understanding of technical objects. Also, the use of AR-technology can significantly reduce time costs.

Recent approaches, methods, and software based on AR-technology were studied, new AR-applications in analytical geometry, mathematical analysis, descriptive geometry, engineering graphics, and computer graphics were developed with the aim of the educational/retraining process of engineers.

Further research will be focused on the improvement of reliability of the developed solutions, expanding technological capabilities and functionality, integration with CAx-systems. AR-applications to the industry are the first-priority step, especially for production planning, assembly, quality control, etc.

References

1. Larkin, J.: Wayray calls for developers to compete in the first true AR SDK challenge and hackathon. Automot. Ind. AI **198**(2) (2018)
2. Antosz, K.: Maintenance – identification and analysis of the competency gap. Eksploatacja i Niezawodnosc – Maintenance Reliab. **20**(3), 484–494 (2018). https://doi.org/10.17531/ein.2018.3.19
3. Dynnyk, O., Denysenko, Y., Zaloga, V., Ivchenko, O., Yashyna, T.: Information support for the quality management system assessment of engineering enterprises. In: Ivanov, V., et al. (eds.) DSMIE 2019. LNME, pp. 65–74. Springer, Cham (2020). https://doi.org/10.1007/978-3-030-22365-6_7
4. Zheng, Y., Kuang, Y., Sugimoto, S.: Revisiting the PnP problem: a fast, general and optimal solution. In: IEEE International Conference on Computer Vision, ICCV 2013, pp. 2344–2351 (2013)
5. Xia, L., Ding, H., Lu, J., Zhang, H.: Development and application of virtual collaborative experiment technology based on Unity platform. In: Proceedings of 2018 IEEE International Conference of Safety Produce Informatization, IICSPI 2018, pp. 546–550 (2019). https://doi.org/10.1109/IICSPI.2018.8690340
6. Rana, K., Patel, B.: Augmented reality engine applications: a survey. In: Proceedings of the 2019 IEEE International Conference on Communication and Signal Processing, ICCSP 2019, pp. 380–384 (2019). https://doi.org/10.1109/ICCSP.2019.8697999
7. Rahman, H.R., Herumurti, D., Kuswardayan, I., Yuniarti, A., Khotimah, W.N., Fauzan, N.B.: Location based augmented reality game using KudanSDK. In: Proceedings of the 11th International Conference on Information and Communication Technology and System, ICTS 2017, pp. 307–310 (2018). https://doi.org/10.1109/ICTS.2017.8265689
8. Wong, P.K., Vong, C.M., Wong, K.I., Ma, Z.: Development of a wireless inspection and notification system with minimum monitoring hardware for real-time vehicle engine health inspection. Transp. Res. Part C: Emerg. Technol. **58**, 29–45 (2015). https://doi.org/10.1016/j.trc.2015.07.001
9. Babak, N.G., Kryukov, A.F.: Mobile application for visualization of the advertising booklet using augmented reality. In: 2018 4th International Conference on Information Technologies in Engineering Education, Inforino 2018 (2018). https://doi.org/10.1109/INFORINO.2018.8581841
10. Makino, R., Yamamoto, K.: Spatiotemporal information system using mixed reality for area-based learning and sightseeing. In: Geertman, S., Zhan, Q., Allan, A., Pettit, C. (eds.) CUPUM 2019. LNGC, pp. 283–302. Springer, Cham (2019). https://doi.org/10.1007/978-3-030-19424-6_16
11. Verner, I., Cuperman, D., Romm, T., Reitman, M., Chong, S.K., Gong, Z.: Intelligent robotics in high school: an educational paradigm for the industry 4.0 era. In: Auer, M.E., Tsiatsos, T. (eds.) ICL 2018. AISC, vol. 916, pp. 824–832. Springer, Cham (2020). https://doi.org/10.1007/978-3-030-11932-4_76
12. Židek, K., Lazorík, P., Piteľ, J., Pavlenko, I., Hošovský, A.: Automated training of convolutional networks by virtual 3D models for parts recognition in assembly process. In: Trojanowska, J., Ciszak, O., Machado, J.M., Pavlenko, I. (eds.) MANUFACTURING 2019. LNME, pp. 287–297. Springer, Cham (2019). https://doi.org/10.1007/978-3-030-18715-6_24
13. Bawa, M., Caganova, D., Szilva, I., Spirkova, D.: Importance of Internet of Things and big data in building smart city and what would be its challenges. In: Leon-Garcia, A., et al. (eds.) SmartCity 360 2015-2016. LNICSSITE, vol. 166, pp. 605–616. Springer, Cham (2016). https://doi.org/10.1007/978-3-319-33681-7_52

14. Caganova, D., Cambal, M., Luptakova, S.W.: Intercultural management - trend of contemporary globalized world. ElektronikaIrElektrotechnika **6**, 51–54 (2010)
15. Sobrino, D.R.D., Kos'al, P., Caganova, D., Cambal, M.: On the possibilities of intelligence implementation in manufacturing: the role of simulation. In: 3rd Central European Conference on Logistics, CECOL 2012, vol. 309, pp. 96–104 (2013). https://doi.org/10.4028/www.scientific.net/AMM.309.96
16. Varela, M.L.R., Putnik, G.D., Manupati, V.K., Rajyalakshmi, G., Trojanowska, J., Machado, J.: Collaborative manufacturing based on cloud, and on other I4.0 oriented principles and technologies: a systematic literature review and reflections. Manag. Prod. Eng. Rev. **9**(3), 90–99 (2018). https://doi.org/10.24425/119538
17. Pavlenko, I., Ivanov, V., Kuric, I., Gusak, O., Liaposhchenko, O.: Ensuring vibration reliability of turbopump units using artificial neural networks. In: Trojanowska, J., Ciszak, O., Machado, J.M., Pavlenko, I. (eds.) MANUFACTURING 2019. LNME, pp. 165–175. Springer, Cham (2019). https://doi.org/10.1007/978-3-030-18715-6_14
18. Pavlenko, I., Neamtu, C., Verbovyi, A., Pitel, J., Ivanov, V., Pop, G.: Using computer modeling and artificial neural networks for ensuring the vibration reliability of rotors. In: 2nd International Workshop on Computer Modeling and Intelligent Systems, CMIS 2019. CEUR Workshop Proceedings, vol. 2353, pp. 702–716 (2019)
19. Ivanov, V., Pavlenko, I., Liaposhchenko, O., Gusak, O., Pavlenko, V.: Determination of contact points between workpiece and fixture elements as a tool for augmented reality in fixture design. Wirel. Netw. **27**(3), 1657–1664 (2019). https://doi.org/10.1007/s11276-019-02026-2
20. Zidek, K., Lazorik, P., Pitel', J., Hosovsky, A.: An automated training of deep learning networks by 3D virtual models for object recognition. Symmetry **11**(4), Article number 496 (2019). https://doi.org/10.3390/sym11040496
21. Adamenko, D., Pluhnau, R., Nagarajah, A.: Case study of model-based definition and mixed reality implementation in product lifecycle. In: Ivanov, V., et al. (eds.) DSMIE 2019. LNME, pp. 3–12. Springer, Cham (2020). https://doi.org/10.1007/978-3-030-22365-6_1
22. Crisan, A.N., Pop, G.M.: Creative, quality oriented rethinking of the assessment strategy at the university level courses. a case study. In: Ivanov, V., et al. (eds.) DSMIE 2019. LNME, pp. 33–42. Springer, Cham (2020). https://doi.org/10.1007/978-3-030-22365-6_4
23. Kudrna, J., Miller, A., Edl, M.: Methods of industrial engineering used in network organizations. Paper Presented at the Creating Global Competitive Economies: A 360-Degree Approach. In: Proceedings of the 17th International Business Information Management Association Conference, IBIMA 2011, vol. 4, pp. 2037–2042 (2011)
24. Maydaniuk, V.P., Arseniuk, I.R., Lishchuk, O.O.: Increasing the speed of fractal image compression using two-dimensional approximating transformations. J. Eng. Sci. **6**(1), E16–E20 (2019). https://doi.org/10.21272/jes.2019.6(1).e3
25. Al Salaimeh, S.: Development of the computer graphics management system using text of natural language. J. Eng. Sci. **5**(2), E7–E9 (2018). https://doi.org/10.21272/jes.2018.5(2).e2
26. Kuric, I., Kandera, M., Klarák, J., Ivanov, V., Więcek, D.: Visual product inspection based on deep learning methods. In: Tonkonogyi, V., et al. (eds.) InterPartner 2019. LNME, pp. 148–156. Springer, Cham (2020). https://doi.org/10.1007/978-3-030-40724-7_15
27. Fesenko, A., Basova, Y., Ivanov, V., Ivanova, M., Yevsiukova, F., Gasanov, M.: Increasing of equipment efficiency by intensification of technological processes. Periodica Polytechnica Mech. Eng. **63**(1), 67–73 (2019). https://doi.org/10.3311/PPme.13198
28. Sokolov, V., Krol, O., Stepanova, O.: Nonlinear simulation of electrohydraulic drive for technological equipment. In: Journal of Physics: Conference Series, vol. 1278, no. 1, p. 012003 (2019). https://doi.org/10.1088/1742-6596/1278/1/012003
29. Adrianto, D., Hidajat, M., Yesmaya, V.: Augmented reality using Vuforia for marketing residence. In: 2016 1st International Conference on Game, Game Art, and Gamification, ICGGAG 2016 (2017). https://doi.org/10.1109/ICGGAG.2016.8052642

30. Blanco-Pons, S., Carrión-Ruiz, B., Lerma, J.L.: Augmented reality application assessment for disseminating rock art. Multimed. Tools Appl. **78**(8), 10265–10286 (2018). https://doi.org/10.1007/s11042-018-6609-x
31. Azkorreta, K.O., Rodríguez, H.O.: Augmented reality applications in the engineering environment. In: Zaphiris, P., Ioannou, A. (eds.) LCT 2014. LNCS, vol. 8524, pp. 83–90. Springer, Cham (2014). https://doi.org/10.1007/978-3-319-07485-6_9
32. Dangelmaier, W., Fischer, M., Gausemeier, J., Grafe, M., Matysczok, C., Mueck, B.: Virtual and augmented reality support for discrete manufacturing system simulation. Comput. Ind. **56**(4), 371–383 (2005). https://doi.org/10.1016/j.compind.2005.01.007
33. Martin-Gutierrez, J., Fernandez, M.D.M.: Applying augmented reality in engineering education to improve academic performance & student motivation. Int. J. Eng. Educ. **30**, 625–635 (2014)
34. Liarokapis, F., et al.: Web3D and augmented reality to support engineering education. World Trans. Eng. Technol. Educ. **3**(1), 11–14 (2004)
35. Chen, H., Feng, K., Mo, C., Cheng, S., Guo, Z., Huang, Y.: Application of augmented reality in engineering graphics education. In: 2011 International Symposium on IT in Medicine and Education, vol. 2, pp. 362–365 (2011). https://doi.org/10.1109/ITiME.2011.6132125
36. Veide, Z., Strozheva, V., Dobelis, M.: Application of augmented reality for teaching descriptive geometry and engineering graphics course to first-year students. In: Joint International Conference on Engineering Education and International Conference on Information Technology, pp. 158–164 (2014)
37. Parmar, D., Pelmahale, K., Kothwade, R., Badguajar, P.: Augmented reality system for engineering graphics. Int. J. Adv. Res. Comput. Commun. Eng. **4**(10), 327–330 (2015)
38. Bun, P., Trojanowska, J., Ivanov, V., Pavlenko, I.: The use of virtual reality training application to increase the effectiveness of workshops in the field of lean manufacturing. In: Bruzzone, A.G., Ginters, E., Mendivil, E.G., et al. (eds.) 4th International Conference of the Virtual and Augmented Reality in Education, VARE 2018, pp. 65–71 (2018)
39. Ivanov, V., Pavlenko, I., Trojanowska, J., Zuban, Y., Samokhvalov, D., Bun, P.: Using the augmented reality for training engineering students. In: Bruzzone, A.G., Ginters, E., Mendivil, E.G., et al. (eds.) 4th International Conference of the Virtual and Augmented Reality in Education, VARE 2018, pp. 57–64 (2018)

A Review in the Use of Artificial Intelligence in Textile Industry

Filipe Pereira[1(✉)], Vítor Carvalho[1,2], Rosa Vasconcelos[3], and Filomena Soares[1]

[1] Algoritmi R&D Centre, University of Minho, Guimarães, Portugal
vcarvalho@ipca.pt, fsoares@dei.uminho.pt
[2] 2Ai Lab, School of Technology, IPCA, Barcelos, Portugal
[3] 2C2T R&D Centre, University of Minho, Guimarães, Portugal
rosa@det.uminho.pt

Abstract. This paper presents an analysis of the state of the art of artificial intelligence applications in the textile industry. A review of the existing literature was performed. This article presents three methods of analyzing textile yarn. Some techniques, used in textile fabric inspection, are presented throughout this paper, as well as the use of artificial intelligence on improving the performance of productive systems using neural networks and artificial vision. The preliminary results demonstrate that the techniques covered are an asset in obtaining defects in textile fabrics at the industrial level. Taking into account the various methods of inspection and analysis of textile yarn, all present pros and cons in applicability in the textile area. In terms of advantages, all allow a better analysis of the textile yarn and defect detection with high quality, but with applicability in more complex systems. As a disadvantage, they present the fact that they do not have an already standardized algorithm that can be used, which makes its use more complex. Some possible future applications are also described.

Keywords: Artificial intelligence (AI) · Textile industry · Industry 4.0 · Yarn evaluation · Machine learning · Image processing

1 Introduction

The textile sector is one of the most important for the quality of life of the entire world population. Clothing is part of one of the basic needs of the human being.

Among the strategies used to stay competitive in the market, one of the most essential for the textile industry is innovation. New practices, use of computer intelligence and development of sustainable materials are increasingly present in the universe of fashion, sport, uniforms and objects. The technological evolution of fabrics brings important functionalities and can be a differentiating factor in this type of industry.

The manufacture of fabrics and clothing has gone through an update process that goes from choosing the raw material to the production methods. Low environmental impact solutions have been the subject of ongoing research and investment in new production machines, energy and methods.

One of the applications that has been widely used at an industrial level is the use of vision systems, which allows the analysis and recognition of patterns in fabrics in the textile industry. One of the companies that uses this technology is COGNEX. For this, it uses a platform called Cognex ViDi that analyzes patterns in textile fabrics, such as finishes. The main objective is to integrate this technology in an industrial textile system in order to obtain predefined images of fabric samples, to carry out two analyzes. The first analysis will be more preliminary to detect more visible defects and, at a later stage, a more detailed analysis of defects that are not visible to the naked eye (Fig. 1) [1].

Fig. 1. Yarn dye plaid image capture system [1]

This renewal meets new market needs in order to add value to the product while rethinking costs. Some biological processes applied in the dyeing phase, for example, can reduce, on average, 30% in water and electricity consumption [2].

The outline of this paper is as follows. In Sect. 2, Artificial intelligence (AI) in textile industry is presented. In Sect. 3, it is described a yarn analysis solution using AI. Section 4 presents the description of the AI used techniques and in Sect. 5 there are presented some future work proposals based on existing identified gaps and challenges. Finally, in Sect. 6, final remarks are enunciated.

2 Artificial Intelligence in Textile Industry

Artificial Intelligence (AI) is a branch of computer science dedicated to the study of computational activities that requires intelligence when performed by humans. "Intelligence" is the computational part of the ability to achieve goals [3].

The main purpose of AI systems is to perform functions that, if supposed to be performed by a human being, would be considered intelligent. It is a very broad concept, and one that receives so many definitions when we give the word intelligence different meanings [3].

Machine Learning is a subfield of Artificial Intelligence (AI) and its main approach is to learn through experience to find patterns in a data set. It involves teaching a computer, through examples, to recognize patterns instead of programming it with specific rules [4].

In Machine Learning there are three main classifications with respect to the algorithms used and the type of learning of the same, namely:

- Supervised Learning
- Unsupervised Learning
- Reinforcement Learning

The choice of which of the three methods will be used will depend on the data set you have and the problem you want to solve.

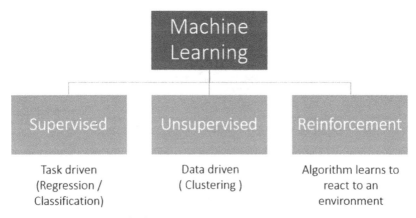

Fig. 2. Types of machine learning [4]

Most machine learning practices use supervised learning. In supervised learning, a set of data is provided and we know what the correct result should be, as there is a relationship between the input and the output, that is, it has the inputs (X) and the outputs (Y) and uses the algorithm to map the learning function that will image the input (X) to the output (Y) (Y = f (X)) [4].

The goal is that, when you have a new input (X), you can predict what the output (Y) is, based on the model produced containing the rules learned with the learning function [4].

In Unsupervised Learning, there is only the input (X) of data and not the corresponding output (Y). The objective in this case is to model the underlying structure or distribution of the data to learn more about it. Unsupervised Learning is called that, because there is no feedback based on the prediction results as in the previous case, in fact there is no way out and there is no teacher. The algorithms are left by their own mechanisms to discover and present the data structure, this allows us to approach problems with little or no idea about the results.

Reinforcement Learning is a trial-and-error method. Their approach is similar to that of a baby who is learning new things. The machine knows nothing at first. She will then try something and, depending on the results, decide whether it was a good decision or

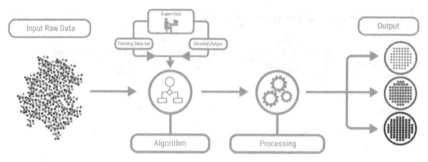

Fig. 3. Supervised learning [4]

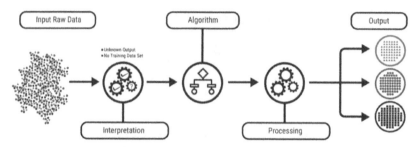

Fig. 4. Unsupervised learning [4]

a bad decision. After countless attempts, if the model is well learned, the machine will become better and better and, eventually, it will only make good decisions. The difficulty lies in the fact that the actions taken by the agent can change the environment, and the agent can obtain his reward long after his action. So, it becomes difficult to know which actions led to which reward.

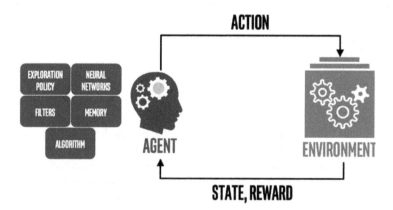

Fig. 5. Reinforcement learning [4]

Several sectors of activity used AI including robotics, artificial vision, comprehension, artificial neural networks and a branch of AI denominated "expert systems". AI has been particularly useful in solving a number of problems in different areas of the textile industry [4]. Expert Systems, Fuzzy Logic and Artificial neural networks will be covered in the following subsections.

2.1 Fuzzy Logic

One of the techniques used in textile industry is the neuro-fuzzy. The neuro-fuzzy hybrid system is a combination of fuzzy and neural network techniques. A neuro-fuzzy system (SNF) is a type of embedded hybrid system consisting of the combination of two well-known modeling techniques such as ANN (Artificial Neural Networks) and FL (Fuzzy Logic). At present, SNFs are of great interest as they bring benefits from both ANN and FL systems, thus eliminating individual disadvantages by combining common features. In addition, different SNF architectures have been researched in various application areas, especially in industrial process control [5–7].

Fan et al. (2001) have employed an adaptive neuro-fuzzy system for the prediction of cotton yarn strength from HVI (High Volume Instrument) fiber properties garment drape and forecasting circular knitting machine parameters. However, the neuro-fuzzy systems are yet to be applied in the field of yarn property prediction [8].

2.2 Artificial Neural Networks

Artificial neural networks are created from algorithms designed for a particular purpose and basically resembles the human brain at two points: knowledge is obtained through learning steps, and synaptic weights are used to store knowledge.

Synapse is the name given to the connection between neurons and the connections are assigned values, which are called synaptic weights. Thus, artificial neural networks have in its constitution a series of artificial (or virtual) neurons that will be connected to each other, forming a network of processing elements [9].

Defects in yarn reduce the value of textile products. In order to try to eliminate this problem, Artificial Intelligence techniques, such as ANN, are applied to identify defects in textile inspection of fabrics. This principle is based on obtaining images that are analyzed in the image acquisition system and saved in the relevant standard format (JPEG, PNG, among others.). Resources are extracted from the acquired image and the image analysis method is used to reduce the feature set dimensionality by creating a new smaller size feature set, which is a combination of older features. [9].

2.3 Expert Systems

This is a technology derived from the family of artificial intelligence, being one of the main practical applications based on knowledge. The specialists seek to apply the knowledge of an expert in the systems and generate actions inherent to the human being to bring innovative solutions and thereby advance in the field or area of application. An expert system is a computational system that tries to imitate human specialists, applying methodologies of reasoning or knowledge about a specific area [10].

The components that make up expert systems are:

- Knowledge base: rules, facts and heuristics corresponding to the knowledge of specialists in the domain on which the system was built;
- Development team: composed of one or more experts who are in charge of the knowledge base, who translate the knowledge described by the experts into a set of production standards;
- Development environment: also known as AI shells, they are user-friendly development environments, which quickly and efficiently generate user interface screens, capture the knowledge base and manage search strategies in the standards base (Fig. 6).

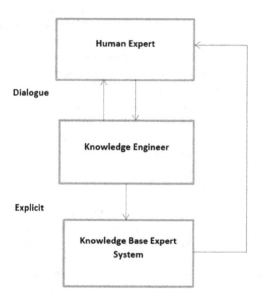

Fig. 6. Expert system concept [10]

One of the applications of the Systems of specialists in the textile industry, is the verification of the coloring in textiles in terms of finish. The work carried out by the systems of specialists in production management is of fabric machines, aided by computer, is evaluated elsewhere.

One of the applications of the expert systems was in the company Sandoz AZ. This company developed an expert system called Wooly. This has been applied in textile dyeing, where it can predict the performance of solidity in a wide range of tests on patterns and colors in textile fabrics and provides corrections to the fabric such as, better dyeing methods, suitable dyes and is able to interact with a color matching system.

In the next chapter, one of the techniques used by expert systems, in color detection, which is the Kubelka-Munk theory, is discussed.

3 Yarn Analysis Solutions Using AI

3.1 Yarn Identification Using Image Processing

This section describes the various methods that exist in the analysis of textile yarn, using artificial intelligence. Three methods using artificial intelligence are described, the first using image processing, the second method identifies patterns using artificial neural networks and the third method verifies the correspondence of textile colors using ANN and the Kubelka-Munk theory.

In the textile industry, there are automatic and manual inspection systems, in which these are the most used in determining the quality of the textile yarn, such as uniformity and impurities. The manual inspection method has several disadvantages, such as solving time-consuming and complex problems. This type of inspection will cause tiredness, lack of attention and errors on the part of the operator, which would cause the unreliability of the products obtained and also the inaccuracy of the results obtained.

The automatic textile thread inspection systems are a very effective method in the continuous improvement of the quality of the textiles, when inserting the image processing and analysis technique. This allowed for greater efficiency in obtaining defects and in the continuous improvement of the industrial process.

The image processing technique allows to transform the data of an image and convert it to mathematical values. This technological solution is very reliable and very efficient. Comparatively with existing solutions, such as human visual analysis, this manages to eliminate all the disadvantages of traditional techniques.

There are more and more studies and research in the identification of defects in textile yarn through image processing that have given very promising results.

There is an ongoing investigation, in which a system was developed that extracts neps images (highly tangled cotton knots) [11] and then classifies it, according to a probabilistic neural network (PNN).

When this test was performed in this investigation, a k-fold cross-validation methodology was used, which is a technique to assess the generalizability of a model, based on a data set. Thus, it was possible to measure what would be the performance of the PNN classifier.

The results obtained through this technique demonstrated that the neps classification, which was carried out through image recognition by the PNN classifier, were very good and validate the technique used in this analysis [12].

Vítor Carvalho et al. (2009) presented a paper on how to apply a textile thread analysis tool using AI and image processing techniques, in order to obtain textile yarn characteristics and prediction models [12].

3.2 Pattern Inspection Using ANN

The importance of product quality inspection is important in companies, as quality defects can significantly reduce the process (for example, in the textile industry, defects can reach 60%). In textile production, online inspection is a slow process due to the slower movement of the fabric out of the weaving machine, making human inspection an inexpensive option [13].

The patterns of the fabrics can have several aspects, such as fabric, knitting, braids, finishes and prints. Changing the visual inspection at the factory level by inspection based on vision systems, can help manufacturers in the textile industry avoid fatigue and human errors in defect detection. Typically, camera-based inspection systems are installed that feature immense defect analysis images with "OK" and "NOT OK" final results [13].

Vision systems to analyze patterns in fabrics, first "learn" the pattern of the fabric, the properties of the textile yarn, the colors and tolerable imperfections from these images. This "teaching" time can be a few weeks, after which the platform can potentially begin to detect defects (such as incorrect knitting patterns) in the final textile fabric, saving the human effort of evaluating hundreds of meters of material. Several challenges are inherent to the inspection of textile fabric patterns, such as its shape, complexity, variability and the large number of types of knitwear [13].

Ajay Kumar et al., (2003) used a feed-forward neural network to segment defects in pattern textile fabric. According to him, "every defect in the textile mesh changes the gray level arrangement of neighboring pixels, and this change is used to segment the defects" [13].

Each pixel acquired from an image is characterized in an information vector. A simplified characterization of the texture of the textile fabric is used based on the connections between the pixels in the neighborhood [13].

The gray-level values of neighboring pixels form the resource vector for each pixel. This information vectors captures certain local textural properties of the fabric, such as thickness, directionality, regularity, patterns, colors, etc. [13].

3.3 Textile Color Matching Using ANN and Kubelka-Munk Theory

The color of a textile fabric can be considered "acceptable" or "not acceptable" or else "very light", "very red" or "very blue". Such analyzes can be done visually or using equipment, based on the difference between an ideal pattern of textile fabric or a sample [14].

Color tolerance is based on numerical color descriptions through "instrumental tolerance systems". As this method presented very wrong results, compared to the visual checks of the fabric, an AI-based system was developed, presenting results of the Pass/Fail (P/F) type in order to improve the precision and efficiency of the instrumental tolerance and errors committed by the analysis by the human eye [14].

To solve this problem of the textile industry related to color patterns, several computer-based methods have been presented in the last few years, using artificial neural networks and Kubelka-Munk (K-M) theory [14].

The color of an opaque material is a function of its absorption and dispersion properties of visible light. While the dispersion changes the direction of the light rays, the absorption converts them into heat [14].

Rocco Furferi et al. (2016) presented a method for accurate estimation of the spectrophotometric response of a textile mixture composed of different colored fibers made of different materials. In particular, the performance of the K-M theory is improved with the introduction of artificial intelligence to determine a more consistent value for the nonlinear function relationship between the mixture and its components. Therefore, a hybrid system composed of K-M + ANN, was developed with the purpose of modeling the color mixing mechanism in order to predict the reflectance values of a mixture [14].

4 Description of the Used AI Techniques

This section describes in a more technical way the systems most used in the analysis of textile yarn. In the first system, the YSQ (Yarn System Quality) project that uses image processing to check the quality of the textile yarn was approached. In the second system, the technique on how to carry out the analysis of fibers and spinning using artificial neural networks was discussed. In the third system, an analysis of how to perform color matching in the textile industry was carried out.

4.1 Yarn Analysis Tool Using Artificial Intelligence and Image Processing

In the last ten years there has been a very significant advance in terms of commercial solutions in obtaining the characteristics of the textile yarn. In this area, USTER Technologies AG has some machines that provide some characteristics but do not obtain all the characteristics of the textile yarn, especially if it is in continuous movement.

There is a prototype, designated as YSQ [15] that obtained some preliminary algorithms in order to obtain a significant amount and variety of characteristics of a textile yarn.

The YSQ prototype [15] consists of an image acquisition and processing system using a microscope attached to a USB camera, which can be subdivided into hardware and software components. At the software level, National Instruments' IMAQ Vision is used to perform image acquisition and processing.

The hardware of this system consists of a USB camera that is in the prototype next to the output plane of an analog microscope. In order to obtain the images, any image processing system needs a light source. In this prototype, a monochromatic light source was used, in order to obtain better contrasts for the relief of the wire geometry (Fig. 1).

This system allows to obtain the following characteristics in textile yarn, namely: the orientation and twist of the fibers, the twisting step and the orientation in folded yarns and the number of cables, spun yarns (single cable) or folded yarn (multiple cables) [15].

4.2 Artificial Neural Network in Fibers and Spinning Analysis

Artificial neural networks are increasingly used to analyze problems in the textile industry. Its use in the textile industry involves the integration of a large number of variables. Due to the high degree of variability of the raw materials, the processing in several stages and the lack of precision in the control of the process parameters, the analysis of the relationship between these variables and the product properties depends on human knowledge [16].

The ability to predict these properties accurately has become a challenge due to the fact that textile materials have a non-linear behavior. As there is a need to forecast some properties or performances of an industrial process in advance, to minimize the cost of installation and time, it becomes important to use the functions performed with

Fig. 7. Yarn image capture system [15]

Artificial Neural Networks, since it does not have a constant understanding, but it can be changed dynamically [17]. An ANN is normally used to predict the composition of the copolymer.

A copolymer is a polymer formed when two (or more) different types of monomers are linked in the same polymeric chain, as opposed to a homopolymer where only one monomer is used. Copolymer refers to a type of polymer that contains two or more distinct repeating units called "monomers". It produces substances of high molecular weight by chemical combination or polymerization of monomers. The final results that an ANN can indicate related to the composition of the copolymer correctly, will depend on the reaction conditions and conversions. Figure 2 shows an intelligent fiber classification system. In this case, ANN is used to identify two types of animal fibers, Merino and Mohair. This ANN model extracts six parameter scales with image processing and the rest using an unsupervised neural network to extract resources automatically, which are determined according to the complexity and accuracy of the obtained model. It can obtain greater precision to the increasing number of colors for learning, since the precision of the ANN depends largely on the selected parameters, this type of algorithm can be used to develop models for previewing and to generate a cotton classification [17].

In [17] it is analyzed the set of elastic properties of the fiber yarn properties. Figure 3 shows an artificial model of neural network for textile thread engineering, developing a "reverse" model of yarn for textile fiber using an ANN.

This one model is totally different from future models, which predict the properties of the final yarn using the properties of the fiber as input elements. The cost reduction of cotton fiber was ensured with the use of linear programming in conjunction with an ANN algorithm (Fig. 4). [17] (Fig. 9).

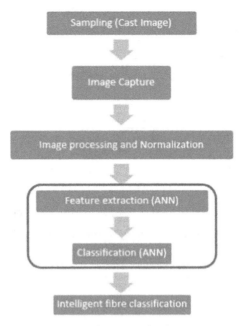

Fig. 8. Intelligent fibre classification system [17]

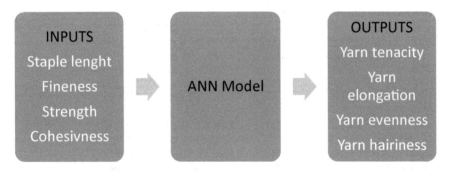

Fig. 9. ANN model for yarn engineering [17]

The Uster fabric scan machine system uses this ANN in the analysis of wrinkled textile fabrics using image processing analysis. Figure 7 shows the artificial neural network model for textile engineering as well as the USTER machine (Fig. 10).

4.3 Artificial Neural Network and Kubelka-Munk (K-M) in Color Matching Analysis

Rocco Furferi et al. (2016), refers that after a mixture of colors is created, an instrumental comparison is always performed against a defined pattern, with the aim of determining the difference between them.

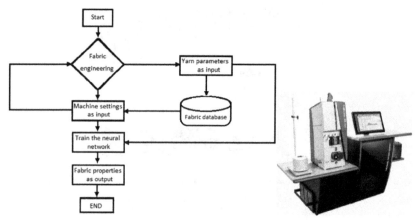

Fig. 10. Artificial neural network for textile engineering (left) and USTER machine (right) [17]

The Kubelka-Munk functions is:

$$F(R) = (1-R)^2/2R = k/s = Ac/s \qquad (1)$$

Where:

R = reflectance; k = absorption coefficient; s = scattering coefficient; c = concentration of the absorbing species; A = absorbance [18].

The comparison is evaluated in various color spaces such as CIELAB under a number of standard illuminants, as shown in Fig. 5 [18].

The hybrid method K-M + ANN used is done in three steps, namely:

1. Definition of the equivalent tissue substrate using the simplified model described in [19].
2. Teach an ANN to infer variations of the recipe in the K-S ratio of the substrate; this proportion is then used to predict the spectral response of the tissue mixture. The sum of steps 1 and 2 will result in a hybrid K-M + ANN system [19].
3. Derive (from the K-M + ANN method) a practical graph that can roughly replace the ANN software in everyday use [19] (Fig. 11).

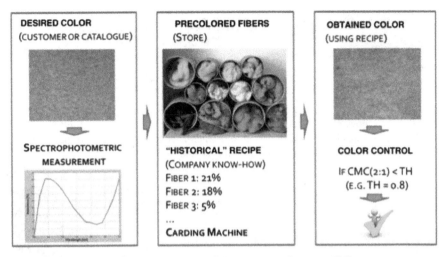

Fig. 11. Recipe-based mixing conceptual process [20].

5 Future Work Proposals Based on Existing Gaps and Challenges Identified

The commercial use of AI in the analysis of pre-production textile processing appears to be limited to only a few applications today, particularly in the identification and classification of textile fibers and yarns.

The identification and classification of the fiber in terms of color, length, uniformity, among others, may be analyzed with the use of AI and with the possibility that it will be more used and developed in the coming years.

The experience of employees who have acquired a lot of know-how over several years cannot be easily automated. This knowledge is usually lost, unless measures are taken to store knowledge and experience in an appropriate mechanism that allows for easy retrieval and later use.

One way to get around this negative aspect of expert systems is to allow a combination of these in domains familiar with programming languages and available shells, in order to build their own expert systems. For that they should need to ally specialists in artificial intelligence.

On the textile production side, AI can be used, for example, to detect visual defects in shirts or collars, or it can be applied to automatically detect and measure wrinkles in the fabric (Fig. 12 and 13). The measurement of the roughness of the textile fabric, based on an artificial vein, can help the textile industries to reduce costs and productivity time in the industrial manufacturing process [21].

Another application and a challenge for the future is the use of artificial intelligence in the design of the textile industry '3D models' of yarn fibers in their designs and prototypes, that can model the properties of yarns and fibers in an automatic and realistic way, without much effort and intervention by the human being [22].

For the future, human 3D modeling experts would have to create a yarn from individual 'virtual fibers' and the objective would also be to use a CT (Computed Tomography)

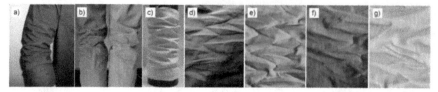

Fig. 12. Comparison of wrinkles in actual wear and those generated by wrinkle simulator; a) wrinkles at the elbow, b) wrinkles at the knee, c) wrinkles in the simulator, d) elbow wrinkles when the coat is spread, e) wrinkles of fabric 3 when it is spread, f) knee wrinkles when the trousers are spread, g) wrinkles of fabric when it is spread [21].

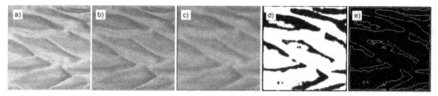

Fig. 13. Image process of wrinkled fabric; a) coloured image, b) grey level image, c) median filtered image, d) binary image, e) edge detected image [22].

scanner, using an AI algorithm in order to convert the data obtained by the CT into a 3D fiber model, as shown in Fig. 8 [23] (Fig. 14).

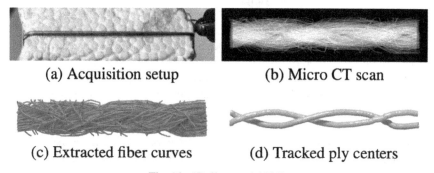

Fig. 14. 3D fiber model [24].

6 Final Remarks

The implementation of AI in the textile sector is currently relatively low compared to other industrial sectors. Currently, most AI applications in textiles involve the use of artificial vision systems on machines to replace humans. These allow the analysis and detection of errors that, even in some cases, is carried out by humans in textile yarn samples.

Currently, the technology presented in this article leaves many promising aspects and allows analyzing the various research and development activities that are being carried out in universities and other institutions.

AI applications in the textile sector are not yet fully in practice. This is due to the simple fact that cutting-edge AI manufacturing applications are more likely to reach larger and more modern sectors of activity.

In conclusion, the use of AI in the textile industry begins to be widely used and in the coming years it will play a major role in the production and final quality of products, taking into account technological advances and market competitiveness.

The implementation of a new project aims to develop a new technological solution, based on image processing and artificial intelligence techniques, for automatic characterization of the main parameters of a textile yarn, with the highest level of parameterization in relation to commercial solutions. These technologies associated with a new development of a solution to be used "online" at an industrial level in the textile area, will contribute to increase the efficiency of the textile industry and the quality of the final products.

Acknowledgements. This work has been supported by FCT – Fundação para a Ciência e Tecnologia within the R&D Units Project Scope: UIDP/04077/2020 and UIDB/04077/2020.

References

1. Textile inspection - industry overview. COGNEX (2019). Accessed 30 Jan 2020
2. Bullon, J., et al.: Manufacturing processes in the textile industry. Expert systems for fabrics production. In: ADCAIJ: Advances in Distributed Computing and Artificial Intelligence Journal, Salamanca, vol. 6, no. 4, pp. 15–23 (2017). ISSN 2255-2863. Accessed 08 Nov 2019
3. Kuo, J., Lee, C.-J., Tsai, C.-C.: Using a neural network to identify fabric defects in dynamic cloth inspection. Text. Res. J. – TEXT. RES. J. **73**, 238–244 (2003)
4. Jeyaraj, P., Nadar, E.R.S.: Computer vision for automatic detection and classification of fabric defect employing deep learning algorithm. Int. J. Cloth. Sci. Technol. (2019)
5. Giarratano, J., Riley, G.: Expert Systems: Principles and Programming, 3rd edn. PWS Publishing, USA (1998)
6. Ahlawat, N., Gautam, A., Sharma, N.: Use of logic gates to make edge avoider robot. Int. J. Inf. Comput. Technol. **4**(6), 630 (2014). ISSN 0974-2239
7. Leão, C.P., et al.: Web-assisted laboratory for control education: remote and virtual environments. In: Uckelmann, D., Scholz-Reiter, B., Rügge, I., Hong, B., Rizzi, A. (eds.) ImViReLL 2012. CCIS, vol. 282, pp. 62–72. Springer, Heidelberg (2012). https://doi.org/10.1007/978-3-642-28816-6_7
8. Veit, D.: Fuzzy logic and its application to textile technology. In Simulation in Textile Technology: Theory and Applications, pp. 112–141 (2012). https://doi.org/10.1533/9780857097088.112
9. Majumdar, A., Majumdar, P.K., Sarkar, B.: Application of an adaptive neuro-fuzzy system for the prediction of cotton yarn strength from HVI fiber properties. J. Text. Inst. **96**(1), 55–60 (2005)
10. Chauhan, N., Yadav, N., Arya, N.: Applications of artificial neural network in textiles. Int. J. Curr. Microbiol. Appl. Sci. **7**(04), 3134–3143 (2018). https://doi.org/10.20546/ijcmas.2018.704.356

11. Shamey, R., Shim, W.S., Joines, J.: Development and application of expert systems in the textile industry (2010)
12. Ghosh, A., Hasnat, A., Halder, S., Das, S.: A proposed system for cotton yarn defects classification using probabilistic neural network. In: Recent Advances and Innovations in Engineering (ICRAIE) (2014)
13. Carvalho, V., Cardoso, P., Belsley, M., Vasconcelos, R., Soares, F.O.: Yarn hairiness characterization using two orthogonal directions. IEEE Trans. Instrum. Meas. **58**(3), 594–601 (2009)
14. Chattopadhyay, R., Guha, A.: Artificial neural networks: applications to textiles. Text. Prog. **35**(1), 1–46 (2004)
15. Shamey, R., Hussain, T.: Artificial intelligence in the colour and textile industry. Rev. Progr. Colorat. **33**, 33–45 (2003)
16. Carvalho, V., Soares, F., Vasconcelos, R.: Artificial intelligence and image processing-based techniques: a tool for yarns parameterization and fabrics prediction, pp.1–4 (2009). https://doi.org/10.1109/ETFA.2009.5347255
17. Zhang, Y., Lu, Z., Li, J.: Fabric defect classification using radial basis function network. Pattern Recogn. Lett. **31**, 2033–2042 (2010). https://doi.org/10.1016/j.patrec.2010.05.030
18. Kumar, A.: Neural network-based detection of local textile defects. Pattern Recogn. **36**, 1645–1659 (2003)
19. Furferi, R., Governi, L., Volpe, Y.: Color matching of fabric blends: hybrid Kubelka-Munk+ artificial neural network-based method. J. Electron. Imag. **25**(6), 061402 (2016)
20. Furferi, R., Governi, L.: Prediction of the spectrophotometric response of a carded fiber composed by different kinds of coloured raw materials: an artificial neural network-based approach. Color Res. Appl. **36**(3), 179–191 (2011)
21. Islam, A., Akhter, S., Mursalin, T.: Automated textile defect recognition system using computer vision and artificial neural networks (2004)
22. Aspland, R., Shanbhag, P.: Comparison of color difference equations for textiles: CMC (2:1) and CIEDE2000. AATCC Rev. **4**(6), 26–30 (2004)
23. Liu, C.: New method of fabric wrinkle measurement based on image processing. Fibres Text. Eastern Eur. **103**, 51–55 (2014)
24. Zhao, S., Luan, F., Bala, K.: Fitting procedural yarn models for realistic cloth rendering. ACM Trans. Graph. **35**(4), 11 (2016). Article 51

HiZeca: A Serious Game for Emotions Recognition

Pedro Santos[1], Vinícius Silva[2], João Sena Esteves[3(✉)], Ana Paula Pereira[4], and Filomena Soares[3]

[1] Department of Industrial Electronics, School of Engineering,
University of Minho, Guimarães, Portugal
a74504@alunos.uminho.pt
[2] R&D Centre Algoritmi, School of Engineering, University of Minho, Guimarães, Portugal
a65312@alunos.uminho.pt
[3] R&D Centre Algoritmi, Department of Industrial Electronics, School of Engineering,
University of Minho, Guimarães, Portugal
{sena,fsoares}@dei.uminho.pt
[4] CIEd, University of Minho, Braga, Portugal
appereira@ie.uminho.pt

Abstract. To comprehend human behavior can be a very difficult task for children with Autism Spectrum Disorder (ASD). These children have difficulties in social interaction, and they manifest repetitive patterns. Furthermore, they present deficits in imitation which can be directly linked to impairments in social interaction skills. Taking this into account, this paper presents the serious game *HiZeca,* in which a virtual agent (ZECA *Avatar*) is able to interact with a child, in order to promote social interaction and training certain facial movements that will be validated, and that will facilitate imitation and recognition of emotions (content, sad, surprised, among others). In order to validate the system, tests were conducted with typically developing children and children with ASD. The results show that, in general, the game was accepted with a positive feedback from the children.

Keywords: Serious game · Facial expressions · Emotions · Human-computer interaction · Autism Spectrum Disorder

1 Introduction

The emotional state is mostly defined by the expression of different non-behavioural cues, such as facial prompts. The learning of such facial cues starts from early age when pupils mirror the facial movements of others [1]. Learning by imitation is fundamental to the development of cognitive and social communication behaviours. Furthermore, it directly influences the cognitive empathy that allows a human being to logically comprehend the emotional states of the others, being paramount for successful social interactions. However, children with Autism Spectrum Disorder (ASD) present deficits in imitation which can be directly linked to impairments in social interaction skills [2].

In order to mitigate these impairments, several works in the literature focus on the use of technological tools such as robots with different configurations (from animal-like to humanoid designs) or virtual agents. The humanoid design has been recently more employed by the researchers in different projects in order to promote social interaction with children with ASD. The humanoid design provides a more realistic approach to explore several facets of emotional skills. One example is project ZECA [3] (Portuguese acronym for *Zeno Engaging Children with Autism*), that uses a humanoid robot developed by *Hanson Robotics*. This robot allows to promote social interaction with the children providing the communication and the development of capacities of understanding the emotions. Another more recent example is the *QTrobot* [4] (launched in 2018), created by *LuxAI* to assist children in interpreting facial expressions and promoting social interaction.

However, the technological robots used to support children with ASD are usually expensive and making them widely available in intervention settings may be impractical. A less expensive alternative is using software applications, which design is an area that has been increasing. These applications have several concepts underlined, such as the association of words with images to teach emotions, used by *Sono Flex* [5], or simply allowing the user to choose an item and producing the audio or video that is linked with that item, used by *Livox* [6]. A more advanced concept is proposed by *TippyTalk* [7], which enables an individual with verbal disabilities to communicate by translating pictures into text messages, allowing a child to communicate and express a desire, need, or feeling. Studies conducted with children [8] showed that significant progress has been made in their behaviour regarding relationships with colleagues and teachers, for example.

As addressed in [1], facial expressions play an important role during a social interaction and children with ASD lack the skills to perceive them. For this reason, the development of applications that are capable of aiding in the learning of emotions and the understanding of states of mind becomes a relevant solution. Furthermore, the design of applications with virtual agents that can infer the user non-behavioural cues may offer a unique opportunity to explore and promote the development of social skills in a wider scale.

This paper presents a serious game called *HiZeca,* used to aid in the learning of emotions and to promote social interaction with children with ASD. The game aims at offering a virtual alternative to the physical robot ZECA for interacting with children with ASD, at a much lower cost. It can be used anywhere with a general purpose computer. The main character of the game is ZECA *Avatar*, a virtual agent that a child interacts with using his facial expressions to imitate the avatar's face. In turn, the avatar mimics the child's expressions. Thus, the child's face will be monitored – using a tool previously presented in [9] – in order to verify the execution of his facial movements and to support their improvement, if they are not executed correctly. ZECA *Avatar* will encourage the improvement of possible incorrect movements. The game has three levels of complexity. It explores and fosters the child's ability to mimic, recognize, and infer the emotional states of ZECA *Avatar*.

This paper is organized in four sections. Section 2 presents the proposed serious game; Sect. 3 shows the preliminary results obtained so far; and the final remarks and future work are addressed in Sect. 4.

2 The Serious Game *HiZeca*

The serious game *HiZeca* is devoted to mimicking facial expressions and emotion recognition. The challenges that are presented to the user will allow to understand the type of emotion and interpret emotional states. For this, the user's face will be monitored in order to verify the execution of the facial movements.

HiZeca has three levels with increased difficulty and running in three different scenarios: Movement Training, Expression Training and Emotion Identification (Fig. 1).

Fig. 1. Game levels

In the first scenario, Movement Training, the purpose is to train some facial movements that will help the user in the execution of emotional expressions. This level also serves to familiarize the user to the main character. The movements made by the user are replicated by ZECA Avatar in an imitation context by model [10]. In this case, the user serves as a model for the avatar to copy. For this, the user is asked to perform a set of basic movements (Fig. 2).

Fig. 2. Level 1 - movement training (in Portuguese).

In the second scenario, Expression Training, various facial expressions associated with five emotions are trained: happiness, sadness, anger, fear and surprise. The user is asked to execute some expressions where the avatar serves as a model. As the user successfully imitates the expression, the avatar moves away becoming smaller, until it disappears remaining only the label of the emotion (Fig. 3). A control bar has been placed strategically so that the user can control the intensity of his/her expression. This scene is an example of learning by deferred imitation [11].

Fig. 3. Level 2 - expression training (in Portuguese).

In the third level, Emotion Identification, the story mode was implemented where the child is invited to listen to a set of 15 different stories and identify the associated emotional state corresponding to one of the previously trained. Figure 4 presents the menu for the emotion selection. In case the answer given by the user is incorrect, the number of lives (options) is decreased and the answer button disappears.

Fig. 4. Level 3 - emotion identification: menu emotion selection.

3 Results

Laboratory tests involving adults were performed in order to test the robustness of the game and validate the implemented methods. Tests were also performed in a school environment, both with typically developed children and children with ASD, in order to verify the acceptance of the tool by children between 6 and 8 years old. Each child was seated strategically, with the face at the level of the camera or webcam.

3.1 Typically Developed Children

The values shown in Table 1 are related to the tests performed for level 1 with three typically developed children. A small flaw in the validation of the mouth opening movement was detected. Only one timeout was recorded, which resulted in a detection rate of 66.67%. For the smile movement, two timeouts were recorded, resulting in a lower validation rate, with a value of 33.33%. For the movement of raising the eyebrows, it was also verified that the validation limits should be high, leading to a timeout for this movement. The movement with the highest average response time was to frown. This movement has a 100% validation rate, but it has a high average time due to the difficulty that the children encountered in executing the movement. As expected, it was also in this movement that a maximum response time was recorded: 39 s. The computation of the average times shown in Table 1 did not include the timeouts that occurred. The small number of children being tested implies that the fluctuation in the values corresponding to the responses of each child greatly affects mean values and validation rates.

Table 2 shows the values related to the tests performed for level 2 with three typically developed children. Each emotion was tested five times by each of the three children, resulting in a total of 15 validations per emotion. In this second level, the happiness and fear emotions were those that registered a maximum validation rate: 100%. For these emotions the average response times were exactly the same: 6.8 s. The value recorded for the emotion of disgust was the lowest of all: 80%. This is an emotion that, although usual, is not one of the most present in everyday life, hence the difficulty in expressing it. Regarding the average times, the lowest value was registered for the emotion of surprise: 6.56 s. On the other hand, the highest value was registered for the emotion of disgust: 8.08 s.

Table 3 shows the values related to the tests performed for level 3 (story mode) with three typically developed children. In this third level, some wrong answers were recorded in some stories, as well as some repetitions, which happened when the children felt the need to listen again and see the story to give their final answer. For stories 4 and 11, two wrong answers were recorded, this being the largest number of wrong answers in all tests. The story with the highest average response time was story 3, related to angry emotion, with a value of 13.67 s. The maximum response time was recorded for story 10. At this level, the children's reasoning ability to quickly associate emotions with the story in question was noted. Of the three levels tested, the third and last level was the one that showed the best performance and acceptance results by these typically developed children.

3.2 Children with ASD

Tests were performed with two children with ASD, who have very different characteristics (named child A and child B).

Child A has a high degree of autonomy. He can speak fluently and read, which promotes the interaction because, at all levels, in addition to sound there are written elements.

Table 1. Summary table with average values of the results of tests in school environment for level 1 with typically developed children.

Movement	Detection	Timeout	Average time (s)	Maximum time (s)
Open the mouth	66,67%	1	5,5	7
Lower the head	100%	0	5	10
Raise the head	100%	0	9,67	13
Tilt head to the left	100%	0	15	21
Tilt head to the right	100%	0	11,33	23
Look to the left	100%	0	4,67	6
Look to the right	100%	0	9,67	16
Raise eyebrows	66,67%	1	8,5	15
Smile	33,33%	2	5	5
Wink	100%	0	10,33	20
Frown	100%	0	16	39
Tilt head to the side	100%	0	5,33	8
Look to the side	100%	0	7,33	11

Table 2. Summary table with average values of the results of tests in school environment for level 2 with typically developed children.

Emotion	Validation	Average time (s)	Maximum time (s)
Happiness	100%	6,80	19
Surprise	93,33%	6,56	22
Fear	100%	6,80	28
Sadness	80%	6,70	18
Anger	86,87%	6,85	20
Disgust	80%	8,08	27

Table 4 summarizes the performance of the child A at level 1. At this level, he just failed to correctly perform two movements: tilting his head to the side and raising his eyebrows. He was unable to perform the movement of raising his eyebrows due to difficulties in controlling each isolated movement of the face. For this reason, the maximum time for executing the movement has been exceeded. Another difficulty felt was the distinction between the right and the left. For this reason, movements that involve the distinction between right and left have high response times, as can be seen in Table 4. For example, the movement of tilting the head to the right has an answering time of 28 s.

Table 3. Summary table with average values of the results of tests in school environment for level 3 with typically developed children.

Story	Wrong answers	Repetitions	Average time (s)	Maximum time (s)
Story 1	0	0	11	22
Story 2	1	1	7	12
Story 3	0	0	13,67	17
Story 4	2	1	11,5	14
Story 5	0	0	6,33	11
Story 6	0	1	7,33	16
Story 7	0	0	7,33	15
Story 8	0	0	2,33	4
Story 9	0	1	2,67	6
Story 10	0	0	8,33	18
Story 11	2	0	7	9
Story 12	0	0	3	5
Story 13	1	0	9,67	13
Story 14	0	1	5,67	8
Story 15	1	0	4,67	7

Table 5 summarizes the performance of child A level 2. At this level, where the ability to express emotions is tested, the child's performance was surprising. In addition to being able to express all emotions correctly, he performed each expression in acceptable times. For the expression of angry, for example, the response times were low (never exceeding 5 s) which translates to an ease in the expression of the emotion. At the last moment only the name of the emotion remains, and the child was able to express each emotion very quickly. The first time the expression associated with disgust was asked, the child had some difficulties, evidenced by the 28 s of response time. But in all the identical requests that followed, response times were low. This means that, after a period of learning, the child was able to easily express the emotion.

Table 6 summarizes the performance of child A at level 3 (story mode). In this third level, there was no story in which the maximum response time was reached or the number of lives exhausted. However, from the recorded data it can be seen that there were three moments in which the repetition of the story was necessary for the child to clarify which emotion was right. For example, in story 5 the child initially selected a wrong answer but, after repeating the story, was able to select the right emotion. As with typically developed children, this third level was the one that aroused the most interest for the child. Since this child said everything he was thinking it was possible to realize that, after hearing the story and having watched the face of the avatar, he associated the expression and context of the story with the right emotion, except in stories 1 and 5. The fact that the child knows how to read also helped when doubts arose regarding the expression or image present, as he was able to associate the emotion with the name of the button.

Child B has many difficulties in concentrating and is a very active child. These complications led to greater difficulty in performing the tests. There were also times when the child fixed his attention and concentration on elements external to the game, resulting in total inattention to the game.

At level 1, the validation of movements was affected by the child's constant movements that caused difficulties in detecting his face. For this reason, the level 1 test was not completed, with only a few movements being tested (Table 7).

Table 4. Summary of the performance of child A at level 1.

Movement	Right answer	Answering time
Open the mouth	Yes	00:00:08
Lower the head	Yes	00:00:10
Raise the head	Yes	00:00:11
Tilt head to the left	Yes	00:00:28
Tilt head to the right	Yes	00:00:28
Look to the left	Yes	00:00:20
Look to the right	Yes	00:00:06
Raise eyebrows	No	TimeOut
Smile	Yes	00:00:06
Wink	Yes	00:00:03
Frown	Yes	00:00:02
Tilt head to the side	No	TimeOut
Look to the side	Yes	00:00:04

Level 2 tests achieved better results. At this level, the child was initially able to express some emotions (Table 8). For the surprise emotion, the child associated the movement of opening the mouth to this emotion having obtained acceptable response times.

At the third and final level this child's behavior changed significantly (Table 9). For the first story presented, two wrong answers were recorded. Then the child was encouraged to listen and see the story again, after which he was able to answer correctly. In the following stories the child's attention and concentration increased significantly. It was necessary to repeat some stories. In these cases, the child could understand the emotion. Response times were better than expected. The acceptance of this third level was also noted and the child's performance was surprising.

After completing all tests with typically developed children and children with ASD, it can be said that the feedback is positive. The acceptance *HiZeca* of was good at all levels. Level 3 was the level that attracted the most interest from all children.

Table 5. Summary of the performance of child A at level 2.

Emotion	Right answer	Answering time
Happiness 1	Yes	00:00:05
Happiness 2	Yes	00:00:10
Happiness 3	Yes	00:00:06
Happiness 4	Yes	00:00:03
Happiness 5	Yes	00:00:03
Sadness 1	Yes	00:00:05
Sadness 2	Yes	00:00:02
Sadness 3	Yes	00:00:05
Sadness 4	Yes	00:00:02
Sadness 5	Yes	00:00:02
Disgust 1	Yes	00:00:28
Disgust 2	Yes	00:00:02
Disgust 3	Yes	00:00:03
Disgust 4	Yes	00:00:03
Disgust 5	Yes	00:00:03
Fear 1	Yes	00:00:04
Fear 2	Yes	00:00:04
Fear 3	Yes	00:00:06
Fear 4	Yes	00:00:11
Fear 5	Yes	00:00:08
Surprise 1	Yes	00:00:04
Surprise 2	Yes	00:00:14
Surprise 3	Yes	00:00:03
Surprise 4	Yes	00:00:02
Surprise 5	Yes	00:00:03
Anger 1	Yes	00:00:04
Anger 2	Yes	00:00:05
Anger 3	Yes	00:00:02
Anger 4	Yes	00:00:02
Anger 5	Yes	00:00:02

Table 6. Summary of the performance of child A at level 3.

Story	Right answer	Answering time
Story 3	Repetition	
Story 3	Yes	00:00:33
Story 6	Yes	00:00:17
Story 9	Repetition	
Story 9	Yes	00:00:04
Story 12	Yes	00:00:10
Story 15	Yes	00:00:11
Story 14	Yes	00:00:32
Story 11	Yes	00:00:09
Story 8	Yes	00:00:07
Story 5	No	
Story 5	Repetition	
Story 5	Yes	00:00:13
Story 2	Yes	00:00:11
Story 1	No	
Story 1	Yes	00:00:59
Story 4	Yes	00:00:07
Story 7	Yes	00:00:15
Story 10	Yes	00:00:12
Story 13	Yes	00:00:25

Table 7. Summary of the performance of the child B at level 1.

Movement	Right answer	Answering time
Lower the head	Yes	00:00:23
Raise the head	Yes	00:00:26
Tilt head to the left	No	TimeOut
Tilt head to the right	No	TimeOut
Smile	Yes	00:00:39
Frown	Yes	00:00:04
Tilt head to the side	No	TimeOut

Table 8. Summary of the performance of child B at level 2.

Emotion	Right answer	Answering time
Happiness 1	No	TimeOut
Happiness 2	No	TimeOut
Happiness 3	No	TimeOut
Happiness 4	No	TimeOut
Happiness 5	No	TimeOut
Sadness 1	Yes	00:00:15
Sadness 2	Yes	00:00:16
Sadness 3	Yes	00:00:02
Sadness 4	Yes	00:00:05
Sadness 5	Yes	00:00:05
Disgust 1	Yes	00:00:08
Disgust 2	Yes	00:00:03
Disgust 3	Yes	00:00:15
Disgust 4	Yes	00:00:12
Disgust 5	Yes	00:00:20
Fear 1	Yes	00:00:06
Fear 2	Yes	00:00:03
Fear 3	Yes	00:00:05
Fear 4	Yes	00:00:07
Fear 5	Yes	00:00:07
Surprise 1	Yes	00:00:05
Surprise 2	Yes	00:00:02
Surprise 3	Yes	00:00:04
Surprise 4	Yes	00:00:16
Surprise 5	Yes	00:00:03
Anger 1	Yes	00:00:28
Anger 2	No	TimeOut
Anger 3	Yes	00:00:19
Anger 4	Yes	00:00:19
Anger 5	Yes	00:00:30

4 Final Remarks

In order to understand the other's emotional states, it is important to be able to interpret their non behavioural cues, for example facial expressions. This ability allows human beings to adjust their behaviour and react suitably. Consequently, understanding facial expressions correctly and extracting the pertinent social information from them is important for social interactions and communication. Individuals with ASD present deficits in imitation which can be directly linked to impairments in social interaction skills.

This paper presented the serious game *HiZeca,* used to aid in the learning of emotions and promote social interaction with children with ASD. It aims at offering a virtual alternative to the physical robot ZECA for interacting with these children, at a much lower cost. It can be used anywhere with a general purpose computer.

Table 9. Summary of the performance of child B at level 3.

Story	Right answer	Answering time
Story 1	No	
Story 1	No	
Story 1	Repetition	
Story 1	Yes	00:00:17
Story 4	No	
Story 4	Repetition	
Story 4	No	
Story 4	Yes	00:00:51
Story 13	Yes	00:00:09
Story 2	Yes	00:00:12
Story 5	Repetition	
Story 5	Yes	00:00:17
Story 3	Yes	00:00:12
Story 6	Yes	00:00:07
Story 14	Repetition	
Story 14	Yes	00:00:02
Story 7	Yes	00:00:05
Story 10	Yes	00:00:05
Story 8	Repetition	
Story 8	Yes	00:00:12
Story 15	Yes	00:00:03
Story 11	Yes	00:00:04
Story 9	Repetition	
Story 9	Yes	00:00:23
Story 12	Yes	00:00:06

The game, with three levels of complexity, explores and fosters the child's ability to mimic, recognize, and infer the emotional states of virtual agent ZECA *Avatar*, the main character.

Preliminary tests were conducted in a school environment, both with typically developed children and children with ASD, in order to verify the acceptance of the tool by children between 6 and 8 years old.

After completing all tests with typically developed children and children with ASD, it can be said that the feedback is positive. The acceptance of *HiZeca* was good at all levels. Level 3 was the level that attracted the most interest from all children.

Future work considers optimizing further the system by improving the automatic recognition of facial expressions, and conducting tests in a school environment with a larger sample of children with ASD, in order to find out if the game can actually be a complementary intervention tool. With this game, children are expected to promote their cognitive abilities and improve their ability to communicate and interact with peers.

Acknowledgments. This work has been supported by FCT – Fundação para a Ciência e Tecnologia within the R&D Units Project Scope: UIDB/00319/2020. Vinicius Silva also thanks FCT for the PhD scholarship SFRH/BD/ SFRH/BD/133314/2017.

References

1. Piaget: Piaget and His School. Springer, Heidelberg (1976). https://doi.org/10.1007/978-3-642-46323-5
2. Ingersoll, B.: The social role of imitation in autism. Infants Young Child. **21**(2), 107–119 (2008)
3. Silva, V., Soares, F., Sena Esteves, J.: Mirroring emotion system – on-line synthesizing facial expressions on a robot face. In: 8th International Congress on Ultra Modern Telecommunications and Control Systems and Workshops (ICUMT), Lisbon, Portugal, 18–20 October 2016
4. LuxAI, QTrobot - LuxAI.com. http://luxai.com/qtrobot/. Accessed 6 Dec 2018
5. Tobii Dynavox logo, Sono Flex for Communicator 5 - Tobii Dynavox. https://www.tobiidynavox.com/software/content/sono-flex-for-communicator-5/. Accessed 7 Dec 2018
6. Livox, Livox: para pessoas com deficiências e transtornos de aprendizagem. http://www.livox.com.br/pt/quem-somos/#sobre. Accessed 6 Dec 2018 (in Portuguese)
7. TippyTalk - Comunicação Instantânea para Pessoas Não Verbais. (n.d.). http://www.tippy-talk.com/. Accessed 16 Jan 2019. (in Portuguese)
8. Ferreira, V.: Utilização das tecnologias em crianças com perturbações do espectro autista em contexto da Prática de Ensino Supervisionada. Universidade Católica Portuguesa, Centro Regional de Braga, Faculdade de Ciências Sociais, Braga (2013).(in Portuguese)
9. Santos, P., Silva, V., Soares, F., Simões, A.: Facial virtual tracking: a system to mirror emotions. In: Moura Oliveira, P., Novais, P., Reis, L.P. (eds.) EPIA 2019. LNCS (LNAI), vol. 11805, pp. 68–79. Springer, Cham (2019). https://doi.org/10.1007/978-3-030-30244-3_7
10. McLeod, S.: Bandura - Social Learning Theory (2016)
11. Moura, M.L., Ribas, A.: Imitação e desenvolvimento inicial: evidências empíricas, explicações e implicações teóricas. Estudos de Psicologia (2002). (in Portuguese)

Portable Bathing System for Bedridden People

M. Leonor Castro-Ribeiro[1], A. A. Vilaça[1], Mariana A. Pires[1], Karolina Bezerra[2], Cândida Vilarinho[2], and Ana Olival[1(✉)]

[1] School of Engineering, University of Minho, 4710-057 Braga, Portugal
[2] METRICs Research Center, University of Minho, 4800-058 Guimarães, Portugal

Abstract. A portable bathing system for bedridden people was designed to help facilitate the bath and to reduce the lack of privacy of such patients. This system comprises a cylindric stainless steel reservoir with an insulation of polyurethane foam, a rectangular low-density polyethylene board and a galvanized steel transport system. This system allows the reproduction of a conventional shower bath at a constant temperature, improving the bedridden hygiene, self-esteem and quality of life, while reducing the physical effort made by the caregiver. The modeling of the three components and the analysis of the transport system was done with Inventor® Software, obtaining for the transport system a maximum stress of 0.011 MPa and a maximum displacement of 3.438×10^{-6} mm, for an applied load of 200 kg. Since the displacement was relatively small, the system can withstand the load.

Keywords: Bedridden · Portable bath · Medical device

1 Introduction

According to the World Health Organization more than one billion people, about 15% of the world's population, have some form of disability and between 110 million and 190 million adults have significant movement difficulties [1]. Consequently, it is necessary to create devices to aid these people and their caregivers [2].

These disability rates are increasing due to population aging and increased chronic health conditions. It is estimated that the world's population over 60 years of age will grow to double in 32 years [2].

Disabled people need help in numerous basic activities, especially in the bath which is essential for a person's health. Hygiene care is based on an initial assessment of the needs and preferences of the user. This evaluation leads to the establishment of the following objectives: comfort and relaxation, circulation stimulation, cleaning, improving self-image and treating the skin [3].

The bedridden people bath can be a full bath in the bed (caregiver totally washes the person using wipes, face towel, bath towels and bath bowls), a partial bath in bed (given by the caregiver who helps cleaning areas where secretions accumulate: face, hands, armpits and perineal area) and self-help bath in bed (for people confined to the bed, but able to wash themselves completely except for the back, legs and feet) [3].

These forms of sanitation compromise the health of the patient, which can cause the appearance of ulcers, bedsores, rashes and urinary infections. In addition, it is a very exhausting procedure for the caregiver who performs the bath [4].

The proposed project consists in the development of a portable shower system with a shower, a waterproof flexible tub-shaped board and a transport system. The main purpose of the project is to enable and facilitate the shower of a bedridden person.

This device was developed to reduce the stress of the caregiver and the patient associated with this activity, offer the bedridden greater thermal comfort and privacy, improve skin hydration and enable a shower bath sensation.

Thus, it is intended that this device could be used domestically, in homes, hospitals and day centers.

2 State of Art

As the number of people with physical limitations increases, it is necessary to develop better devices and techniques to assist caregivers. In this context, technological solutions have been developed to assist the bed bath, improving the quality and efficacy of the bedridden people bath.

Some examples of these devices are a portable bathtub, a water reservoir and a hygiene apparatus.

The portable bathtub was created with the aim of innovating and improving the quality of nursing care, allowing patient mobility without causing much discomfort. This design includes a vertical stand that allows the side areas of the bed to have different heights, allowing it to accommodate multiple beds. The tub is made of a flexible polyvinyl chloride (PVC) laminate. This bed is also attached to a silicone tube and to the shower in the bathroom. This tube at one end has the showerhead and at the other end a water drain valve [5].

GADE Hospital has developed FlexCare, a portable system for bathing, to maintain a constant temperature, throughout the procedure providing greater comfort for the patient. The system is made by a digital panel with temperature control and an ergonomically designed trolley that has swivel handles and casters [6].

Hygienic apparatuses have also been developed to support bathing in bed, including a hygienic bath system designed to provide comfort, stimulate blood circulation and hydrate the skin of bedridden patients, especially in the intensive care unit. This device is portable and has a rotating massager, moved by compressed air pressure [4].

In addition to the devices presented there are several patented systems that have the purpose of assisting the bathing of people in bed, as we can see in Table 1.

US6802088B1 and WO2011138660A1 are patents of devices that can be used as bathtubs in beds, which are comfortable and improve the bedridden bath.

The device of the Comfort bed bath patent consists of a portable bed with a horizontal bottom surrounded by vertical walls and an opening at the top thus having the shape of a bath. The tub has a removable side panel incorporated into at least one side wall with a latching mechanism. This is used to attach the removable side panel to the tub. This device has an adjustable backrest/footrest. The bathtub has a flexible inlet conduit and a

Table 1. Patents of devices related to bathing systems for bedridden patients.

Patent	Inventor	Name
US3778848A [8]	T Lyytinen	Mobile hospital bathing unit
US5054136A [9]	Jitsuo Inagaki	Bed with a bathtub
US5729848A [10]	Yukigaya Precision Industry Co., Ltd	Bed with a wash tub for pubes
US20020125179A1 [11]	Chi-An Chang	Air cushion bed with water-draining device
WO2002078491A1 [12]	George Khait	Universal service bed
US6802088B1 [7]	Daniel M Gruner	Comfort bed bath
WO2008069767A1 [13]	Mehmet Akdemir	Bed system performing automatic cleaning of bedridden patients
WO2011138660A1 [14]	Minurri Donato	Bathing system bedridden patients

flexible outlet conduit for introduction of water. Water can be supplied by a conventional showerhead. This tub is designed for use on beds or stretchers [7].

The Bathing system for bedridden patients of patent WO2011138660A1 is a device composed of an inflatable PVC air tube. This device has an inflatable and ergonomic main central body, a user support headrest and a tubular air wall [14]. This system protects the user against accidental falls, prevents overflow of water during the bath and channels the waste water into the drain hole. An important aspect of this system is that it allows the patient's elevated position in relation to the waste water, guaranteeing a better hygiene of the body [14].

The devices and patents presented are examples that need improvement in order to be applied. Despite improving the bed bath experience when compared to the conventional methods there are some components that should be improved. Some of these devices are difficult to transport and not straightforward to use, while others rely on the existence of water sources. Moreover, the patient may need to be moved and their privacy cannot be ensured.

3 Methodology

In the elaboration of the project it was necessary to follow a certain methodology that guarantees the introduction of rational procedures. This methodology consisted of Brainstorming, an Objectives Tree, Interviews, elaboration of a House of Quality and the establishment of the project components.

The brainstorming performed allowed the drafting of possible future designs of the device, insertion of new elements, improvement of existing products, clarification of some of the objectives of this project and the need to carry out future interviews and surveys.

The first idea of this system was to be constituted by a reservoir of water for the bath, a reservoir for the residual water and a waterproof plate to put in the bed.

An Objectives Tree was built, consisting of the main objectives of the project, presented in Fig. 1.

This proposal has as its main function "Bathe people in bed" and in the line of development of this objective were defined a series of main objectives concerning characteristics that the device must have to satisfy the caregiver and bedridden needs. Is also take in consideration that the project must be achievable and ecologic.

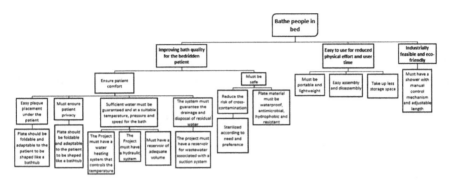

Fig. 1. Objective tree of the portable bathing system

The interviews allowed a better understanding of the opinion of patients, caregivers and health professionals. The interviewees were a quadriplegic patient, one nurse, a caregiver and a technical director of Retirement Home Nossa Senhora da Misericórdia, Braga, Portugal.

In a general way, it was given greater importance to following aspects: mechanism of the plate, the portability of the device, the reduction of the physical effort, the thermal comfort and the cost/benefit relation. The interviewees also suggested possible applications for this device in houses, retirement homes, Palliative Care Units and in the provision of Home Care.

A survey was carried out, Portable Bathing System - Consumer Requirements, to realize the most important requirements and their respective weight for buyers, caregivers and users.

With the results of the surveys answered by 54 people, a House of Quality was built. The design requirements that were given more importance were the following: design of the reservoir; design of the plate; and plate material: antimicrobial, foldable and resistant.

3.1 Project Components

This system is composed of a reservoir with two compartments, one for warm water and other for residual water. This reservoir needs a heating system and a pumping system. The diagram of the overall system is presented in Fig. 2.

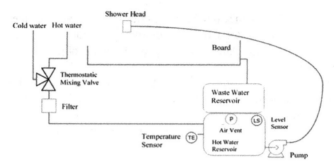

Fig. 2. Overall diagram of the system.

Heating System

To ensure a safe and comfortable bath for the user, temperature is an essential factor. According to the Association of Manufacturers of Thermostatic Mixing Valves the temperature cannot exceed 42 °C, nor be below 38 °C [15].

This solution would require users to make a small change to the faucet to be used to fill the reservoir, avoiding problems with other heating solutions that required shutdown of the device.

This change consists on the placement of quick couplings between the water source and the faucet. After the water enters the system, it passes through the thermostatic valve and then through the filter, accumulating in the hot water reservoir. In this tank there are a temperature sensor, an air trap and a level sensor. The water exits the reservoir and is propelled into the shower through a suction pump. After going through the shower, the water flows down the board to the place where it is collected by a tube that, due to gravity, takes it to a reservoir of residual water.

Pumping System

The main purpose of the pumping system is to ensure the pumping of water from the reservoir to the shower.

The water flow chosen was 30 L in 10 min i.e. 5×10^{-3} m^3/s. Thus, to ensure that the water flow has this value, it was necessary to dimension the water circulation ducts [16].

For the flow value 5×10^{-3} m^3/s and for a maximum velocity of 1.5 m/s the minimum diameter of the selected pipe is 0.0065 m. The commercial selected tubes are shown in Table 2.

Table 2. Pipes main characteristics and estimated head loss (ΔHc)

Parts	Material	Diameter (m)	Length (m)	J (m m^{-1})	ΔHc(m)
Tube 1 [17]	PPR	0.0144	0.1	1.01E−02	1.01E−03
Tube 2 [18]	Stainless steel	0.008	1,5	2.29E−01	3.44E−01

J – Head loss estimated with the empirical Hazen-Williams equation.

To choose the most appropriate hydraulic pump for this project, the maximum desired water flow rate, manometric head and operating voltage of the pump has been considered [19]. The desired water flow rate, as mention before, is 3 L/min. In this system the maximum height that the pump must pump the water is 1.5 m corresponding to a manometric head of 3.19 bar. Therefore, the pump chosen for this system should allow this pressure to be selected for the desired water flow rate.

Morphologic Map

To select the individual necessary components, a morphological map, shown in Fig. 3, was constructed.

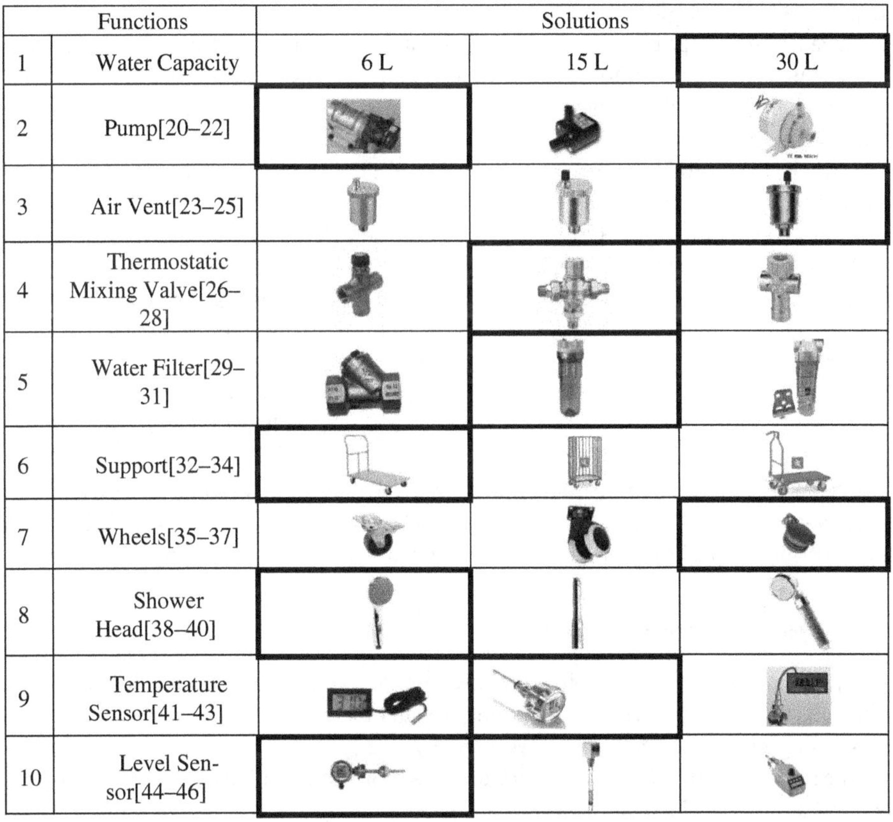

	Functions	Solutions		
1	Water Capacity	6 L	15 L	30 L
2	Pump[20–22]			
3	Air Vent[23–25]			
4	Thermostatic Mixing Valve[26–28]			
5	Water Filter[29–31]			
6	Support[32–34]			
7	Wheels[35–37]			
8	Shower Head[38–40]			
9	Temperature Sensor[41–43]			
10	Level Sensor[44–46]			

Fig. 3. Morphological map of the project.

Initially, it was considered that a water capacity between 6 and 10 L. would be necessary. However, after the interviews with professionals involved in the bedding process, we chose a capacity of 30 L.

The pump chosen for this system was the SHURflo diaphragm pump model 2088-474-144 which, with a pressure switch, allows to select the pressure between 2,1 and 3,4 bar to obtain a certain flow rate. Thus, with this pump, the approximate pressure of

3,2 bar would be selected to obtain an approximate flow rate of 3 L/min. This model is a pump with nominal voltage of 24 V and pumps liquids of temperatures up to 54 °C. These characteristics prove that it is adequate and safe for this system [20].

This design includes an air vent to prevent negative phenomena that could impair the system reservoir's durability and performance, such as corrosive processes caused by oxygen, air pockets and cavitation. The selected air trap was VALCONTROL air cleaner, with a ½" diameter [25].

After selecting the heating system, it was necessary to select the thermostatic valve that corresponds to the requirements of the design with respect to the diameter of the same, respective minimum flow to ensure a stable temperature and good resistance to temperature and wear. A Thermostatic Mixer with Caleffi replaceable cartridge, with a diameter of ½", was selected [28].

Because the water in the network may contain impurities and high concentrations of chlorine, it is important to use a water filter in the tubes to ensure that the water that contacts with the bedridden is adequate. The selected filter was a Pentek 158623 Clear Housing with ½" diameter [30].

To make the entire system portable, a four-wheeled car with an elongated handle was selected to carry both tanks. Thus, 4 polypropylene wheels with steel brakes, a 80 mm diameter and capable of holding 75 kg each were selected [37].

The shower head selected should allow water-saving, to respect the ecological needs. In this way, an ECCO Sensea Shower was selected from Leroy Merlin. This device has a Stop feature, which allows the control of the water flow through the shower [38].

For the user to confirm the reservoir water temperature, Baumer's CombiTemp ™ TFRx temperature sensor with a platinum resistance, which is applicable for liquid measurement, was selected. This sensor has at one end a digital display showing the measured temperature [43].

On the other hand, for the user to know when the tank is full or, for some reason, with lower water level than expected a Kobold level sensor was selected [44]. This sensor must measure water levels of 330 mm or more and have a maximum length of 250 mm.

3.2 Developed Reservoir Prototype

The system must have a reservoir that allows the storage of water. This tank must have a sufficient volume for the required amount of water and for the placement of the system elements (temperature sensor and level sensor).

The reservoir consists of two cylindrical reservoirs: one for heated water and another for residual water, which simplifies its design.

For the system to be portable, a 4-wheel structure was also coupled to the reservoir to support the weight of the reservoir.

Materials

The selected material was 316L stainless steel because it has low thermal conductivity, is resistant to corrosion in several chemical environments and steel is commonly used for the construction of several different types of water reservoirs [47, 48]. Since this material also has a high modulus of elasticity it was selected for the outer constitution of the reservoir.

This material was also selected for the constitution of the residual water reservoir.

An insulation was selected with the objective of minimizing heat transfer from the interior of the reservoir to the environment. The material selected was polyurethane foam because it has low thermal conductivity and density.

Design

As mentioned before the reservoirs for residual water and hot water that make up the reservoir must have a capacity of 30 L. Stainless steel hot water containers should have a thickness between 1.4 mm and 2 mm. In this project the thickness chosen was 2 mm [15].

In this design the vessel was sized to prevent water from cooling 1 °C in an hour. This way it was chosen an insulation thickness of 25 mm. Factors taken in account were: heat transfer values, the final dimensions of the hot water reservoir, the existence of heat losses on the flat faces of the reservoir and that polyurethane foam is not very dense.

The residual water tank does not require insulation and the thickness of stainless-steel chosen was 1,4 mm.

Reservoir Prototype

A 3D model of the prototype was developed in Inventor Software considering the dimensions necessary to incorporate all the electronics and meet the full set of requirements. This prototype was created to be easily used by caregivers, healthcare professionals or common citizens. The system can be divided, as mentioned before in a warm water reservoir and a residual water reservoir. Both reservoirs are cylindric and can be separate and have removable caps for future cleaning.

The warm water reservoir, seen in Fig. 4, has a structure where the second reservoir fits and it is suggested the use of screws to fixate the two reservoirs.

Fig. 4. Warm water reservoir prototype

The dimensions include a base with a diameter of 418 mm and a height of 860.4 mm. system. In Fig. 5 it can be seen the reservoir and the entrances for the different components included in the reservoir. The entrance number 1 refers to the entrance of residual water coming from the bed board, 2 is the entrance of warm water close to the level sensor, 4 is the place of the temperature sensor and 5 is the pump entrance.

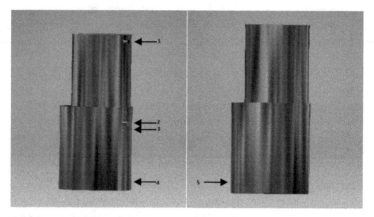

Fig. 5. Reservoir Prototype in which at the left 1 represents the entrance of residual water, 2 the entrance of warm water, 3 the entrance of the level sensor, 4 the entrance of the temperature sensor and at the right 5 represents the entrance of the pump.

The technical drawing of the reservoir can be seen in Fig. 6.

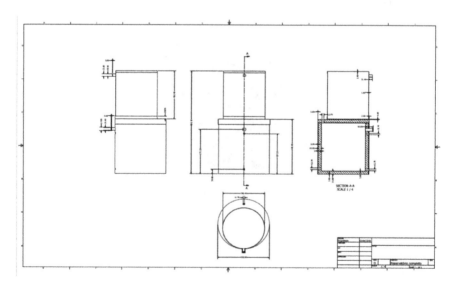

Fig. 6. Technical drawing of the reservoir developed.

3.3 Developed Plaque Prototype

The board serves as an interface between the water and the bed of the bedridden people. In this way, the board needs to be: waterproof (to allow it to not wet the bed of the bedridden), foldable (to fold and take the form of a bathtub and facilitate its transport and storage); resistant (not to break); and antimicrobial (to facilitate cleaning).

Considering that privacy is important for the bedridden person, the board has the sides high enough to ensure that the bedridden person feels comfortable regarding their privacy. The raised margins are also important to ensure that water does not overflow during bathing in bed. The chosen method to connect the margins was the use of a common zipper.

The board needs to connect with the water waste reservoir. This way it was chosen a nipple DIN8063 of PVC, to avoid outflow of water from the adapter it is necessary to cover the inner part in contact with the board with polyethylene plastic.

Materials

The material chosen was low density polyethylene (LDPE). This material is inert compared to most common chemicals because of its paraffinic nature, high molecular weight and partially crystalline structure, and ethylene polymers are non-toxic [49]. This material has a unique combination of properties like toughness, high impact strength, high flexibility, good processability, stability and remarkable electrical properties [49].

Antimicrobial additives can be added to the polyethylene, so the material has antimicrobial properties [50].

Dimensions

The dimensions of the board should conform to the general measurements of most hospital beds. From a bibliographical research, it was verified that there are several models of hospital beds with different dimensions [51, 52].

In this way, the dimensions chosen were for the central board a length of 200 cm, width of 90 cm and for the margins that will rise and give the plate the shape of a bath 20 cm high.

As the chosen material is very malleable it has been defined that the central body of the board should be 5 mm thick and the areas at the tips of the board 1 mm thick. The zipper is installed in the remaining 0.4 cm where materials of different thicknesses are not connected.

Plaque Prototype

The board was modeled in 3D in two positions: extended and closed. The prototype of the board extended can be seen in Fig. 7.

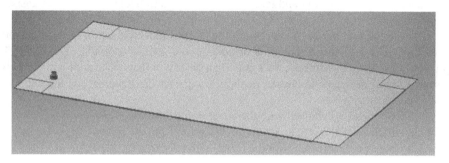

Fig. 7. 3D model of the board extended.

The board closed with a tub format can be seen in Fig. 8.

Fig. 8. 3D model of the folded board.

The technical drawings of the plate are in Fig. 9.

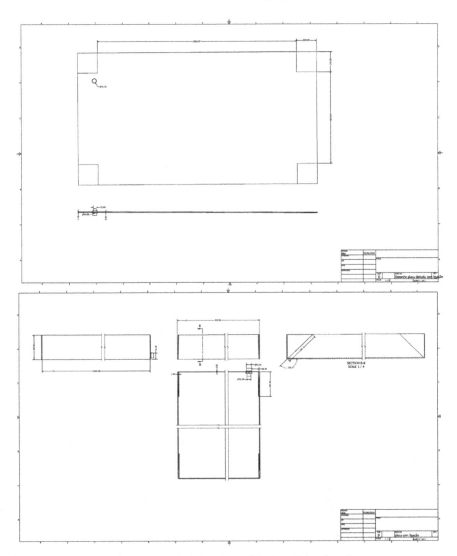

Fig. 9. Technical drawings of the board developed.

3.4 Developed Support System Prototype

Modelling

The support selected to carry the bath system was modeled in 3D using Inventor® software and it is shown in Fig. 10. This support is integrally made of galvanized steel, suitable due to its mechanical strength properties. To allow the whole structure to be movable, as mentioned before, 4 wheels were placed under the support, the constitution of which includes stainless steel on the metal parts and polypropylene on the wheels.

The support includes a region that allows the user to push the system, another for accommodating the reservoir and another to accommodate the pump.

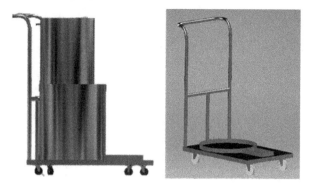

Fig. 10. 3D model of the transport system.

Analysis

As the support will have to support all components of the bath system, a finite element analysis was performed, using the Inventor® software.

This analysis was done by assigning a bearing restriction to each of the wheels of the support and a restraint restriction to the connection of the wheels to the support.

The applied force was of 1960 N to predict if the support would be able to withstand a load of 200 kg.

The stress distribution, according to the von Misses criterion, is represented in Fig. 11, showing a greater accumulation of stress in the place where the force is applied, representing the reservoir, with a maximum stress 0.011 MPa.

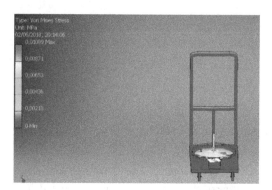

Fig. 11. Analysis of the stress distribution in the transport system.

In addition to the stress, the strain distribution was also obtained along the support, shown in Fig. 12. It was verified that the support bars accumulated more deformation, mainly in the "handle", being the maximum displacement obtained 3.438×10^{-6} mm.

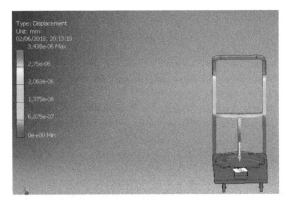

Fig. 12. Analysis of the displacement distribution in the transport system.

Since the displacement that occurred during the test was relatively low, it can be deduced that the system can withstand the applied load. Control part [53] is not detailed in this study.

4 Conclusions and Future Perspective

This work consisted in the development of a portable bathing system for bedridden patients that helps to facilitate the bath and to reduce the lack of privacy. Thus, a reservoir, a board and a transport system were designed, and the materials were chosen to correspond to the initial objectives.

The system aims to facilitate the work of any user compared to existing traditional methods, as it reduces the time spent in bathing, relieves stress of the caregiver and reduces repetitive efforts. This system benefits the bedridden person because it is possible to reproduce a conventional shower bath at a constant temperature and with greater hygiene increasing their self-esteem and quality of life.

In the future, this work may be studied in greater depth to improve some of its characteristics and the concept needs to be validated.

As polyurethane foam is very porous, in future studies, the thickness of the polyurethane could be changed to 40 mm or higher and the thickness of the stainless steel changed to 1,4 mm. This would change the design, weight and heat transfers of the reservoir. These modifications should be analysed to see if they would bring improvements to this project.

5 Responsibility Notice

The authors are the only responsible for the printed material included in this paper.

Acknowledgments. This work has been supported by FCT – Fundação para a Ciência e Tecnologia within the R&D Units Project Scope: UIDP/04077/2020.

The authors are grateful for the contribution of Dr. Isabel Rocha, Director of Santa Casa da Misericordia in Braga, Portugal.

References

1. WHO: World Report on Disability. World Health Organization, Malta (2011)
2. Disability and Health. http://www.who.int. Accessed 10 May 2018
3. Bolander, V.B.: Sorensen e Luckmann enfermagem fundamental: uma abordagem psicofisiológica, 1st edn. Lusodidacta, Lisboa (1998)
4. Honna, S.H., Ferreira, A.C.M., Curimbaba, R.G.: Desenvolvimento de um aparelho para higienização de pacientes acamados. In: III Encontro Científico do GEPro Grupo de Estudo de Produção, p. 12 (2014)
5. Backes, D.S., Gomes, C.A., Pereira, S.B., Teles, N.F., Backes, M.T.S.: Portable bathtub: technology for bed bath in bedridden patients. Rev. Bras. Enferm. **70**, 364–369 (2017)
6. FlexCare. Comprovação Laboratorial. http://gadehospitalar.com.br. Accessed 16 May 2018
7. Gruner, D.M.: Comfort bed bath (U.S. Patent No. US6802088B1) (2004). https://patents.google.com/patent/US6802088
8. Lyytinen, T.: Mobile hospital bathing unit (U.S. Patent No. US3778848A) (1971). https://patents.google.com/patent/US3778848A/en
9. Inagaki, J.: Bed with a bath-tub (U.S. Patent No. US5054136A) (1989). https://patents.google.com/patent/US5054136A/en
10. Yamagish, T.: Bed with a wash tub for pubes (U.S. Patent No. US5729848A) (1994). https://patents.google.com/patent/US5729848
11. Chang, C.-A.: Air cushion bed with water-draining device (U.S. Patent No. US20020125179A1) (2002). https://patents.google.com/patent/US20020125179
12. Khait, G.: Universal service bed (WO Patent No. WO2002078491A1) (2002). https://patents.google.com/patent/WO2002078491A1/en
13. Akdemir, M.: Bed system performing automatic cleaning of bedridden patients (WO Patent No. WO2008069767A1) (2008). https://patents.google.com/patent/WO2008069767A1/pt
14. Donato, M.: Bathing system for assisted and non-sufficient bedridden patients (WO Patent No. WO2011138660A1) (2011). https://patents.google.com/patent/WO2011138660A1/en
15. Bezerra, K., et al.: Bath-ambience—a mechatronic system for assisting the caregivers of bedridden people. Sensors **17**, 1156 (2017)
16. Brater, E.F., King, H.W., Lindell, J.E., Wei, C.Y.: Handbook of Hydraulics, for the Solution of Hydraulic Engineering Problem, 7th edn. McGraw-Hill, New York (1996)
17. Ficha Técnica: alfa PP-H.Alfa Tubo Engineering Pipes. http://www.alfatubo.pt. Accessed 22 Apr 2018
18. Stainless steel uncoated. http://kottmann.eu. Accessed 22 Apr 2018
19. Quintela, A.: Hidráulica, 12th edn. Fundação Calouste Gulbenkian, Lisboa (2000)
20. SHURflo. Product Data Sheet. Model: 2088-474-144. http://shurflo.com/. Accessed 22 May 2018
21. 3.6L/Min Mini DC Brushless Submersible Water Pump. https://www.lightobject.com. Accessed 22 May 2018
22. TL-A02. http://www.topsflo.com. Accessed 22 May 2018
23. 5020 MINICAL - Purgador de ar automático 3/8″ M e 1/2″M. https://www.caleffi.com. Accessed 13 May 2018
24. 5022 VALCAL - Purgador de ar automático 1/4″M, 3/8″ e 1/2″M. Cromado. https://www.caleffi.com. Accessed 13 May 2018
25. VALCONTROL. Purgador de ar. http://www.valcontrol.pt/. Accessed 13 May 2018
26. Válvula Misturadora Termostática AQS. https://www.zantia.com/. Accessed 13 May 2018
27. Válvula Misturador Termostática. http://www.lusosol.com. Accessed 13 May 2018
28. 520 Misturadora termostática regulável. https://www.caleffi.com. Accessed 13 May 2018
29. Filtro água LATÃO Y PN16 1/2″. http://www.leroymerlin.pt. Accessed 13 May 2018

30. Pentek 158623 1/2″ 3G Slim Line 10″ Clear Housing 12-Pack. https://www.filtersfast.com. Accessed 13 May 2018
31. Aquafilter 10″ Water/Biodiesel Filter Housing 1″ 3/4″ 1/2″ BSP PRV. https://www.ebay.com. Accessed 13 May 2018
32. Carro Plataforma em Aço com Capacidade de 300 Kg Tm52 Marcon. https://www.ferramentaskennedy.com.br/. Accessed 13 Apr 2018
33. Carrinho de carga manual com posição horizontal/de movimentação/de aço. http://www.directindustry.com/. Accessed 13 Apr 2018
34. Plataforma. http://www.directindustry.com/. Accessed 13 Apr 2018
35. Roda pivotante industrial rodízio de metais com freio 75 mm. http://www.cablematic.pt. Accessed 13 Apr 2018
36. Roda sem travão D48MM 35KG. http://www.leroymerlin.pt. Accessed 13 Apr 2018
37. Roda com travão D80MM 75KG. http://www.leroymerlin.pt. Accessed 13 Apr 2018
38. Chuveiro ECCO SENSEA. http://www.leroymerlin.pt. Accessed 10 Apr 2018
39. Chuveiro VITALIO GET. http://www.leroymerlin.pt. Accessed 10 Apr 2018
40. Bathroom Booster SPA Anion Water-saving Handheld Rain Shower Head Nozzle. https://www.banggood.com. Accessed 10 Apr 2018
41. Termômetro Digital LCD com Leitura Instantânea Detector Impermeável. http://www.directindustry.com/pt. Accessed 10 Apr 2018
42. Termômetro Digital LCD com Leitura Instantânea Detector Impermeável – PRETO. https://pt.gearbest.com. Accessed 10 Apr 2018
43. Sensor de temperatura com sensor PT100/de resistência de platina/com rosca/com carcaça metalica. http://www.directindustry.com/pt/. Accessed 10 Apr 2018
44. KOBOLD. Level Sensors. Model D/R. https://kobold.com. Accessed 15 Apr 2018
45. Sensor de nível para liquidos/com sensor de temperatura/programável/analógico. http://www.directindustry.com/pt. Accessed 15 Apr 2018
46. Sensor de nível LFP Cubic TDR – Sick. https://www.ffonseca.com/pt/. Accessed 15 Apr 2018
47. Stainless Steel - Grade 316L - Properties, Fabrication and Applications (UNS S31603). https://www.azom.com/. Accessed 22 Apr 2018
48. Votruba, L., Broža, V.: Water management in reservoirs. In: Developments in Water Science, 1st edn. Elsevier Science (1989)
49. Coutinho, F.M.B., Mello, I.L., de Santa Maria, L.C.: Polietileno: principais tipos, propriedades e aplicações. Polímeros **13**, 01–13 (2003)
50. BioCote. Antimicrobial Technology. https://www.biocote.com. Accessed 24 Apr 2018
51. ARTIFOFO Camas. http://www.artifofo.pt. Accessed 25 Apr 2018
52. Camas EUPHORIA. http://www.orthosxxi.com. Accessed 25 Apr 2018
53. Barros, C., Leão, C.P., Soares, F., Minas, G., Machado, J.: RePhyS: a multidisciplinary experience in remote physiological systems laboratory. Int. J. Online Eng. **9**(SPL.ISSUE5), 21–24 (2013). https://doi.org/10.3991/ijoe.v9iS5.2756

Manufacture of Facial Orthosis in ABS by the Additive Manufacturing Process: A Customized Application in High Performance Sports

Anna Kellssya Leite Filgueira[✉], Isabella Diniz Gallardo, Ketinlly Yasmyne Nascimento Martins, Rodolfo Ramos Castelo Branco, Karolina Celi Tavares Bezerra, and Misael Elias de Morais

State University of Paraíba (UEPB), Campina Grande, Paraíba, Brazil
{yasmyne.martins,rodolfo.ramos}@nutes.uepb.br

Abstract. The aim of the present study is to manufacture a customized facial orthosis with Acrylonitrile butadiene styrene (ABS) material using the Additive Manufacturing process with fused deposition modeling technology (FDM) and to present its benefits when applied in high performance sports. The facial orthosis manufacture was possible due to the combination of three-dimensional technologies and medical images from computed tomography (CT). The development process was carried out in six stages: 1. Capture of the individual's digital image; 2. Manipulation of CAD (Computer-Aided Design) software - which runs through the following substeps - geometric mesh treatment, combination of the desired bone region, mirror of the region (offset) of interest for orthosis and device's digital three-dimensional modeling; 3. Simulation by finite elements; 4. Device additive manufacturing; 5. Post-processing of Additive Manufacturing; 6. Device validation. The design is totally directed towards the individual's anatomy perfect coupling, having two lateral holes for the wrap fitting that will fix the device in the cranial box during use. The choice of ABS material provided a thickness of 3 mm in the device, maintaining the characteristic of resistance to impact, ensuring lightness, excellent visual aspect, good finish, which guaranteed the maintenance of the athlete's performance. The FDM process powered these benefits by adding characteristics such as low cost and customization to the device. The study provides an alternative for the development of automated and fully customized orthoses, in addition to demonstrating a good applicability of the ABS material, increasing the scientific subsidies on the subject and its correlations.

Keywords: Orthopedic devices · Three-dimensional printing · Manufacturing

1 Introduction

The term additive manufacturing (AM) refers to a manufacturing process through the addition of material through successive and flat layers [1]. The great advantage of using AM is the possibility of producing a physical object of any complexity in a relatively short period of time [2]. A simple way to classify the processes related to it is through the state or raw materials initial form used in the manufacture (liquid, solid and powder). For this, the ISO/ASTM 52900:2015(E) [3] proposes framing in seven categories: Vat Polymerization; Material Extrusion; Material Jetting; Binder Jetting; Powder Bed Fusion; Sheet Lamination; Directed Energy Deposition.

Fused deposition modeling (FDM) technology, in the category of material extrusion technology, has the ability to build robust, durable and dimensionally stable parts through specialized three-dimensional (3D) printers, providing better accuracy and repeatability when compared to any other three-dimensional printing technology available [4]. FDM is the most used method among all AM techniques for the manufacture of pure plastic parts, as they receive characteristics such as low cost, minimum waste and great material variability [5, 6].

In this process, the construction material is heated in an extrusion head where the temperature is controlled. In addition, there is a controlled chamber temperature. The semi-liquid material in filament form is extracted from the extrusion head and is deposited layer by layer [7]. Of all material possibilities, those commonly processed include Polycarbonate (PC), Polyphenylsulfone (PPSU), Acrylonitrile Butadiene Styrene (ABS) and Lactic Polyacid (PLA) [8, 9]. ABS, in particular, is an industrial thermoplastic widely used in FDM technology. It is a resistant and ductile material, which has heat tolerance and wear resistance, maintaining an extrusion temperature in a range of approximately 230 ºC to 240 ºC. Besides that, it has specific properties such as good impact, tensile and abrasion resistance [10].

The combination of these characteristics that unite the FDM process and the ABS material when associated with digital modeling software enables the construction of a light, resistant, durable, comfortable and fully planned device according to the individual's needs. In this context, a wide range of applicability is emphasized, especially with regard to the area of medical devices, especially the development of orthoses - a device indicated for resting, immobilizing, protecting, providing feedback, correcting and promoting tissue healing [11].

One of the types of orthoses in which the manufacture using AM has a particularity are the facial orthoses applied to the sports area. These aim to provide the athlete's early return safely, protecting the injured area from a new impact and avoiding aggravation situation from simple to complex. The increase in three-dimensional technology, in these cases, is a personalization feature related to the process allowing the body structure's perfect coupling [12].

Therefore, the objective of the present study is to manufacture a customized facial orthosis with ABS material by the AM process using FDM technology and to present its referred benefits when applied in high performance sports.

2 Material and Methods

The present study was developed in the period of November 2020, at the Laboratory of Three-dimensional Technologies (LT3D) of the Center for Strategic Technologies in Health (NUTES), located at the State University of Paraíba (UEPB). Refers to a scientific research of an applied nature and qualitative approach that was submitted and approved by the ethics committee under the CAAE 36304820.0.0000.5187.

The manufacture of the facial orthosis was possible due to the combination of three-dimensional technologies and medical images from computed tomography (CT). The manufacturing process comprises six steps, expressed in flowchart 1 (Fig. 1).

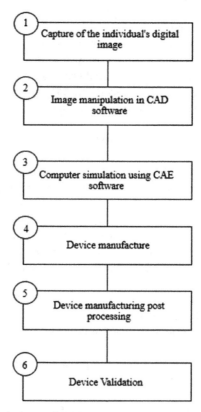

Fig. 1. Methodological steps flowchart (Source: Elaborated by the author, 2021).

The step 1 refers to capturing the individual's digital image. This was performed by means of CT, which allowed obtaining the reliable anatomical contour.

Step 2 is marked by the manipulation of CAD (Computer-Aided Design) software for medical images. It is through these softwares that it will be possible to carry out, respectively, the following substeps: Geometric mesh's treatment; Segmentation of the necessary bone region; Mirroring of the region (offset) of interest for orthosis; And finally, a digital three-dimensional modeling of the device. In this phase, the file, bulletin in.DICOM, will be converted to.STL (Standard Triangle Language) format that refers to a standard language of triangles to form the surface mesh. The digital modeling of the device was performed based on the anatomical contour and requirements regarding the clinical case and the use applicability.

The Step 3 deals with finite element simulation using CAE (Computer-Aided Engineering) software. This phase has significant relevance for the digital model material structure behavior analysis, carried out through stress and deformation parameters of the produced device, making it possible to make necessary corrections before the manufacturing step itself.

Step 4 is the manufacture of the device. To accomplish this step, a 3D printer with FDM technology and ABS was used. The choice of material was based on the mechanical properties that were desired to be achieved in the device.

Step 5 refers to the post-processing of additive manufacturing, where all the support material is removed and the device finishes superficially.

Step 6 indicates device validation. Some points of the device applicability were analyzed, such as: Material resistance during use, Individual's comfort, Correlation between device thickness and field of view, Limitations provided and Device adjustment.

3 Theoretical Reference

The AM, defined as the process of adding layers by layers, is intended for the construction of parts used directly as final products or components, in a fast, automated and totally flexible way [1, 8]. This method offers several advantages in many applications when compared to classic manufacturing processes based on material removal, such as milling or turning, allowing, in most cases, that the desired objects are produced in a reduced time and at a lower cost [13].

There are several sectors that can benefit from the use of AM technologies, being currently widespread in the aerospace, automobile, bioengineering (medicine and dentistry), electrical products (household appliances), electronic products in general, and in the jewelery, arts, civil engineering and architecture sectors [1].

In health sciences, in barriers that go beyond the medical field, AM has taken on large proportions, with the potential to revolutionize the future of this area. The most expanded application, in this sense, currently, is the use of biomodels for surgical planning, first reported by Paul D'Urso. The method consists of using medical image acquisition to produce reliable physical models of human anatomy that guide the performance of surgeries [1]. The process is reported in the diagram in Fig. 2.

For Volpato [1], this process can also be applied in the assistive technology (AT) development, another area that has emerged with the advances of AM. The possibility of expanding the functional capacity of people with physical disabilities, disability or reduced mobility, with personalized assistance in the creation of products adapted to anatomy and according to the specific needs of each patient quickly and with low cost-effectiveness makes the effectiveness of the process increase [14].

The creation through 3D printing of tools, utensils and other medical equipment, represents speed and cost reduction [5]. It is estimated that the time and cost savings provided by the application of AM techniques in the construction of models is between 70% and 90% [13].

Another great advantage of AM is that it is not restricted to just one type of technology, but it achieves numerous alternatives of variant processes between the states of liquid, solid and powder, when analyzing the initial conditions of the raw material [15]. With regard to information on the addition and adhesion mechanisms involved, the variants are classified as: Vat Polymerization; Material Extrusion; Material Jetting; Binder Jetting; Powder Bed Fusion; Sheet Lamination; Directed Energy Deposition.

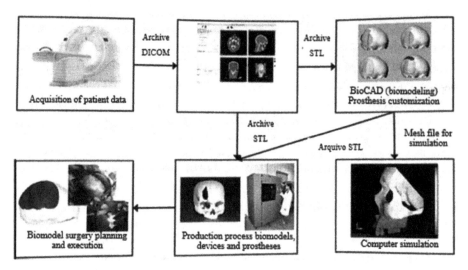

Fig. 2. Processing and manufacturing of biomodels diagram (Source: VOLPATO, 2017)

3.1 Fused Deposition Modeling Technology (FDM)

This technology is characterized by the deposition of extruded material continuously in the form of a wire, heating it to a semi-liquid or pasty point. The table, consisting of a lift mechanism, positioned on the Z axis, receives the material through the movements of the extrusion head on the xy axes, where, to expel the material through a calibrated nozzle, the filament traction itself works as a plunger to beginning of the process system. Thus, when the extruded thin filament material comes in contact with the part surface material, it solidifies and adheres to the previous layer (Fig. 3) [1].

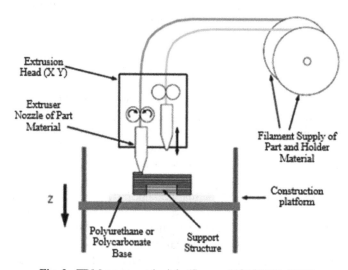

Fig. 3. FDM process principle (Source: VOLPATO, 2017)

In the machine, the liquefactor head plays an important role, which is the key to the success of the molten deposition: The material in the form of filament is pulled or pushed with the help of the driving wheels that are connected to the electric motors and then enters the heating chamber. The material flows through the tube blender and is deposited through an extrusion tip [7].

The printer specialized in FDM features an additional exclusive extruder nozzle for adding the supports, since it is necessary to create a structure so that the parts that do not articulate with the first layers can be connected. In the absence of support, the nozzle deposits material in an empty space causing deformation in the part [1].

In most commercial 3D printers it is not possible to control the material's deposition trajectory, but the positioning of the part on the printing table can interfere with the characteristics desired by the product. This positioning is related to the printing parameters that are determined in CAM (Computer Aided Manufacturing) software [16].

CAM software, defined as computer-aided manufacturing, has a main function aimed at converting a component geometry to be manufactured in a machine programming language, tool path performing calculation, from the geometric representation of the part, treats therefore, specific software for printing parameters [17].

In FDM processing, the monofilaments used are generally made of engineering thermoplastic polymers, such as ABS, polycarbonate and polycarbonate blends [18]. In this process, due to the large number of parameters and the interaction over them, it is often difficult to determine the actual parameter levels in order to achieve the best component properties manufactured by FDM [4].

3.2 Acrylonitrile Butadiene Styrene (ABS)

ABS plastic is a petroleum-derived thermoplastic resin consisting of the copolymerization of three monomers (Fig. 4): acrylonitrile (synthetic monomer produced from propylene hydrocarbon and ammonia - from 15% to 30%), butadiene (alkene obtained from dehydrogenation of butane - from 5% to 15%) and styrene (produced from the dehydrogenation of ethylbenzene - from 40% to 60%).

With regard to the manufacturing process for this material, a study compared the mechanical properties of two types of ABS prototypes made by two different techniques [19], where one is the removal of material made by a CNC machining center (milling machine) and another by PR, by adding material in the FDM process and concluded that the specimens printed by FDM showed a better performance compared to the machined ones in the tests, obtaining better results in comparison of tensile strength.

A specific study carried out with the ABS material, analyzed the layer height, scanning angle and density filling [4], understood as its mechanical properties when processed by FDM. In this, in turn, it is reported that the ideal manufacturing parameters appear to be 55° for raster angle (angle of the deposited beads in relation to the x axis of the construction table), 0.5 mm layer thickness and a percentage of 80% fill. However, they also point out that when the percentage of filling is increased, the properties of final traction, strength, modulus of elasticity and resistance to flow are also increased.

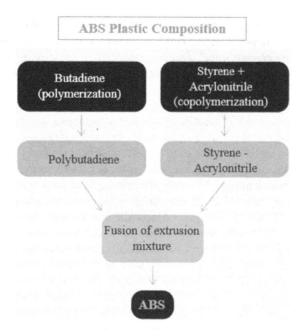

Fig. 4. ABS plastic composition (Source: MaisPolímeros, 2021).

4 Results and Discussion

The orthosis was developed for a male soccer player, who was selected as a midfielder of a semi-professional team from Paraíba, who had a clinical picture of nasal fracture. The capture of images, obtained by the CT examination, when processed by the computational software of reading .DICOM, allowed to create a virtual three-dimensional model of the individual (Fig. 5).

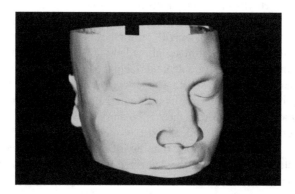

Fig. 5. Individual three-dimensional model (Source: Research data, 2020).

From this representation, the 3D model was imported into CAD software for medical images, InVersalius 3, which is free software for reconstructing images from CT or

MRI (magnetic resonance image) equipment, for the correction and smoothing of the triangular mesh, generating a digital file of images in .STL format, which represents a triangular mesh characterized by the surface of the analyzed region.

With the digital model finalized, an offset (mirror) of the region of interest was created to start the device modeling phase. For this, aspects such as device applicability, device's functionality correlation with the need to maintain the athlete's complete field of view, comfort, usability and lightness were taken into account.

The key point for the device development was the thickness - finally defined at 3 mm. The reach of this parameter was only possible due to the realization of a direct correlation with the material of manufacture, since the clinical case required minimum thickness so that there was no athlete's field of vision interruption, at the same time that there is a need that the device maintains good impact absorption due to the direct confrontation generated by the sport.

In this regard, among all the possibilities, ABS was defined as a viable option for providing lightness, excellent visual aspect, good finish, good resistance to impact, traction, abrasion, heat and wear [10, 20]. In addition, according to studies by Davanço and collaborators [19], the characteristics of this material are enhanced when manufactured by deposition.

Another important aspect to be reported that directly influenced the material's choice was the characteristics of compressive strength. The study of Martinez and colleagues (2019), compared the axial compression data between ABS and PLA through a specimen with internal structural mesh. This study, for the ABS samples, the graphics tension-deformation showed similar inclinations to the same fiber density and the elasticity modulus proved to be superior to the PLA [21].

The roughness control provided by the temperature control characteristic in the chamber of this type of material also adds a potential benefit for biomedical use with regard to the accuracy of the device.

In an intermediate phase, after modeling and before manufacture, the device digital model (Fig. 6) was tested for stress and strain parameters through finite element simulation using the CAE software. This step allowed the product performance to be verified in terms of strength, stiffness, fatigue, temperature and dynamics, before the physical model production, providing an increase in the process economy.

The device design is totally directed towards the perfect coupling of the individual's anatomy, having two lateral holes for the wrap's fitting that will fix the device in the cranial box during use. With the virtual modeling stage closed and the material defined, the manufacturing process itself begins. For this, the FDM type AM process was used, with the Raise 3D printer, and the CAM system embedded in the machine itself, in which the parameters for printing are planned and defined. After the printing parameters were determined, the G-code (Geometric Code) file was generated, which stores all associated parameters.

The entire manufacturing process lasted four hours, ending in printed material with associated support (Fig. 7). According to Volpato [1], FDM technology needs support to connect the parts that are not associated with the previous layer, without prejudice to the final production of the device, making it possible to build parts that differ from blocks.

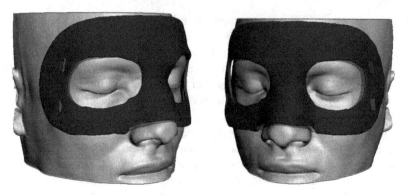

Fig. 6. Device digital model produced (Source: Research data, 2020).

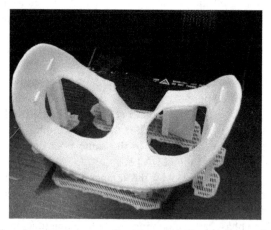

Fig. 7. Device after complete manufacturing step. (Source: Research data, 2020).

After these steps, the device is directed to the only step of post-processing, which is the support removall. Figure 8 shows the finished device.

For validation, the athlete used the device during a complete soccer match, corresponding to 90 min. After finishing, he demonstrated satisfaction with the feedback usability, reporting that in no time did the mask interrupt its performance during the game, including that it allowed direct head shocks with the ball and the other players, without any damage. In addition, he emphasized the perfect maintenance of his field of vision, the comfort and the lightness that the device provides.

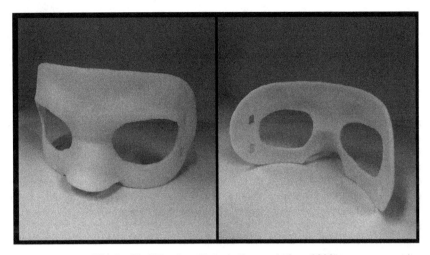

Fig. 8. Final Device. (Source: Research data, 2020).

5 Conclusions

The study proposal was successfully achieved once it provided an alternative for the development of facial orthoses in a fast, automated and fully customized way. One of the factors to be highlighted was the use of 3D technologies in the device manufacturing process, which characterizes it as an innovative process in the use of health technologies.

The ABS type material used in the manufacture, provided to the developed device with a minimum thickness, without causing damage to the athlete's field of vision, but maintaining this material impact resistance property. In addition, it allowed comfort and lightness that maintained the athlete's complete performance, due to its characteristics of good resistance to compression, impact, traction, abrasion, heat and wear.

The FDM process, in turn, enhanced these benefits by adding resources such as low cost and the device customization. In this case, post-processing is not necessary to finish the device, only to remove the support material that is necessary for structuring it in the printing process.

Further studies are needed in different types of materials with application in athletes to offer greater diversity of data, besides that to encourage the fomentation of quantitative research being this the study limitation point.

References

1. Volpato, N.: Manufatura aditiva: tecnologias e aplicações da impressão 3D. São Paulo, Blucher (2017)
2. Chua, C.K., Leong, K.F., Lim, C.S.: Rapid Prototyping, Principles and Applications. World Scientific Publishing (2003)
3. ISO/ASTM 52900:2015(E): Standard terminology for additive manufacturing – general principles – terminology. Genève, West Conshohocken (2016)

4. Samykano, M., Selvamani, S.K., Kadirgama, K., Ngui, W.K., Kanagaraj, G., Sudhakar, K.: Mechanical property of FDM printed ABS: influence of printing parameters. Int. J. Adv. Manuf. Technol. **102**(9–12), 2779–2796 (2019). https://doi.org/10.1007/s00170-019-03313-0
5. Stratasys Homepage. http://www.stratasys.com/~/media/Main/Secure/Technical_Application_Guides. Accessed 12 Dec 2020
6. Kai, C.C, Fai, L.K., Sing, L.C.: Rapid Prototyping: Principles and Applications, 2nd edn, 448p (2003)
7. Arivazhagan, A., Masood, S.H.: Dynamic mechanical properties of ABS material processed by fused deposition modelling. Int. J. Eng. Res. Aplicativos (IJERA) **2**, 2009–2014 (2012)
8. Hopkinson, N., Hague, R., Dickens, P.: Rapid Manufacturing: An Industrial Revolution for the Digital Age. Chichester, London (2006)
9. Gibson, I., Rosen, D.W., Stucker, B.: Generalized: additive manufacturing process chain. In: GIBSON, I. et al (ed.). Additive Manufacturing Technologies: Rapid Prototyping to Direct Digital Manufacturing. Springer, Heidelberg (2010). https://doi.org/10.1007/978-1-4419-1120-9_3
10. Filho, E.G.: Desenvolvimento e fabricação de moldes flexíveis (tpu) de baixo custo por manufatura aditiva para produção de próteses cranianas de PMMA. Dissertação. Universidade Tecnológica Federal do Paraná, Curitiba (2019)
11. Carvalho, J.A.: Amputações de Membros Inferiores: em busca da plena reabilitação, 23nd ed. São Paulo, Manole (2003)
12. Ghoseiri, K., Ghoseiri, G., Bavi, A.: Ghoseiri, R: Face-protective orthosis in sport-related injuries. Prosthet. Orthot. Int. **37**(4), 329–331 (2013)
13. Garcia, L.H.T.: Desenvolvimento e fabricação de uma mini-impressora 3D para cerâmica. Dissertação. Escola de Engenharia de São Carlos (2010)
14. Rodrigues, J.: Cruz Lms; Sarmanho APS: Impressora 3D no Desenvolvimento de Pesquisas com Próteses. Rev. Interinst. Bras. Ter. Ocup. v2(2), pp. 398–413 (2018)
15. Chua, C.K., Leong, K.F., Lim, C.S.: Rapid prototyping: principles and applications. 3edn. Singapore (2010)
16. Bellini, A., Güçeri, S.: Mechanical characterization of parts fabricated using fused deposition modeling. Rapid Prototyping J. **9**(4), 252–264 (2003)
17. Holzmann, H.A.: Uso do software CAM em auxílio a estimativa de custos de fabricação na fase inicial do desenvolvimento de produtos poliméricos injetados. Universidade Tecnológica Federal do Paraná, Dissertação (2014)
18. Nikzard, M., Masood, S.H., Sbarski, I.: Thermo-mechanical properties of highly filled polymeric composites for fused deposition modeling. Mater. Des. **32**, 3448–3456 (2011)
19. Davanço, L. Furhmann, J.G. Chuan, C.T.: Estudo comparativo das propriedades mecânicas entre plástico ABS prototipado por deposição x Usinado em cnc. Congresso Técnico Científico da Engenharia e da Agronomia. Belém, Pará (2017)
20. Mais Polimeros Homepage. http://www.maispolimeros.com.br/2018/09/24/plastico-abs-e-suas-principais-caracteristicas-eaplicacoes/#:~:text=Det%C3%A9m%20propriedades%20especíﬁcas%20como%20boa,C%20a%2080%C2%B0C. Accessed 12 Jan 2021
21. Martinez, A.C.P., Souza, D.L., Santos, D.M., Pedroti, L.G., Carlo, J.C., Martins, M.A.D.: Avaliação do comportamento mecânico dos polímeros ABS e PLA em impressão 3D visando simulação de desempenho estrutural. Gestão e Tecnologia de Projetos **14**(1), 125–141 (2019)

E-Health in IDPs Health Projects in Pakistan

M. Irfanullah Arfeen[1(✉)][iD], Adil Ali Shah[2], and Demetrios Sarantis[3]

[1] QASMS, Quaid-i-Azam University, Islamabad, Pakistan
m.arfeen@qau.edu.pk
[2] Center for Advanced Studies in Engineering (CASE), Islamabad, Pakistan
[3] United Nations University Operating Unit on Policy-Driven Electronic Governance, Guimarães, Portugal
sarantis@unu.edu

Abstract. Almost all the developed countries have adopted E-health system and Pakistan is also one of these countries where E-health is now being partially implemented. In addition to this implementation the need of the day is to implement E-health in Health Projects such as Internally Displaced People (IDP) Camps to make the projects even more successful and result oriented. Not many studies have been conducted to find the rationale behind E-Health system not being adopted in IDPs Health Projects in Pakistan. This research is an attempt to fill the gap and identify the challenges due to which E-health is not yet adopted in Health Projects. The data gathered through the questionnaire was then structured and regression analysis with significant test was conducted to identify the factors that contribute in non implementation of E-health system in IDPs Health Projects. The result shows that the healthcare staffs are willing to adopt E-health system but due to lack of ICT at the Project Implementation Unit level such E-health system could not be implemented. Therefore in order to make Health Projects more successful using E-health system basic ICT facilities should be provided at Project Implementation Unit.

Keywords: ICT · e-health · IDP

1 Introduction

Information Communication Technology (ICT) has always been a source of information for healthcare not only for public but also for doctors. According to the survey 74% people use internet to search health related issues and problems ("Internet World Stats" 2008). Not only general public but doctors and physicians use Internet to gather information about health related issues [1]. E-Health was first launched in 1999 when health information was available via internet [2]. E-health has positive as well as negative impacts on doctors and patients. Using technology many patients now tend to use E-health for information gathering and to avoid going to doctors [3] which has been a drawback and disappointment for some [4].

E-health can also be defined as a healthcare application that improve the services, effectiveness, efficiency and quality of health sector and hence the health of patient [5].

Health sector around the world has witness major changes since year 2000. Developed countries in the world are consistently trying to make improvements in health sector, such improvements demand not only reducing the cost but also improving the quality of healthcare easier and accessible [6].

The information regarding when exactly the first time health was used electronically is not know as there is very less literature available, however according to [7] about 8 million people search health related information on net in year 1996, which increased to 23 million in 1999 and about 100 million in 2001. According to [8] more than 70% internet users search health related information through credible sources. Before the introduction of E-health in developed countries traditional paper based healthcare approach was used to maintain the health record of patients in hospitals and clinics. This practice is used in many developing countries like Pakistan where the book keeping procedure is used to record data. The main disadvantage of this method is that paper based record keeping is decentralized method because of which the information of patients cannot be shared.

Since past few decades, E-Health has been used for various you tubes purposes in the developed countries that include recording and maintaining of patients details, sometimes for research purposes and trend analysis. The main advantage of using E-health is that the data and record is become centralized and so can be shared anywhere when required. In the modern era E-health is also been used to circulate health resources which enables to use these resources efficiently and effectively.

1.1 E-Health in Pakistan

Healthcare in any country is important in building a prosperous society. Pakistan is one of the developing countries where health facilities are not up to the standard and SOPs defined by World Health Organization [9]. In a recent report it has been observed that only 29% of people in Pakistan enjoy full health services. Pakistan has the lowest budgets allocated to health in the entire region [9]. On the other hand the population is increasing on a rapid scale. In year 2000 the population of Pakistan was around 142 million which has increased to 198 million in 2014, but the budget allocated to health sector has not increased [10]. The recent survey has shown the doctor to patient ratio in Pakistan in around 1:1,378 while the nurse patient ratio is around 1:22,346 [11]. To overcome the doctor patient ratio and provide a basic facility to people, use of ICT in health sector was introduced which helped in improving the quality of health services around the world. This phenomenon helped in working beyond the conventional method. The idea of E-health was given in 1994 during a World Telecommunication Development Conference. In 2005 E-health was also recognized by World Health Organization (WHO) in order to achieve cost effective and improved quality of healthcare [12].

In November 2008, a Non Profit Organization in the name of E-HAP was established in Pakistan in order to enhance and spread the knowledge of E-health among the people of Pakistan. The main vision of E-HAP was to improve the health sector and provide quality and improved healthcare to people especially in rural area of the country by implementing E-health Practice in Pakistan [13].

In Pakistan around 62% of population is living in rural area [14]. Basic health units (BHUs) are established in parts of rural area in Pakistan to provide healthcare to people. But the question arises that despite so many advantages of E-health system, why it is not yet adopted and implemented in Pakistani BHUs especially in IDPs camps where the basic health facilities are not up to the standard and the serious patients had to visit travel to major cities for health services.

1.2 IDP Camps in Pakistan

Many international NGOs have developed their own definition of IDP mainly focus on their own perspective which reflects their concern for people, but Cohen and Deng in 1998 expressed the IDPs in more appropriate way. According to them;

IDPs are people who are forced to leave their place and homes and to migrate to other place within a country without their will in order to avoid violence or arm conflict or due to a natural or human made disaster [14].

According to [14] millions of people are displaced in the world due to conflicts annually. Thousands of them find refuge with their relatives while many of them are forced to live in camps where they become victim of physical and mental stress which destroy their health [15]. In 2014 more than youtube32 million people are displaced due to conflicts around the world of which mostly are the victim of conflicts in Syria and Congo [15]. According to survey around 2.6 million people were displaced in North Africa and Middle East in 2009 with Sahara Africa region considered to place the largest number of IDPs which is around 10.5 million [14].

Fata is an integral part of Pakistan that comprise of seven agencies. The Population of FATA is approximately four million [16]. Since ten years this part of the country is affected by challenges and instability as it is considered as the hub of terrorism in Pakistan [16]. The conflict among the Taliban and Pak Army across the FATA has seriously affected the lives of local people which forced them to move to camps in other part of the Province.

In year 2009, an operation was carried out against the militants in FATA that forced to displace more than two million people [17]. According to an estimate about 0.55 million people have shifted from FATA to KPK Province. Another 0.9 million people were shifted and registered as IDPs in parts of KPK [18]. According to report 90% of people were adjusted in KPK Province as IDPs [14].

The Government of Pakistan in collaboration with many INGOs including United Nations and others established several camps for displaced people of FATA in Nowshera, Hangu, Peshawar, D.I.Khan, Tank and parts of Mohmand Agencies [14]. According to the survey by FATA Disaster Management Authority (FDMA), about 0.4 million families are still registered and living in IDP camps in parts of KPK Province.

According to Fata Disaster Management Authority and UNHCR latest report in Khyber Pakthoonkhwa Province there are three main camps; Jalozai, New Durani and Togh Serai camp and several small camps with Jalozai Camp located in Nowshera is the biggest camp in terms of population i-e 12,400 families followed by New Durrani and Togh serai camp with 11,000 and 110 families respectively. Following table shows the total number of families living in these camps.

The total registered IDP families were about 0.38 million that are registered with NADRA Pakistan of which 12,302 are living in Jalozai Camp, 10,400 families in New Durrani Togh and 86 families hosted by Serai Camp [18]. According to [14] 54% of the

IDPs Population consists of men while 46% comprise of women. 9% of IDPs comprise of age between 5 to 11 and 7% between 0–5 years (Table 1).

Table 1. No of displaced families

Agency	Displaced families	Returned families	Total balance
Bajaur	86,406	66,200	155,606
FR Tank	2,260	960	3,220
Khyber	86,400	37,210	123,610
Khuram	48,650	18,606	67,256
Mohmand	45,950	16,23	47,573
Orakzai	48,990	23,121	72,111
SWA	56,296	21,363	77,659
Total Balance	3,74,952	1,69,083	5,47,035

A startling finding about the IDPs is that healthcare awareness among them is generally very low. According to ([19] hygiene and health are the major issues among the people of Tribal area mainly because they are uneducated and lack knowledge about the healthcare because of which majority of IDPs from FATA are facing severe health issues. However it has been observed that the guidance and knowledge provided by the LHVs to females in IDP Camps has help them a lot in regard to hygiene [19]. In 2012 a study was done by WHO which highlighted that 60% of the IDPs suffer from different diseases and health problems including malaria, scabies, diarrhea etc.s because of lack of knowledge about health.

1.3 IDP Camps in Pakistan

The Statement of Research Problem.

The research problem are specifically focused on the issues and reasons behind non implementation of E-health in health Projects in IDP Camps in Pakistan. For this, research problems have been identified to evaluate its reasons.

1.4 IDP Camps in Pakistan

Research Problem 1
To investigate the variety of perceptions among health care staff about E-Health and to systematically survey the availability of organizational support at basic information and communication (ICT) level.

Solution Methodology: A systematic survey will be done to know the perception of health care staff work in IDPs health Projects and basic ICT resources available.

1.5 IDP Camps in Pakistan

Research Problem 2

Extracting "significant factor's" to identify the reasons behind the failure of adoption and implementation of E-Health in health projects for IDPs in Pakistan.

1.6 IDP Camps in Pakistan

Solution Methodology: A detailed questionnaire was the research tool developed to identify the significant factors that are contributing in non implementation of E-Health in Health Projects. The data gathered would be then analyzed using SPSS.

2 Literature Review

The literature related to E-Health is adequate as E-health is being used around the world since decade. However literature related to E-health in Pakistan is scarce. The intensity is much more if we try to pursue literature related to E-health in rural areas of Pakistan and that in IDP Camps in Pakistan. The author was unable to search literature related to E-health in IDP Camps in Pakistan. However literatures related to E-health in rural areas of developing and under developed countries were available. Parts of literature were found but it was hardly in the interest of author and related to thesis.

Relevant researches were identified and collected from different sources and were thoroughly studied and analyzed to check the current status of E-health around the world. The studied researches reveled that majority of the studies were conducted to investigate the benefits of implementing E-health in BHU's, issues in implementing E-health and willingness of stakeholders for the implementation E-health in BHU's were studied. The relevant researches were grouped in accordance to benefits; issues in implementing E-Health and willingness to implementing E-health discussed below were divided into categories.

The first category of reviewed literature that were obtained; specifically focused on the benefits of using E-health in health sector. The information obtained was generally from book written by famous authors about the health and its benefits were obtained. Literature also includes journals published recently, articles related to E-health, and online articles about E-health and its benefits of using E-health.

The second category of literature that was obtained; specifically focused on 'issues in implementation E-health Literature related to this literature was easily available in form of researches that includes journal papers and conferences; online articles and from other reliable sources including web pages.

The third category of literature focused on willingness to use of E-health by organizations stakeholders. Data for this literature was collected from Journals, Online reports and articles, and case studies.

2.1 E-health and Its Benefits

This section of chapter will determine the benefits of using E-health in healthcare sector. Majority of the studies were of view that E-health had more benefits than having draw backs. The main aim of E-health is to offer benefits to end users by providing better health facilities. E-health is specifically helpful in rural areas where people can receive expert medical support from distant health units with easy access to patient medical history. [20] in their study described E-health as a tool take enable us to make more effective decision making and hence improve the quality of health care. Having E-health not only improves the quality and but also is more cost and time effective for patients who live in areas where health care support is not easily available. [21] in his book illustrate e-health double face as it not only improves the quality of healthcare but also it enables to reduce the overall time and cost of healthcare. [22] also in a study examined that implementation of e-health reduces the overall medical expense of the end users significantly. Related researched have also been done in past which highlights the benefits of E-health application. [23] in his study observed positive outcomes using E-health in chronic disease management. [24] in their study also concluded that health information system has a positive outcome when it comes to quality, efficiency and satisfaction. They also found that more than 92% of the recent studies were positive about the benefits of using health information system. [25] in his research stated that E-health could be very useful for mental health patients and advance collaborative management. A famous butler project studied the satisfaction and acceptance level of E-health through connection of patients with different E-health applications. The results showed that the patients had a positive impact and shoed great improvement in their daily activities.

Studies show that there are various benefits of using E-health application subject to successful implementation and acceptance by the organization and stakeholders. However its adaptation is also influenced by the culture, skills, policies and individual perception.

2.2 Issues in Implementation, Acceptance and Adoption of E-Health

Implementation of E-health has always been difficult as many factors need to be considered such as social, cultural issues. Organization issues also play its part in use of technology in the field of health. These issues need to be addressed in order to assure implementation of Technology in health sector. Before the implementation of any system, proper planning and research must be done in order to assure its success. Implementations of Technology in health sector also require a great deal of research and planning to make it successful in that part of the world. South Korea, in this regard was the most successful country where E-health application was promoted in 2003 by the Government and by the end of year 2008, E-health was implemented and adopted in most part of the country.

[26] in their study described introduction of Technology in healthcare as a dynamic process, that require certain planning at organizational level before its implementation can be done. [27] in his study reported the challenges and issues that were faced during the implementation of E-health in European countries. The authors observed the use of E-health especially telemedicine is the future and were of view that implementation

of E-health can be useful in different aspects including history of patient data to any specialist and doctor. Other uses of E-health could be telenursing, teleconsulting and online inspection of patient's data with full patient history and records which could be very useful and convenient for patients as well as doctors. The author is in a view that using E-health health call centers could also be established to provide online health facilities to patients in future. Patients with chronic diseases could be easily paid more attention using online health services. The author emphasized that the use of E-health application can be beneficial for both end users which can improve the issues and reduce the cost, time and improve the standard of life and health for patients especially in rural areas where basic health facilities are not up to the desired standard. The authors were also of a view that that implementation of E-health in would be difficult initially as acceptance level of end users will low due the change of health system but slowly and gradually the patients and health care staff will accept and adopt the new style of health treatment. [28] designed an online questionnaire to examine the level of acceptance of E-health among the patients of different culture. The research consisted of three models of IT acceptance among patients who registered for E-health services. The results showed that the acceptance level of all tested IT models were positive which proved that the patients were intended to use E-health system in future. In [29] in their research examined Clinical Communications Implementation (ECCI) its adoption and use on among healthcare staff in different hospitals and private clinics in Scotland. ECCI initially found to be useful for end users however different issues found like inconsistency and data duplicity and reliability made it less effective than expected.

In past decade several researches have been conducted to examine use and benefits of E-health. In spite of several benefits of electronic health it is many countries in the world are still reluctant to adopt e-health system facility among its rural areas. The adaptation and implementation issues are more found in under develop and developing countries. The main reason could be because the trend of acceptance of new technology in these countries are less as compared to developed countries where they are more interested in new innovations and adoption of latest technologies and systems. Other reasons such as implementation and its maintenance issues, unawareness among the users and acceptance by stakeholder, language and culture barriers could also play its part in non-implementation of E-health system in some countries. Therefore in order to ensure successful implementation of E-health system it is important that the stakeholders and government and NGOs should play its part for adoption and implementation of E-health system.

2.3 Willingness to Use E-Health

In order to assure successful execution of a system, the most significant step is to take stakeholders into confidence to prevent the system from failure. Hence various researches have been conducted to assess the willingness of use of E-health system around the world. [30] in his research designed a tool to assess the willingness to use E-health system in under developed countries health departments. The tool was developed with an aim to examine the willingness of doctors and managers to use E-health system in their institutes. Two different tools were designed by the authors one of which was for the managers and doctors while other tool was designed to cover the healthcare staff.

The tools consisted of three categories with each category represent determinant to use and implementation of E-health system. These tools contained both qualitative and quantitative questions and was interviewed based. These tools were later applied on trial basis in the hospitals in Pakistan.

[31] examined E-Health in Hospitals in Pakistan as a test case. They found E-Health to be very useful and found that majority of the healthcare staff had knowledge about E-Health system and its use. [32] examined the willingness of end user E-Health system using grid-based model. Through this framework the author defined levels of preparedness as high, medium and low and identifies four main requirements. These include pre defined policies, government interest and willingness and proper IT structure in order to ensure that the E-health system is proper implemented and the users are willing to use it. On the other hand some impact needs to be address before doing so that include knowledge about ICT and culture issues to be solved in order to achieve the desired objectives. The framework was very useful to examine the essentials that are needed to implement the E-health system in an organization. The framework designed in their study was very useful to recognize the preparedness of using E-health system, but it had certain limitations that were later addressed by Chattopadhyay in his study. [33] recommended an abstract framework theory and integrated approach to assess the willingness of organizations to use ICT in health department. The framework designed was applied at different levels to identify and measure the willingness of use of E-health system at each level including Government and Non Government organization level. The framework was initially implemented in Cardiogly health department. The work done was later extended by Li et al. in his research in 2009 by assessing the willingness level specifically focusing on Electronic Health Record in E-health system. Both the studies conducted had certain limitation. One of the major drawbacks of the studies was that both the frameworks were theoretical in nature and needed empirical validation to real assumptions. [34] found E-Health to be very useful and were optimistic about the implementation of E-Health around the continent soon.

The available literature discussed above related to implementation, adoption and willingness to use E-health system was generally related to developed countries and it seems that no real study related to E-health in Pakistani perspective has been conducted. However many researchers have been done by Pakistani researches related to E-health and some scattered attempts have been found but again no real research have been found realted to implementation, adaptation and willingness to use E-health system in health departments in rural areas specifically in IDPs camp in KPK and other provinces. To overcome this gap, the current research is conducted in Pakistani perspective. The research tries to examine the awareness of E-health system among health staff and willingness to use E-health system specifically in IDPs health projects in Pakistan. To do so, following research questions were developed to know the real reasons.

To investigate the variety of perceptions among health care staff about E-Health and to systematically survey the availability of organizational support at basic information and communication (ICT) level.

Extracting "significant factor's" to identify the reasons behind the failure of adoption and implementation of E-Health in health projects for IDPs in Pakistan.

3 Analysis and Discussion

Data collected through questionnaire was later structured into nine identified factors with eight independent variables and one dependent variable and were accessed using SPSS for extracting signification factors to know the reasons why E-health has not been adopted in IDPs Health Projects in Pakistan (Table 2).

Table 2. Factors/attributes in 3 point scale. Factors identified (One Dependent & Eight Independent Variables).

S. no	Factors	Three point scales
1	Basic Knowledge of Computer	Advance (2.0); Basic (1.0); No (0)
2	Are you willing to use Computer at BHU	Yes (2.0); Not Sure (1.0); No (0)
3	Knowledge of Electronic Health Record (EHR)	Advance (2.0); Basic (1.0); No (0)
4	Use of Electronic Health Record at Work Place	Yes (2.0); Not Sure (1.0); No (0)
5	Availability of Electricity at Work Place	Yes (2.0); Sometimes (1.0); No (0)
6	Availability of Computer with Internet Connection at Work Place	Yes (2.0); Sometimes (1.0); No (0)
7	Total Computers for Official Use at Work Place	<2 (2.0); 1–2(1.0); 0 (0)
8	Use of Computer at Work Place	Office Use(2.0); Personal & Official (1.0); Personal (0)
9	Are you Satisfied with existing Paper based Health Record System?	Extremely Satisfied (2.0); Not Sure (1.0); Not Satisfied (0)

Factors 1 to 8 are Independent Variables while Factor 9 is dependent variable.

4 Perception of Healthcare Staff

On the basis questionnaire based research tool that was developed to know the perception of Healthcare Staff following response were received from the respondents (Fig. 1).

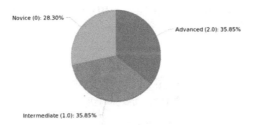

Fig. 1. Knowledge of computer

Basic Knowledge of Computer?

The above figure shows that the respondents with advance and intermediate knowledge of computer were almost the same with very few respondents had no or very little knowledge of computer i-e 28% only while 36% respondents had advance and intermediate knowledge of computer (Fig. 2).

Willingness to have a computer at BHU?

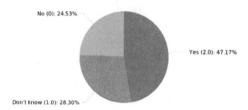

Fig. 2. Willingness to have computer

The graph shows that almost half of the respondents were willing to use computer for electronic health record while 28& of the respondents were not sure about using computer at BHUs and 24% were not in favor of using computer to record patients health data electronically (Fig. 3).

Knowledge of Electronic Health record (EHR).

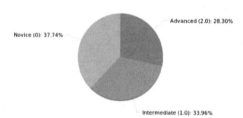

Fig. 3. EHR knowledge

The level of knowledge of Electronic health Record (EHR) among all the respondents is almost of equal ratio. 28% of respondents had advance knowledge of EHR while 34% and 38% of the respondents had intermediate or no knowledge of Electronic Health Record (Fig. 4).

Use of Electronic Health Record (EHR) in BHUs and its training.

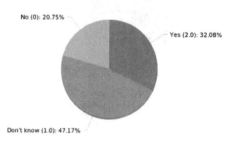

Fig. 4. Use of EHR

The level of knowledge of Electronic health Record (EHR) among all the respondents is almost of equal ratio. 28% of respondents had advance knowledge of EHR while 34% and 38% of the respondents had intermediate or no knowledge of Electronic Health Record (Fig. 5).

Satisfaction using current paper-based Personal Health Record (PHR) system?

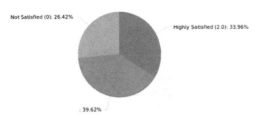

Fig. 5. Satisfaction using PHR

The satisfaction level of respondents using Paper based record is high compare to non satisfaction which is around 26% while majority of the respondents were neither highly satisfied not satisfied with the current PHR system (Fig. 6).

Availability of steady electricity in BHUs?

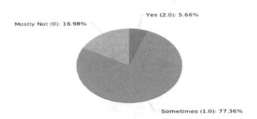

Fig. 6. Availability of electricity

One of the most important factors in implementing E-Health in IDPs Health Project is availability of Electricity. The above graph shows that the electricity availability to 77% of the respondents is sometimes while 17% has no facility of electricity at their BHUs and only 6% of BHUs ha continuous supply or availability of steady electricity (Fig. 7).

3.2.7 Availability of Computer and Steady Internet Connection?

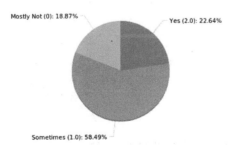

Fig. 7. Availability of steady internet

Another important factor that contributes in the implementation of E-health is availability of steady internet connection. Unfortunately only 23% of the respondents working on health projects have availability of computer with steady internet connection with 19% of the respondents have no facility of internet. Another 59% of the respondents sometimes have the facility of internet at their BHUs (Fig. 8).

Available Computers with software in BHUs?

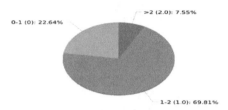

Fig. 8. Available computers with software's

The above figure shows that majority of the respondents have one to two computers only with software that are essential in implementing E-Health while 23% do not have any computers at BHUs and only 8% have got two or more than two computers with software's (Fig. 9).

Nature of use of computers at BHU?

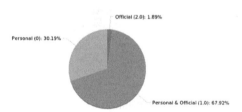

Fig. 9. Use of computer

Interestingly according to the above figure 30% of the healthcare staff use computer for their personal use in Health Project while only 2% of the staff use computers for official work while 68% of the healthcare staff use computer for both personal and official purpose at their work place (Fig. 10).

Perception of Healthcare Staff about Use of E-Health System.

Can E-Health system be applied to improve quality and reduce the cost of healthcare?

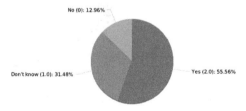

Fig. 10. Can E-Health improve Quality & Cost of health

Majority of the respondents were of a view that using E-Health system can improve the quality and cost of healthcare while 31% of respondents were not sure and 13% of a view that implementing E-Health will not have effect on cost and quality of healthcare (Fig. 11).

Can E-health improve the quality of services in Health Projects?

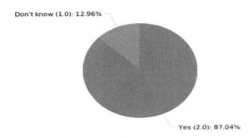

Fig. 11. Can E-Health improve the Quality of Service

The above graph shows that majority of the respondents believe that using E-health technology can improve the quality of service that are provided to IDPs in BHUs (Fig. 12).

How often patients are referred to another healthcare center or hospital other than BHUs for treatment?

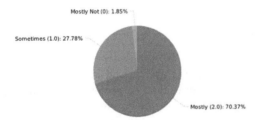

Fig. 12. No. of Patients referred to other Hospitals

The above result shows that due to lack of resources 70% of the times the patients in IDPs camp are referred to other hospital. The main reason is lack of resources which can overcome with the help of technology. Therefore it is essential to use E-health system in rural areas like IDP Camps.

5 Statistical Data Using SPSS

The data collected through questionnaire were structured and multiple regression and significant tests were performed using SPSS to identify the correlation among the variables and significant factors which are important to identify in order to know the factors due to which E-health is yet to be implemented in IDPs Health Project in Pakistan. In total nine factor were checked with eight independent variables and one dependent variable.

6 Findings, Conclusions and Recommendations

The data was analyzed using SPSS to know the significant factors. The results shows that at personal level the healthcare staffs working in 11 NGOs/INGOs in IDP camps in Pakistan are willing to adopt and use E-Health system (refer to significant factor "A") but due to unavailability of Information and Communication Technology (ICT) (refer to significant factor's B, C, D) that is essential in implementation of E-Health system. In other words we can say that the stakeholders know the benefits of E-health system but cannot implement due to basic ICT facilities in project units. These include unavailability of electricity, steady internet facility and because of which the organizations have not provided staff with computers, printers, fax machines and other ICT infrastructure. However, these ICT infrastructures can easily be improved and provided by the organizations and others stakeholders by providing health units with generators for continuous supply of electricity; steady internet can also be provided as many internet wifi devices are easily available now a days; and provide computers or laptops to staff with accessories including printers, fax and photo copier machines, and scanners with important software including EHR and other renowned software's. Through ICT connections can easily be established with the province major hospitals, specialists, surgeons in order to share data of patients and provide quality and cost effective health services to people of camps.

7 Recommendations

On the basis of results it can easily be inferred that lack of basic ICT facilities at Health Project Unit is the reason why paper based health record (PHR) approach is still used in this modern era. Before the start of research it was believed that may be the healthcare staff are reluctant to use E-Health system and are happy with the PHR system but the research results are opposite to that. At personal level majority of the staff is not highly satisfied with PHR system and they want E-Health system to be implemented but due to lack of facilities the system is yet to be implemented. Hence organization and other stakeholders that are implementing health projects in IDP Camps are to be blamed for lack of ICT infrastructure. It is therefore the responsibility of provincial health department who heads the overall projects to compel all the organizations to provide basic ICT infrastructure in health units. The health department should assure before the start of project that health units have all the facilities that are required for electronic health record (EHR) of patients. It must be assured that the entire patient's medical history is recorded with and patients who require surgeons and specialist advice can make connection via teleconference and medical history of patient is shared with particular specialist so that proper health treatment can be advised rather than referring patients to hospital in other cities which is costly, time consuming and difficult for a patient. Therefore in order to provide quality health service with cost and time effective it is important to adopt E-Health system in IDPs health projects too, which not be contribute in project success but can have all the medical history of patients intact that can be helpful in providing patients more effective healthcare facilities.

8 Future Work

The research can be enhanced in more depth.

The primary focus in this research was health projects that are implemented in IDPs Camps in KPK only. In future other health projects around the country can also be focused with more detailed analysis.

The study can form a baseline for the needs assessment and the actual need of the healthcare organizations in the rural and remote areas of Pakistan as well as IDP camps.

References

1. Bennett, N.L., Casebeer, L.L., Zheng, S., Kristofco, R.: Information-seeking behaviors and reflective practice. J. Contin. Educ. Heal. Prof. **26**(2), 120–127 (2006)
2. Eysenbach, G.: What is e-health? J. Med. Internet Res. **3**(2), e20 (2001)
3. Chew, F., Grant, W., Tote, R.: Doctors online: using diffusion of innovations theory to understand internet use. Family Med.-Kansas City **36**, 645–650 (2004)
4. Broom, A.: Virtually he@lthy: the impact of internet use on disease experience and the doctor-patient relationship. Qual. Health Res. **15**(3), 325–345 (2005)
5. HIMSS SIG develops proposed e-health definition. HIMSS News **13**(7), 12 (2003)
6. Docteur, E., Oxley, H.: Health-care systems: lessons from the reform experience (2003)
7. Tustin, N.: The role of patient satisfaction in online health information seeking. J. Health Commun. **15**(1), 3–17 (2010)
8. Hesse, B.W., et al.: Trust and sources of health information: the impact of the internet and its implications for health care providers: findings from the first Health information national trends survey. Arch. Int. Med. **165**(22), 2618–2624 (2005)
9. WHO. http://www.who.int/countries/pak/en/ (2012)
10. Qureshi, Q.A., Ahmad, I., Nawaz, A.: Readiness for e-health in the developing countries like Pakistan. Gomal J. Med. Sci. **10**(1) (2012)
11. Nasrullah, M., Muazzam, S., Bhutta, Z.A., Raj, A.: Girl child marriage and its effect on fertility in Pakistan: findings from Pakistan demographic and health survey, 2006–2007. Matern. Child Health J. **18**(3), 534–543 (2014)
12. http://www.drzaki.org/associations/ehap/. Accessed 22 Oct 2014
13. Powell, S., Rosner, R., Butollo, W., Tedeschi, R.G., Calhoun, L.G.: Posttraumatic growth after war: a study with former refugees and displaced people in Sarajevo. J. Clin. Psychol. **59**(1), 71–83 (2003)
14. http://www.unhcr.org/pages/49c3646c146.html. Accessed 6 Nov 2016
15. http://www.internal-displacement.org/publications/2013/global-overview-2012-people-internally-displaced-by-conflict-and-violence/. Accessed 24 Mar 2015
16. Chaudhry, S.A.: A report by FATA Research Centre on crisis of IDPs in FATA: issues, challenges and way forward (2014)
17. Said, S.: The unrealized crisis of FATA IDPs, 1 November 2013. [http://frc.com.pk/articles/the-unrealized-crisis-offata-idps/] (2012)
18. Khan, Z.A.: Military operations in FATA and PATA: implications for Pakistan. Strateg. Stud. **31**, 136–145 (2012)
19. Khan, S.: Health assistance to internally displaced persons of South Waziristan Agency in camps and host community (2014)
20. Black, A.D., et al.: The impact of eHealth on the quality and safety of health care: a systematic overview. PLoS Med. **8**(1), 188 (2011)

21. Gerkin, D.G.: E-health can be a two-edged sword for the medical doctor. Tenn. Med. **102**, 7–8 (2009)
22. Akematsu, Y., Tsuji, M.: An empirical analysis of the reduction in medical expenditure by e-health users. J. Telemed. Telecare **15**(3), 109–111 (2009)
23. Marziali, E.: E-health program for patients with chronic disease. Telemed. J. E Health **15**, 176–181 (2009)
24. Buntin, M.B., Burke, M.F., Hoaglin, M.C., Blumenthal, D.: The benefits of health information technology: a review of the recent literature shows predominantly positive results. Health Affairs **30**(3), 464–471 (2011)
25. Cornish, P.A., Church, E., Callanan, T., Bethune, C., Robbins, C., Miller, R.: Rural interdisciplinary mental health team building via satellite: a demonstration project. Telemed. J. E Health **9**(1), 63–71 (2003)
26. Cresswell, K.M., Worth, A., Sheikh, A.: Actor-network theory and its role in understanding the implementation of information technology developments in healthcare. BMC Med. Inform. Decis. Making **10**(1), 67 (2010)
27. Vitacca, M., Mazzù, M., Scalvini, S.: Socio-technical and organizational challenges to wider e-Health implementation. Chron. Respir. Dis. **6**(2), 91–97 (2009)
28. Wilson, E.V., Lankton, N.K.: Interdisciplinary research and publication opportunites in information systems and health care. Commun. Assoc. Inf. Syst. **14**(1), 51 (2004)
29. Pagliari, C., Donnan, P., Morrison, J., Ricketts, I., Gregor, P., Sullivan, F.: Adoption and perception of electronic clinical communications in Scotland. Inf. Prim. Care **13**(2), 97–104 (2005)
30. Khoja, S., Scott, R.E., Casebeer, A.L., Mohsin, M., Ishaq, A.F.M., Gilani, S.: e-Health readiness assessment tools for healthcare institutions in developing countries. Telemed. e-Health **13**(4), 425–432 (2007)
31. Shoaib, S.F., Mirza, S., Murad, F., Malik, A.Z.: Current status of e-health awareness among healthcare professionals in teaching hospitals of Rawalpindi: a survey. Telemed. e-Health **15**(4), 347–352 (2009)
32. Wickramasinghe, N.S., Fadlalla, A.M., Geisler, E., Schaffer, J.L.: A framework for assessing e-health preparedness. Int. J. Electron. Healthc. **1**(3), 316–334 (2005)
33. Chattopadhyay, S., Li, J., Land, L., Ray, P.: A framework for assessing ICT preparedness for e-health implementations. In: 10th International Conference on e-health Networking, Applications and Services, 2008. HealthCom 2008, pp. 124–129 (2008)
34. Li, J., Land, L.P.W., Ray, P., Chattopadhyaya, S.: E-Health readiness framework from electronic health records perspective. Int. J. Internet Enterprise Manage. **6**(4), 326–348 (2010)
35. Durrani, H., Khoja, S.: A systematic review of the use of telehealth in Asian countries. J. Telemed. Telecare **15**(4), 175–218 (2009)

Environmental Parameters Monitoring System with an Application Interface for Smartphone

Jorge Ramos[1], André Teixeira[2], Carlos Arantes[3], Sérgio Lopes[4], and João Sena Esteves[4(✉)]

[1] Smart Grid Division, Efacec, Porto, Portugal
jorge.silva.ramos@efacec.com
[2] Department of Industrial Electronics, School of Engineering, University of Minho, Guimarães, Portugal
a71913@alunos.uminho.pt
[3] Protection, Automation and Control Division, Efacec, Porto, Portugal
carlos.arantes@efacec.com
[4] R&D Centre Algoritmi, Department of Industrial Electronics, School of Engineering, University of Minho, Guimarães, Portugal
{sergio.lopes,sena}@dei.uminho.pt

Abstract. Air pollution has become a major cause of health problems and mortality. Monitoring systems based on fixed stations are not able to adequately characterize air pollution on a large area such as city. This paper describes the study, development, implementation and validation of a mobile system for monitoring environ-mental parameters. The system consists of a sensor network, a communications system, a web application and a smartphone application. The sensor network is scalable as it has a bus topology and can collect data about nitrogen dioxide and particulate matter. The communications system aggregates the collected data and sends it to a remote web server using the HTTP protocol and a 3G mobile network. The web and the smartphone applications provide ways for viewing the data stored on the remote web server. The Google Maps API was used to represent the obtained records on a map. All system components were tested individually and, at a later stage, in an integrated manner. The results obtained in tests with the developed system were compared with the results from fixed stations. The measurement errors were less than 10%.

Keywords: Environmental monitoring · Mobile monitoring system · HTTP protocol · Smartphone application

1 Introduction

The daily life of society has been affected by the increase in environmental pollution, either at atmospheric level or at sound level. According to a report by the European Environment Agency in 2016, air pollution was responsible for 467 000 premature deaths in Europe [1]. In order to reduce these numbers, systems have been implemented to monitor ambient air. These systems allow the creation of air quality indicators and, consequently, the determination of critical points and the implementation of air pollution prevention and reduction measures. Monitoring consists of a "series of observations and measurements of some physical, chemical or biological variables, which make it possible to understand and predict some environmental changes" [2]. The main polluting substances monitored are carbon monoxide, nitrogen dioxide, sulfur dioxide, ozone and particulate material [1, 3]. This last substance consists of solid or liquid material that is suspended on the air in the form of particles [4].

Traditionally, air monitoring is done from networks of stationary stations. These networks (still used today) are made up of stations with air quality sensors, installed in strategic areas of the cities. From the data fusion from each station, it is possible to measure air pollution in respective city spots. To characterize the air pollution in the city, the number of stations that make up the network should be proportional to the area to be monitored. For example, in 2017 a monitoring system of this type was introduced in the Philippines. This system consists of 4 new large fixed stations, which allow the monitoring of particulate material, sulfur dioxide, ozone and nitrogen dioxide in real time. This system constantly measures the levels of concentrations of polluting substances and generates, on an hourly basis, the average for each one [3].

Despite their high dissemination, these networks of fixed stations require high installation and maintenance costs. Thus, most cities contain a reduced number of stations for the area to be covered, making the data collected inefficient and the characterization of the environment incomplete [5]. For example, the city of Beijing, with about 16.800 km^2, is covered by only 35 monitoring stations [6]. In this way, areas outside the reach of stations are estimated using mathematical methods, making it impossible to detect local changes in air quality, such as sporadic emissions and heavy traffic [8–10].

The main objectives of the work described in this paper were the study, development, implementation and validation of a mobile system for monitoring environmental parameters. The goals of the project were:

a) flexibility to install a variable number of sensors to measure different environmental parameters;
b) georeferentiation of every sensor reading, enabling to provide detailed maps of air quality on given areas;
c) to provide the data through multiples platforms, web and mobile, in real-time.

2 Related Work

This chapter presents two examples of existing air quality monitoring stations, one fixed and one mobile, with the latter being implemented in vehicles. The main characteristics, advantages and drawbacks of these two types of stations are summarized.

2.1 Fixed Monitoring Station

The CitySense project, implemented in Cambridge, is an example of the implementation of fixed stations [11]. Each station is made up of a set of sensors capable of measuring temperature, relative humidity, wind speed and atmospheric pressure. In addition, it also consists of a box of reduced dimensions (about 8 × 8 cm), containing carbon dioxide sensors and all the necessary hardware for data communication via GSM (Global System for Mobile Communications). The carbon dioxide sensors used are Vaisala GMP343. These sensors are characterized by having low energy consumption, good accuracy and a total cost of around 2500 € [11, 12].

From this example, it may be concluded that the use of small stations scattered around cities is a possibility for air quality monitoring systems. Since the cost and size of these stations is significantly less than that of conventional stations, it is possible to introduce a larger number across cities. In this way, more values are obtained for the area to be covered, improving the characterization of air quality [6]. Although this is just one example, it has the characteristics common to other fixed systems described in [6]. In Table 1, the advantages and disadvantages related to monitoring from fixed stations are pointed out [6, 7].

Table 1. Main advantages and disadvantages of a fixed monitoring station

Advantages	Disadvantages
Use of high-quality hardware, allowing the acquisition of accurate and reliable data	Careful placement of monitoring stations due to the lack of air pollution standard
Lack of localization systems	Placement of a number of stations proportional to the area to be covered
Use of several sensors without constraints regarding the physical dimensions and total weight of the system	Use of mathematical methods in order to predict air quality in unmonitored areas (extrapolation)
Easy system maintenance	Limited measurement area
	High cost of implementation

2.2 Mobile Monitoring Stations

Google, in partnership with Aclima, a company in the field of air quality, has implemented three street view cars systems that collect data about air quality. The main polluting substances monitored were nitrogen monoxide, nitrogen dioxide and black carbon particles. Data collection was done for one year across all streets within the residential, industrial and commercial areas of Oakland, California. After that year, all the data obtained were analyzed, concluding that this type of monitoring reveals a "remarkable and stable heterogeneity in the daily concentrations of some pollutants" [13]. In addition, it was found that some areas within the city have a greater air pollution, for example, in the vicinity of a metal scrap the levels of nitrogen monoxide and dioxide were higher than the levels of the rest of the street. The same happened near highways, that is, as the collection was carried out farther from the highways, the concentrations of the parameters decreased [13].

This monitoring system consists of an air collection system on the top of the car and a measurement system implemented in the trunk of the car. The collection system enables the ambient air to be sent to the measurement system, ensuring that it is constantly changed. Regarding the measurement system, it consists of high cost laboratory analyzers, which allows to acquire precision and accuracy in the measurements made. Whenever new measurements are collected, they are sent to the Aclima cloud where they will be further processed [13]. Table 2 shows the advantages and disadvantages related to monitoring from mobile stations [14].

Table 2. Main advantages and disadvantages of a mobile monitoring station.

Advantages	Disadvantages
Data acquisition in different locations	Data redundancy directly related to the number of vehicles traveling in an area
Visualization of the values collected in different locations in real time	Introduction of a localization module to geo-reference the values obtained
Lower cost in the implementation and acquisition of the system	Introduction of a communication module for sending data from the respective collection site. Thus, it will be necessary to use mobile data, leading to an increase in the cost of use

3 System Architecture

This section describes the components and procedures of the proposed monitoring system. The developed system architecture is represented in Fig. 1.

Each Monitoring Unit is based on a microcontroller connected with sensors. These units are responsible for collecting data related to the environmental parameters and sending them to the local Central Unit through a LIN bus. The LIN bus implements a sensor network that offers: (1) scalability, to increase the number and variety of sensors, for example to monitor also cabin air parameters such as CO2; (2) easier deployment in vehicles, by minimizing the cables necessary to interconnect exterior/engine and cabin units and facilitating the installation of sensors in different vehicle locations. The sensors used consist of one MiCS-4514 to measure the Nitrogen Dioxide (NO2) and a HPMA115S0-XXX to measure the particulate material, namely PM2.5 and PM10. In addition, a DHT 11 sensor was implemented to measure temperature and relative humidity measurement. These sensors have the best cost-benefit ratio for the proposed system.

The Central Unit consists of a microcontroller, a GPS module and a 3G communications module. This unit receives the data collected from Monitoring Units through LIN BUS protocol, groups it and sends it to the Cloud Service along with geographic coordinates.

The Cloud Service saves the data received from central units and makes it available to client applications, namely a web application and an Android application that allows users to easily access real-time air quality data for the location they are at.

In the android application, the AQI (Air Quality Index) values are computed using the data read from the Cloud and represented on a map. This is explained in more detail in subsection below.

Environmental Parameters Monitoring System   453

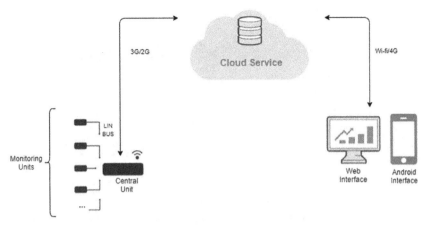

Fig. 1. Block diagram of the monitoring system, adapted from [15].

3.1 Monitoring Units and Central Unit

The Monitoring Unit architecture is presented side by side with the respective PCB in Fig. 2. As shown in the figure, the monitoring units are equipped with a UART, I2C and ADC interfaces that allow the choice of parameters and sensors to be used by the end user. In addition, a connector for gas sensors of the MQ series was introduced as they fit into low-cost sensors for measuring a wide variety of gases. The unit is powered by the 12 V voltage of the Central Unit and LIN protocol is used for communication between all units.

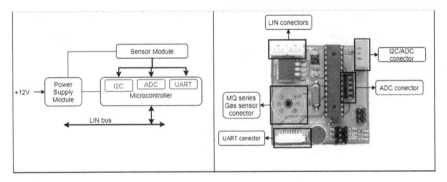

Fig. 2. Monitoring Unit architecture on the left and the respective PCB on the right.

Figure 3 presents the architecture and the PCB of the Central Unit. The board has a 3G modem, a GPS module and connectors for the LIN and CAN communication protocol. The 3G module was chosen considering the available budget. The CAN Bus is used to enable communication with the car's CAN network. This unit is powered by a +12 V power supply, such as a car battery or cigarette lighter adapter.

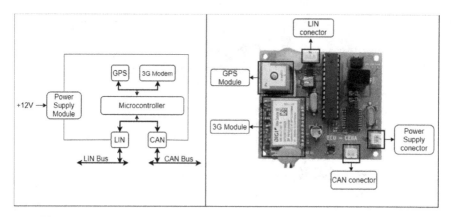

Fig. 3. Central unit architecture on the left and the respective PCB on the right.

4 Data Transmission and Storage

The Cloud Service is based on the ThingSpeak IoT platform. This platform provides storage, analysis and real time data visualization. It stores data in channels, each one with a maximum of 8 data fields. In the developed Cloud Service, each channel corresponds to a Central Unit, and each field is a measured parameter (temperature, humidity, PM2.5, PM10, NO2, latitude, longitude).

The communication with Central Units as well as client applications is implemented using HTTP protocol. For sending data, an HTTP POST method is used by Central Units in which, the data frame sent contains the sensor measured values and the location at which they were obtained. A message is sent periodically every 10 s.

The HTTP POST message structure is presented in Fig. 4 and displays the fields for each monitored parameter. It is important to note that field 6 is not defined since it is used for testing purposes. After receiving the message, the Cloud Services assigns this dataset an identifier (entry_ID).

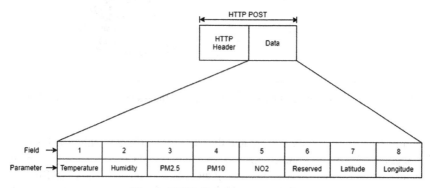

Fig. 4. HTTP Post frame structure.

Client applications request data from the Cloud Service using HTTP GET method. The desired channel is defined in the URL and the reading api key required for private channels is added as query parameters. The response contains the last saved data in a json format as shown below:

{"created_at": "2020-11-29T12:13:23Z", "entry_id":142, "Temperature": "15", "Humidity": "60", "PM2.5": "1.3", "PM10": "6.0", "NO2": "25.1", "Latitude": "41.407067", "Longitude": " − 8.520499"}

The first key/value pair corresponds to the time the data was recorded in the Cloud. The keys field 1 to field 5 and their respective values represent the environmental parameters sent. Finally, the last two key/value pairs correspond to the location (latitude and longitude respectively).

5 Mobile Application

A mobile application was developed to present the data collected by the mobile environmental station. It is an Android app that computes the Air Quality Index (AQI) for a central unit and shows their data on a map. The AQI consists of a color classification for the different levels of the air quality. The color assignment is related to an AQI value calculated from the pollutant concentrations. The Android app was writ-ten in Java and it uses the Google Maps service API. Its interface consists of a unique Activity with a map where colored markers are displayed according to the calculated AQI value. The markers also show the individuals sensor values obtained from the Cloud Service.

The app flowchart is represented in Fig. 5. First, an HTTP GET is made to obtain the json data that contains the latest sensor readings. The entry_ID value is compared with the last entry_ID to ensure that a new marker is only drawn when a new dataset is uploaded to the Cloud. If the entry_IDs does not match, the AQI calculation for each pollutant is performed by comparing the values of concentration with a reference table values, using the following mathematical model:

$$I_p = \frac{I_{Hi} - I_{Lo}}{BP_{Hi} - BP_{Lo}}(C_p - BP_{Lo}) + I_{Lo} \tag{1}$$

Where:

- I_p is the index for pollutant p;
- C_p is the concentration of the pollutant p;
- BP_{Hi} is the concentration breakpoint that is greater than or equal to C_p (Table 3);
- BP_{Lo} is the concentration breakpoint that is less than or equal to C_p (Table 3);
- I_{Hi} is the AQI value for the upper limit BP_{Hi} (Table 3);
- I_{Lo} is the AQI value for the lower limit BP_{Lo} (Table 3).

Then, the final AQI value is obtained by the maximum value of the AQI calculated for each pollutant. Finally, a custom marker with the corresponding AQI color is drawn on the map. When pressing the marker, pollutant information, temperature and humidity and occurrence date are shown. All previous markers remain on the map.

This algorithm is executed every 5 s, so that the Cloud data is periodically read.

All these procedures are executed in background to avoid freezing the map interface.

Table 3. AQI table [16]

Pollutant / Classification	PM_{10} (μg/m³)		$PM_{2.5}$ (μg/m³)		NO_2 (μg/m³)	
	Min	Max	Min	Max	Min	Max
Very good 0-25	0	15	0	10	0	50
Good 25-50	15	30	10	20	50	100
Fair 50-75	30	50	20	30	100	200
Bad 75-100	50	100	30	60	200	400
Very Bad >10	100	-	60	-	401	-

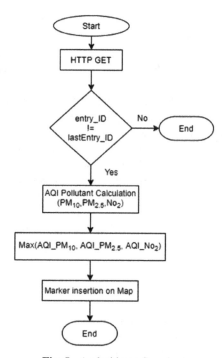

Fig. 5. Android app flowchart

6 Results

The results obtained from experimental tests carried out on the Android application, which include communication with the Cloud, are presented below in order to vali-date the system implementation.

Whenever the application starts-up, a Google map is created and the marker corresponding to the last data transmission is added. As a new marker is added the focus of the map is redirected to that point (see Fig. 6). The added marker shows the color corresponding to the AQI. The marker info window was customized in order to display the detailed information in a table format. To access this window, it is only necessary to click on the desired marker. On top of the info window the AQI classification obtained with the appropriate color is presented. The parameters, the measurements and the AQI value calculated for each one appear below. At the end, the time and date when the measurement took place are displayed.

The application also allows switching the map view between normal, satellite, or hybrid.

Fig. 6. Android Interface.

7 Conclusions and Future Work

The use of fixed monitoring stations has several limitations in terms of environmental monitoring. The need to estimate data for an area based on measurements made in a few places was a problem for this type of system, as it makes it impossible to create reliable air quality profiles that follow the change and the existing standards in a given area. The use of mobile monitoring stations, namely systems implemented on mobile platforms, makes it possible to take advantage of the movements of these platforms to collect data from different zones. In this way, measurements are taken in more places, achieving a better characterization of the monitored areas.

A fully functional prototype of a monitoring system was developed to be implemented in a car, which allows the monitoring of some of the main polluting substances: particulate matter and nitrogen dioxide. This system has a sensors network that make it scalable, enabling the addition of new units for measurement both inside and out-side the vehicle. In addition, the expandability of the sensor network is also guaranteed by the modularity of the monitoring units – the existence of several interfaces facilitates the replacement of the sensors. The implemented system makes possible to send data to a remote server or to the car's CAN network. This last feature enables, for example, to generate an alert whenever the air quality of the passenger compartment is outside a defined range. The data sent to the Cloud can be available online, allowing it to be viewed by the population, as well as the definition of long-term policies to improve air quality. The smartphone application is responsible for viewing the data as it allows viewing the records on a virtual map provided by the Google Maps service and also for the treatment and processing of the data obtained. The prototype developed is a low cost system that easily provides valuable data for responsible authorities and for citizens through both web and mobile platforms.

The achieved results fulfill the proposed objectives. In the first tests, the values obtained with the new system were compared with the results from the fixed stations, obtaining measurement errors of less than 10%. In the remaining tests carried out, the system collected data in several areas, taking advantage of the car's movement. Using the system, it was found that the air quality in the city center of Vila Nova de Famalicão, Portugal, is lower than the air quality in its periphery. In order to reach this conclusion, a large number of measurements were made in each of the zones under evaluation.

Regarding future work, there are some aspects of the current system that could be improved. The following changes are suggested:

- Use of electrochemical sensors in the monitoring units to reduce the error of the measurements, obtaining more accurate values;
- Addition of an SPI interface allowing the introduction of more sensors;
- Provide wider coverage and reduce communications cost by replacing the 3G module with one that supports NB-Iot;
- Implementation of an interpolation of the obtained data to be able to provide a continuous map of air quality.

Acknowledgements. This work has been supported by FCT – Fundação para a Ciência e Tecnologia within the R&D Units Project Scope: UIDB/00319/2020. The authors are also grateful to CEIIA – Centre of Engineering and Product Development. CEIIA provided a very significant support in several parts of this work, namely the ones related with hardware development and software communication.

References

1. Guerreiro, C., González Ortiz, A., Leeuw, F., Viana, M., Horálek, J.: Air quality in Europe - 2016 report. European Environment Agency (EEA) (2016)
2. Lovett, G., et al.: Who needs environmental monitoring? Front. Ecol. Environ. **5**, 253–260 (2007)
3. GMA NEWS. https://www.gmanetwork.com/news/scitech/technology/612877/denr-launches-eu-backed-air-quality-monitoring-system/story. Accessed 01 Jun 2018
4. EEA: "EEA Signals 2013: Every breath we take - Improving air quality in Europe" EEA Signals 2013 (2013)
5. Lo Re, G., Peri, D., Vassallo, S.D.: Urban air quality Monitoring using vehicular sensor networks. In: Gaglio, S., Lo Re, G. (eds.) Advances onto the Internet of Things. AISC, vol. 260, pp. 311–323. Springer, Cham (2014). https://doi.org/10.1007/978-3-319-03992-3_22
6. Yi, W., Lo, K., Mak, T., Leung, K., Leung, Y., Meng, M.: A survey of wireless sensor network based air pollution monitoring systems. Sensors **15**, 31392–31427 (2015)
7. Dhingra, S., Madda, R., Gandomi, A., Patan, R., Daneshmand, M.: Internet of Things mobile-air pollution monitoring system (IoT-Mobair). IEEE Internet Things J. **6**, 5577–5584 (2019)
8. Völgyesi, P., Nádas, A., Koutsoukos, X., Lédeczi, Á.: Air quality monitoring with SensorMap. In: 2008 International Conference on Information Processing in Sensor Networks (ipsn 2008) (2008)
9. Aeroqual. https://www.aeroqual.com/mobile-air-quality-monitoring. Accessed 11 Dec 2017
10. Katulski, R., Stefański, J., Sadowski, J., Ambroziak, S., Namieśnik, J., Wardencki, W.: Mobile monitoring system for control of atmospheric air quality. Polish J. Environ. Stud. **20**, 677–681 (2011)
11. Murty, R., Mainland, G., Rose, I., Chowdhury, A., Gosain, A., Bers, J., Welsh, M.: City-Sense: an urban-scale wireless sensor network and testbed. In: 2008 IEEE Conference on Technologies for Homeland Security, pp. 583–588. Waltham, MA, USA (2008)
12. Vaisala. https://www.vaisala.com/pt/products/instrumentos-sensores-e-outros-dispositivos-de-medicao/instrumentos-para-medicoes-industriais/gmp343. Accessed 15 Dec 2018
13. Apte, J., et al.: High-resolution air pollution mapping with google street view cars: exploiting big data. Environ. Sci. Technol. **51**, 6999–7008 (2017)
14. Silva, L., Mendes, B., Rodrigues, D., Ribeiro, P., Mendes, J.: A mobile environmental monitoring station for sustainable cities. Int. J. Sustain. Dev. Plan. **11**, 949–958 (2016)
15. ThingSpeak. https://thingspeak.com/pages/commercial_learn_more. Accessed 3 Nov 2020
16. https://www.airqualitynow.eu/about_indices_definition.php?fbclid=IwAR3o2V5Gpgnfn mKc-hNKsIFk5vaZavWJoMheK_evsBkWHU0_09-HbdjCVis. Accessed 30 Nov 2020

Mechatronic Design of a Wall-Climbing Drone for the Inspection of Structures and Infrastructure

Erika Ottaviano[1], Pierluigi Rea[2(✉)], Massimo Cavacece[1], and Giorgio Figliolini[1]

[1] Department of Civil and Mechanical Engineering, University of Cassino and Southern Lazio, Cassino, Italy
ottaviano@unicas.it
[2] Department of Mechanical, Chemical and Materials Engineering, University of Cagliari, Via Marengo, 2, 09123 Cagliari, CA, Italy
pierluigi.rea@dimcm.unica.it

Abstract. The inspection of structures and infrastructure is nowadays a relevant problem in Europe, and especially in Italy, after the recent collapses of bridges. Large research activity has been therefore devoted towards the inspection for further management and maintenance of large areas. For those infrastructures difficult to access UAVs, mobile, and climbing robots are the key solution for carrying instrumentation and sensors. UAVs in most of cases are an efficient solution, but they suffer of some drawbacks, among all the security distance to maintain during the flight operation, avoiding the robot to get next to, or in contact to the surface to inspect. Additionally, when dealing with structures exposed to harsh environmental conditions such as strong wind, the possibility of having collisions between to the robot and the surface is quite high. Therefore, to limit possible crashes to surfaces or walls for indoor and even outdoor inspection, wall-climbing drones have been recently conceived being able to fly next to and climb vertical surfaces. In this paper, we present a novel wall-climbing drone based on a multirotor having legs and passive wheels. First preliminary design simulation is reported together with a built prototype.

Keywords: Robotics · Mechatronics · Wall-climbing drone · Robotic inspection

1 Introduction

The integrity of infrastructure and structures, such as bridges, skyscrapers, wind turbines and large aircraft, is closely related to safety issues. Nowadays, due to their aging and potential concerns about their damage or even their collapse, the interest in Structural Health Monitoring (SHM) has increased worldwide, [1, 2]. Although there is a great deal of research on inspecting large structures difficult to access using Unmanned Aerial Vehicles (UAVs) or mobile robotics [3–9], in most of cases the inspection still require great human efforts, installation of additional structures or the use magnet-based technology or vacuum adhesion, and it is time consuming and high-cost. In addition, in most of cases, it requires the stop of the traffic causing big problems to the viability. The new trend in robotics inspections made by UAVs addresses the development of a new concept of a wall-climbing drone, which allows approaching to any type of structure by flying and adhering to the target using a perching mechanism. Furthermore, it does not require the installation of any additional infrastructure and which offers maximum mobility and safety such as the robot to wall. These robots have greater mobility than existing wall robots because they can fly. Furthermore, the robot can also adhere to the surface, therefore, it can perform a close inspection and eventually the maintenance operation of the structure [10]. An example is the SCAMP (Stanford Climbing and Aerial Maneuvering Platform) project [11], which is able to fly, passively perch, climb, and take off. Small drones are generally perfect for resisting to small impacts such as collisions with a wall, they are extremely quick in movements and changes in direction, and can reach higher adhesive forces than larger drones. SCAMP is able to fly, climb, immediately recover stability in the event of a fall and slip into those tunnels where other larger drones cannot access. For its construction, carbon fiber materials were used, drawbacks are related to small dimension (and duration) of the battery. Another example is the VOLIRO [12], which is a novel aerial platform that incorporates the advantages of existing multi-rotor systems with the agility of omnidirectional controllable platforms. It consists of a hexa-copter with tiltable rotors allowing the system to decouple the control of position and orientation. Nowadays, multirotor drones have reached a high level of popularity. Although drones with four or six rotors are common and particularly fast, they are not known for their flexibility due to their fixed rotors [13].

The novelty of VOLIRO is the use of six rotors that operate independently and can rotate 360° achieving interesting features, such as flying on the side, flying upside down and in an upright position. Having in mind the above-mentioned technical solutions, in this paper, we present simulation results and a first prototype of a wall-climbing drone, which can be used for in indoor and outdoor inspections.

2 Mechatronic Design of a Wall-Climbing Drone

The wall-climbing drone should have the characteristics here described: it must fly and move close or attached to walls maintaining the contact or staying at given distance; it must be equipped by suitable sensors, e.g., a camera; it should be low-cost and easy operation. The wall-climbing drone is designed for the inspection of structures and infrastructures. Wall-climbing robots are classified as wall-sticking mechanisms, tilt-rotor-based drones, or multi oriented rotors, as it was recalled in the previous section.

In this context, we have developed a wall-climbing drone composed by four fixed rotors, which are used for the flying mode. The novel idea is to redesign the mechanical architecture by using two additional rotors with fixed axes that are mounted at 90° with respect to the first four ones. To be able to get close to a wall keeping a fixed distance, two limbs are used having passive wheels at the end tips. The two rotors produce a specific propulsion needed to keep the contact to a vertical surface, the passive wheels are used to reduce the friction when climbing the surface.

Figure 1 shows the mechanical design of the wall-climbing drone.

The two legs are equipped with lightweight wheels for maintaining the contact with any vertical surface. Suitable springs allow the damping and reduce the contact impact with the wall. Figure 2 shows the design of the spring whose main characteristics are K

Fig. 1. 3D view of the novel wall-climbing drone.

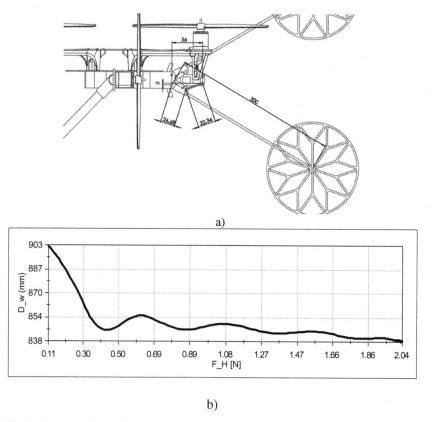

Fig. 2. Design of the spring a); b) distance from the wall D_w VS propulsion force F_H.

= 0.5 N/mm; C = 0.01 N/(mm/s); L = 32.34 mm; initial elongation is $L_0 = 30$ mm. The legs' maximum range is 60–140°. Mechanical specifications are reported in Table 1.

Table 1. Specifications.

Component	Characteristics	Quantity
Frame	Diagonal wheelbase 450 mm Frame mass 282 g Takeoff mass 800 g–1600 g	2 (top and bottom)
ESC	OPTO current 30 A Signal frequency 30 Hz–450 Hz Power pack 3S–4S LiPo	4 + 2
Motor	Stator dim. 22 × 12 mm - KV 920 rpm/V Propeller 10 × 3.8 in; 8 × 4.5 in	4 + 2
KK multi-controller V.5.5	4 rotors control Integrated circuit Atmega IC	1
Propeller pairs	10 in	6
Battery	11.1 V–5000 mAh – mass 440 g	1
Wheel	Dimensions: 8 × 8 × 178 mm Mass 9.07 g	2
Spring	K 0.5 N/mm; C = 0.01 N/(mm/s); L_0 30 mm	2

3 Simulation Results

Simulations of the wall-climbing drone have been developed firstly for sizing the actuation and springs allowing then the system to stick and climb a vertical wall.

A simulation has been carried out considering motion sequence of the approach and contact to a vertical surface, as it is shown in Fig. 3. For the reported numerical results, the horizontal propulsion is considered, while simulation results for the four propellers with vertical axis are not shown. The motion laws used to drive the climbing drone toward the wall is shown in Fig. 4a. In particular, it is worth noting that the horizontal

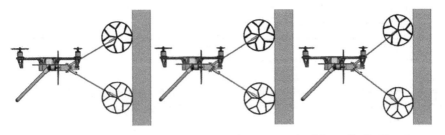

Fig. 3. A sequence for the simulation of the surface approach of the wall-climbing drone.

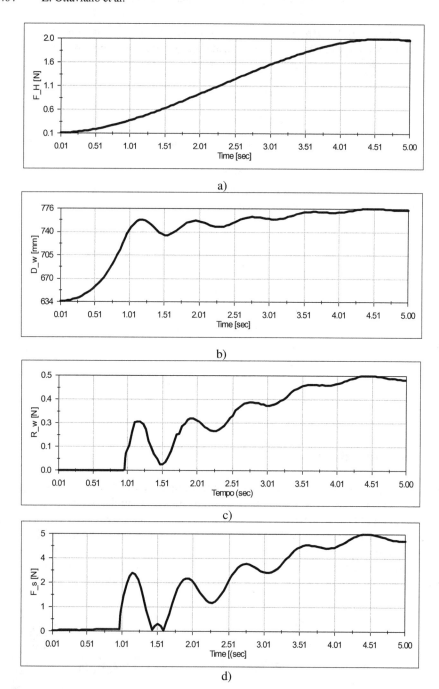

Fig. 4. Numerical results of the simulation in Fig. 3: a) propulsion force F_H; b) wall distance D_w; c) wheel reaction force against the wall R_w; d) spring force F_s.

propulsion force increases in order to get the drone closer to the wall. The distance from the wall is shown in Fig. 4b. In particular, the force of the propellers has a suitable initial value, because a starting threshold of the force is required for getting close to the wall. Subsequently, when the wheels get in contact with the surface this force grows.

4 Mechatronic Design

Figure 5 shows the built prototype, which is currently under testing to reduce masses and static and dynamic balancing. A scheme for the control of the propellers is reported in Fig. 6. It is worth noting that this solution allows decoupling the control of the propellers with vertical axis and those with horizontal axis (labeled as motor 5 and 6 in Fig. 6), the latter used only when approaching to a surface. The idea of decoupled control strategies has been also used in [14–16] for the mechatronic design of multibody systems. The synthesis procedures for the mechanics are described in [17].

The Electronic Speed Controller (ESC) for the drone is a robust, powerful component that connects the flight controller to the motors. Given that each brushless motor requires an ESC, a quadcopter will require 4 ESCs. The ESC takes the signal from the flight controller and power from the battery and makes the brushless motor spin.

The ESC communicates with the control board using a Pulse Position Modulation PPM digital modulation technique. In particular, the FC (Flight Control) sends a sequence of square wave pulses to the ESC which, interpreting their position and/or duration (τ), varies the angular speed of the motors, Fig. 6. The FC sends a pulse of period T and the duration τ in which the signal is high is the factor that determines the final angular speed of the motor. Having chosen motors with a maximum absorption of 10.6 A, an ESC of at least 20 A must be used. ESC can be programmed by specific software. The used KK multi-controller is a flight control board for remote control of multi-copters with 2, 3, 4 and 6 rotors, which is used to stabilize the drone during flight. It takes signals coming from the 3 gyroscopes on the board (roll, pitch, yaw) and sends the information to the

Fig. 5. Built prototype.

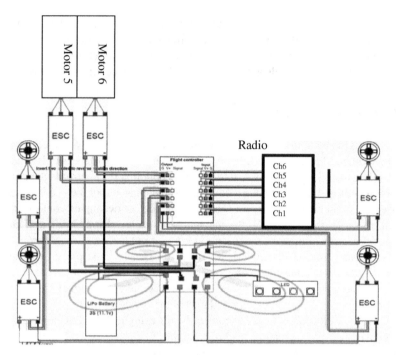

Fig. 6. A scheme of the control for the wall-climbing-drone.

Atmega IC circuit. The latter processes the information according the KK software and sends in output a control signal to the ESCs, which are plugged onto the board and connected to the actuators. Based on the signal from the IC, the ESCs will modify the speed of actuators to control the flight.

5 Conclusion

In this paper, the design and first prototype are presented for a wall-climbing drone that can be used for inspection of structures and infrastructure. Simulation results of the system approaching a vertical surface are reported, they are used both in design stage, for sizing the legs and springs, and for defining the operations during the experiments. A first prototype has been built for experimental tests and it is currently under testing.

Acknowledgments. The proposed paper is part of a project, which has received funds from the Research Fund for Coal and Steel RFCS under grant agreement No 800687.

References

1. Yeum, C.M., Dyke, S.J.: Vision-based automated crack detection for bridge inspection. Comput. Aided Civ. Infrastruct. Eng. **30**, 759–770 (2015)
2. Lin, Y.Z., Nie, Z.H., Ma, H.W.: Structural damage detection with automatic feature-extraction through deep learning. Comput. Aided Civ. Infrastruct. Eng. **32**(12), 1025–1046 (2017)

3. Ottaviano, E., Rea, P., Castelli, G.: THROO: a tracked hybrid rover to overpass obstacles. Adv. Robot. **28**(10), 683–694 (2014). https://doi.org/10.1080/01691864.2014.891949
4. Rea, P., Pelliccio, A., Ottaviano, E., Saccucci, M.: The heritage management and preservation using the mechatronic survey. Int. J. Archit. Heritage **11**(8), 1121–1132 (2017). https://doi.org/10.1080/15583058.2017.1338790
5. Rea, P., Ottaviano, E.: Design and development of an inspection robotic system for indoor applications. Rob. Comput. Integr. Manuf. **49**, 143–151 (2018)
6. Figliolini, G., Ceccarelli, M., Di Gioia, M.: Descending stairs with EP-WAR3 biped robot. In: IEEE/ASME International Conference on Advanced Intelligent Mechatronics (AIM), vol. 2, Paper number: 1225436, pp. 747–752 (2003)
7. Figliolini, G., Ceccarelli, M.: EP-WAR3 biped robot for climbing and descending stairs. Robotica **22**(4), 405–417 (2004)
8. Figliolini, G., Rea, P., Conte, M.: Mechanical design of a novel biped climbing and walking robot. In: Parenti Castelli, V., Schiehlen, W. (eds.) ROMANSY 18 Robot Design, Dynamics and Control. CICMS, vol. 524, pp. 199–206. Springer, Vienna (2010). https://doi.org/10.1007/978-3-7091-0277-0_23
9. Ottaviano, E., Rea, P.: Design and operation of a 2-DOF leg-wheel hybrid robot. Robotica **31**(8), 1319–1325 (2013)
10. PhyOrg: Development of a wall-climbing drone (2016). https://phys.org/news/2016-03-wall-climbing-drone.html
11. SCAMP, webpage (2021). http://bdml.stanford.edu/Main/MultiModalRobots
12. Mina, S.K., Verling, S., Elkhatib, O., Gilitschenski, I., Gilitschenski, I.: Voliro: an omnidirectional hexacopter with tiltable rotors. arXiv, Cornell University (2018)
13. Alkalla, M.G., Fanni, M.A., Abdelfatah, M.M.: A novel propeller–type climbing robot for vessels inspection. In: IEEE International Conference on Advanced Intelligent Mechatronics (AIM) (2018)
14. Rea, P., Ottaviano, E., Castelli, G.: A procedure for the design of novel assisting devices for the sit-to-stand. J. Bionic Eng. **10**(4), 488–496 (2013)
15. Ceccarelli, M., Ottaviano, E., Galvagno, M.: A 3-DOF parallel manipulator as earthquake motion simulator. In: Proceedings of the 7th International Conference on Control, Automation, Robotics and Vision, ICARCV 2002, pp. 944–949 (2002)
16. Rea, P., Ottaviano, E., Conte, M., D'Aguanno, A.D., Carolis, D.: The design of a novel tilt seat for inversion therapy. Int. J. Imaging Rob. **11**(3), 1–10 (2013)
17. Figliolini, G., Rea, P., Angeles, J.: The synthesis of the axodes of RCCC linkages. J. Mech. Robot **8** (2016). https://doi.org/10.1115/1.4031950. ISSN 19424302

Influence of Magnitude of Interaction on Control in Decentralized Adaptive Control of Two Input Two Output Systems

Karel Perutka

Tomas Bata University, Nam. T.G.M. 5555, 76001 Zlin, Czech Republic
kperutka@utb.cz

Abstract. This paper presents an analysis of the influence of interactions on the course of the control process using the chosen method of decentralized adaptive control of two input two output systems. In the paper, after the general introduction to decentralized and adaptive control, the necessary theoretical framework is presented. In this chapter, there is described the method used for decentralized control, the method of identifying the system parameters and the method of calculating the parameters of the controller. In the examples, the effect of the change in the gain of the interaction transfer function in one branch and the effect of the change in the gain of the interaction transfer functions in both branches were tested. Furthermore, it was verified that the change of the pole value in the interaction in one branch influences the control, and the effect of the change of the pole value in the transfer functions of the interaction in both branches. The results of these verifications are presented on selected examples and finally their evaluation is performed.

Keywords: Adaptive control · Self-tuning controller · decentralized control · suboptimal linear quadratic tracking · least squares method · Interaction · Gain · poles

1 Introduction

Most systems in nature and engineering are multi-input and multi-output systems. The simplest example is two-input and two-output systems. These systems are often controlled by a multivariable or decentralized controller. The decentralized control is widely used in practice. Decentralized control could be used for a class of two-time-scale interconnected networks with unknown slow dynamics. The network in this case consists of *m* lower-level subsystems interconnected through an upper-level main system with unknown dynamics [1]. Decentralized control can employ fuzzy control such as the case of the adaptive decentralized fuzzy dynamic surface control scheme for a class of nonlinear large-scale systems with input and interconnection delays. It was proven that this method can ensure that all signals of the closed-loop large-scale systems in the presence of both interconnection and input delays [2]. Frequency control is another area of application, for example, decentralized and discretized control of storage systems offering primary frequency control. It was shown that primary frequency control is a suitable and economically viable service for small-scale energy storage (ES) systems [3]. Even PI control can be used, such as adaptive decentralized output feedback PI tracking control design for uncertain interconnected nonlinear systems with input quantization. By designing an input-driven filter, the unknown states are estimated and then an adaptive decentralized output feedback PI tracking controller is constructed via the backstepping method and neural network technique. The stability of the closed-loop system is addressed based on the Lyapunov function technique plus graph theory, and all the signals in the closed-loop system are uniformly ultimately bounded [4]. The decentralized control is also used in robotics, for example as recursive decentralized control for robotic manipulators, where decentralized robust control is designed under the assumption that communications are allowed in the decentralized controllers. The communicated information includes the relative motion states at the joints, the geometry and mass parameters, and the control torques. These communications enable to recursively construct the nominal part of the physical interactions for compensation [5]. Or it is used as torque sensorless decentralized neuro-optimal control for modular and reconfigurable robots with uncertain environments [6]. Backstepping can be also used to design controller, for example backstepping-based decentralized tracking control for a class of interconnected stochastic nonlinear systems coupled via a directed graph, using quartic Lyapunov functions and graph theory, where the signals of the resulting closed-loop system are globally bounded in probability, and the tracking errors converge to a compact set, whose radius can be adjusted by choosing different controller design parameters [7]. In practice, decentralized control is used for example in aircraft [8], heating loads [9], electric vehicle aggregator [10], or in chillers [11].

Adaptive control extends the possibility of use in more systems. One of the possible algorithms is the use of self-tuning controllers, which combine the calculation of the online identification of the parameters of the controlled system together with the calculation of the controller's parameters by the chosen method. For example, the least-squares identification method can be used and the linear-quadratic suboptimal tracking method can be used to calculate the parameters. From theoretical point of view, nice paper about adaptive control is, for example, adaptive output feedback control for a class of nonlinear time-varying delay systems. It discusses the adaptive control problem for a class of nonlinear time-varying delay systems with triangular structure, whose nonlinearities satisfy uncertain homogeneous conditions [12]. We can find a lot of applications of adaptive control in practice, such as in rotor systems [13], at control of autonomous ships [14], at space-crafts [15], or at aircraft [16].

In this paper, the method of decentralized self-tuning control is used. It includes recursive least squares identification method and linear-quadratic suboptimal tracking method with exponential forgetting employing regression polynomials. It discusses the effect of the change of the gain and/or the poles of the interaction transfer function(s) at the given example of a two-input two-output system. The paper is organized in the following way. Firstly, theoretical background of the used methods is given. It is followed by examples of the influences of the changes in the gains and poles of interaction transfer functions. Finally, the conclusion is given.

2 Theoretical Background

All theoretical background is based on the work of perutka. [17] this paper is referred to the articles, on which the theoretical background is based on.

2.1 Decentralized Approach

Using the decentralized approach, the control is divided into a set of sub-tasks that are matched by simple controllers. These partial tasks will then give us the overall course of control. The main advantages of decentralized control are primarily that a more complex system is divided into a set of simple tasks and the resulting controller is more flexible.

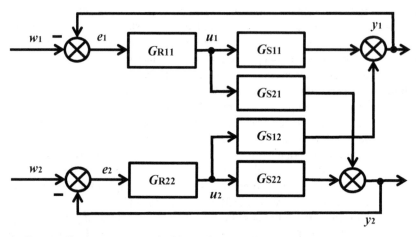

Fig. 1. Decentralized system control with two inputs and two outputs, the so-called P structure

A special example of multidimensional systems is a system with two inputs and two outputs. This can be realized by the so-called P structure, see Fig. 1. In this case, the inputs to the systems describing the interactions are the values of the action signals of the SISO controllers and their outputs are added to the opposite outputs of the main diagonal systems. From this figure, we get the transfer function equations of the model in the form

$$G_{S11} = \frac{G_{S12}G_{S21}G_{R22}}{1 + G_{R22}G_{S22}} \tag{1}$$

$$G_{S22} = \frac{G_{S21}G_{S12}G_{R11}}{1 + G_{R11}G_{S11}} \tag{2}$$

2.2 Recursive Identification Using Approximation Polynomials

A prerequisite for good control is the most accurate description of the regulated system. Identification is the procedure by which the mathematical model of a system is obtained.

The beginnings of identification based on continuous models date back to the middle of the 20th century. For continuous-time identification, the identified model is in the form of the differential equations. Differential equations contain expressions with derivatives over time that are not measurable. It is possible to replace the segment by an approximation polynomial whose derivatives can be calculated analytically in advance and then calculated numerically, see Fig. 2. This approach was for example used by Perutka [20].

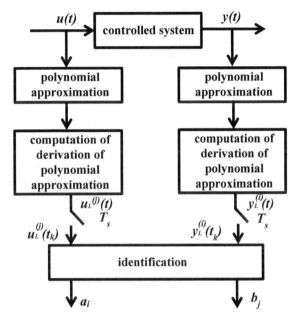

Fig. 2. Identification scheme for continuous-time systems

Least Squares Method with Exponential Forgetting

The estimation of model parameters is computed as

$$\hat{\Theta}(k) = \hat{\Theta}(k-1) + \mathbf{K}(k)\hat{e}(k) \tag{3}$$

The gain vector is calculated as

$$\mathbf{K}(k) = \frac{\mathbf{C}(k-1)\boldsymbol{\phi}(k)}{1 + \boldsymbol{\phi}^T(k)\mathbf{C}(k-1)\boldsymbol{\phi}(k)} \tag{4}$$

and covariance matrix

$$\mathbf{C}(k) = \mathbf{C}(k-1) - \frac{\mathbf{C}(k-1)\boldsymbol{\phi}(k)\boldsymbol{\phi}^T(k)\mathbf{C}(k-1)}{1 + \boldsymbol{\phi}^T(k)\mathbf{C}(k-1)\boldsymbol{\phi}(k)} \tag{5}$$

The following applies to the calculation of the prediction error

$$\hat{e}(k) = y(k) - \boldsymbol{\phi}^T(k)\hat{\Theta}(k-1) \tag{6}$$

In the case of exponential forgetting, the criterion of identification is

$$J = \sum_{i=k_0}^{k} \left(\varphi^{k-i} e(i)\right)^2 \tag{7}$$

where the exponential forgetting factor is chosen in the range of 0 to 1, the most common near 1. If

$$\boldsymbol{\phi}^T(k)\mathbf{C}(k-1)\boldsymbol{\phi}(k) > 0 \tag{8}$$

then

$$\mathbf{C}(k) = \mathbf{C}(k-1) - \frac{\mathbf{C}(k-1)\boldsymbol{\phi}(k)\boldsymbol{\phi}^T(k)\mathbf{C}(k-1)}{\eta^{-1} + \boldsymbol{\phi}^T(k)\mathbf{C}(k-1)\boldsymbol{\phi}(k)} \tag{9}$$

where

$$\eta(k) = \varphi(k) - \frac{1-\varphi(k)}{\xi(k)} \tag{10}$$

If

$$\boldsymbol{\phi}^T(k)\mathbf{C}(k-1)\boldsymbol{\phi}(k) = 0 \tag{11}$$

then

$$\mathbf{C}(k) = \mathbf{C}(k-1) \tag{12}$$

Furthermore

$$\varphi(k) = \left\{1 + (1+\rho)[\ln(1+\xi(k-1))] + \left[\frac{(v(k-1)+1)\eta(k-1)}{1+\xi(k-1)+\eta(k-1)} - 1\right]\frac{\xi(k-1)}{1+\xi(k-1)}\right\}^{-1} \tag{13}$$

$$\eta(k) = \frac{\hat{e}^2(k)}{\lambda(k)} \tag{14}$$

$$v(k) = \varphi(k)[v(k-1)+1] \tag{15}$$

$$\lambda(k) = \varphi(k)\left[\lambda(k-1) + \frac{\hat{e}^2(k)}{1+\xi(k-1)}\right] \tag{16}$$

$$\xi(k) = \boldsymbol{\phi}^T(k)\mathbf{C}(k-1)\boldsymbol{\phi}(k) \tag{17}$$

The parameters estimation vector is in the form

$$\hat{\Theta}^T(k) = \left(\hat{a}_0, \hat{a}_1, \ldots, \hat{a}_{\deg(a)}, \hat{b}_0, \hat{b}_1, \ldots, \hat{b}_{\deg(b)}, d\right) \tag{18}$$

and regressor

$$\boldsymbol{\phi}^T(k) = \left(-y(t_k), \ldots, -y_L^{(n-1)}(t_k), u(t_k), \ldots, u_L^{(m)}(t_k), 1\right) \tag{19}$$

2.3 Self-tuning Controller

The main reason for using adaptive control is that the systems change over time or the characteristics of the controlled system are unknown. the basic principle of adaptive systems is to change the characteristics of the controller based on the characteristics of the controlled process [19] the general scheme of the self-tuning controller is shown in Fig. 3.

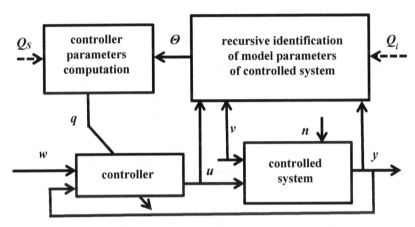

Fig. 3. The general scheme of the self-tuning controller

2.4 Suboptimal Linear Quadratic Tracking Controller

The method was introduced by Dostál [18]. If the system of Fig. 4 is considered

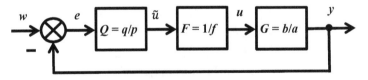

Fig. 4. The system with a feedback controller

Let's minimize a quadratic function with two penalty constants

$$J = \int_0^\infty \left\{ \mu e^2(t) + \varphi \tilde{u}^2(t) \right\} dt \tag{20}$$

The Laplace image of the setpoint holds

$$w(s) = \frac{h_w(s)}{s\tilde{f}_w(s)} \tag{21}$$

It holds for degrees of polynomials

$$\deg(h_w) \leq \deg(f_w), f_w(0) \neq 0 \tag{22}$$

We calculate stable polynomials g and n as results of spectral factorizations

$$(as)^*\varphi as + b^*\mu b = g^*g, \; n^*n = a^*a \tag{23}$$

we solve the following diophantine equation

$$asp + bq = gn \tag{24}$$

Considering the transfer function of the system

$$G(s) = \frac{b_0}{s^2 + a_1 s + a_0} \tag{25}$$

then the controller is

$$F(s)Q(s) = \frac{q_2 s^2 + q_1 s + q_0}{s(p_2 s^2 + p_1 s + p_0)} \tag{26}$$

In this case, the polynomials have the form

$$g(s) = g_3 s^3 + g_2 s^2 + g_1 s + g_0 \tag{27}$$

$$n(s) = s^2 + n_1 s + n_0 \tag{28}$$

and to calculate their coefficients obtained by spectral factorization

$$g_0 = \sqrt{\mu b_0^2} \tag{29}$$

$$g_1 = \sqrt{2 g_2 g_0 + \varphi a_0^2} \tag{30}$$

$$g_2 = \sqrt{2 g_3 g_1 + \varphi(a_1^2 - 2 a_0)} \tag{31}$$

$$g_3 = \sqrt{\varphi} \tag{32}$$

$$n_0 = \sqrt{a_2^0} \tag{33}$$

$$n_1 = \sqrt{2 n_0 - a_1^2 - 2 a_0} \tag{34}$$

2.5 Calculation of Derivatives Using Approximation Functions

To calculate the derivatives, we approximate the closest neighborhood for a given time by the approximation function. For example, we will use the Lagrange polynomial in the form

$$P_2(x) = \frac{(x-b)(x-c)}{(a-b)(a-c)}f(a) + \frac{(x-a)(x-c)}{(b-a)(b-c)}f(b) + \frac{(x-b)(x-b)}{(c-a)(c-b)}f(ac) \quad (35)$$

whose first derivative is

$$f'(x) \cong P_2^-{}'(x) = \frac{2x-(b+c)}{(a-b)(a-c)}f(a) + \frac{2x-(a+c)}{(b-a)(b-c)}f(b) + \frac{2x-(a+b)}{(c-a)(c-b)}f(c) \quad (36)$$

and the second derivative is

$$f''(x) \cong P_2^{//}(x) = \frac{2f(a)}{(a-b)(a-c)} + \frac{2f(b)}{(b-a)(b-c)} + \frac{2f(c)}{(c-a)(c-b)} \quad (37)$$

2.6 Numerical Solution of Differential Equations

The chosen method of numerical solution of differential equations was used for realization, namely the Runge-Kutta-Fehlberg method, which can be described by the following equations

$$y_{n+1} = y_n + \left(\frac{16}{135}k_1 + \frac{6656}{12825}k_3 + \frac{28561}{56430}k_4 - \frac{9}{50}k_5 + \frac{2}{55}k_6\right) \quad (38)$$

$$k_1 = \Delta t f(t_n, y_n) \quad (39)$$

$$k_2 = \Delta t f\left(t_n + \frac{1}{4}h, y_n + \frac{1}{4}k_1\right) \quad (40)$$

$$k_3 = \Delta t f\left(t_n + \frac{3}{8}h, y_n + \frac{3}{32}k_1 + \frac{9}{32}k_2\right) \quad (41)$$

$$k_4 = \Delta t f\left(t_n + \frac{12}{13}h, y_n + \frac{1932}{2197}k_1 - \frac{7200}{2197}k_2 + \frac{7296}{2197}k_3\right) \quad (42)$$

$$k_5 = \Delta t f\left(t_n + h, y_n + \frac{439}{216}k_1 - 8k_2 + \frac{3680}{513}k_3 - \frac{845}{4104}k_4\right) \quad (43)$$

The reason for using this method in the computational algorithm is the higher speed of computation than the implicit ones.

$$k_6 = \Delta t f\left(t_n + \frac{1}{2}h, y_n - \frac{8}{27}k_1 + 2k_2 - \frac{3544}{2565}k_3 + \frac{1859}{4104}k_4 - \frac{11}{40}k_5\right) \quad (44)$$

3 Results

The system is described as
$$G_{11}(s) = \frac{2}{s^2+3s+2}, G_{22}(s) = \frac{1.5}{s^2+2.4s+3.1}, G_{21}(s) = \frac{k_1}{T_1 s+1}, G_{12}(s) = \frac{k_2}{T_2 s+1}$$

3.1 Increase of First Interaction Gain

See Figs. 5 and 6.

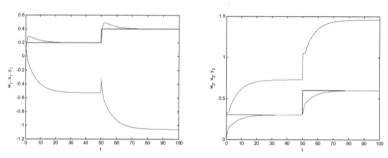

Fig. 5. Output of control – 1st subsystem (left), 2nd (right), for $k_1 = 0.1, T_1 = 0.1, k_2 = 1, T_2 = 1$

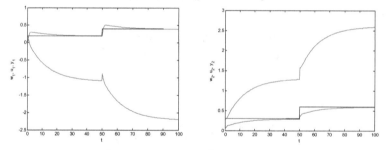

Fig. 6. Output of control – 1st subsystem (left), 2nd (right), for $k_1 = 0.3, T_1 = 0.1, k_2 = 1, T_2 = 1$

3.2 Increase of Both Interaction Gains

See Fig. 7.

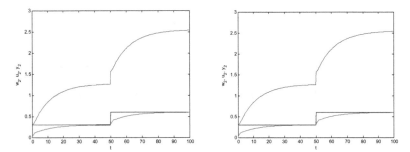

Fig. 7. Output of control – 1st subsystem (left), 2nd (right), for $k_1 = 0.2, T_1 = 0.1, k_2 = 1.4, T_2 = 1$

3.3 Increase of First Interaction Pole

See Fig. 8.

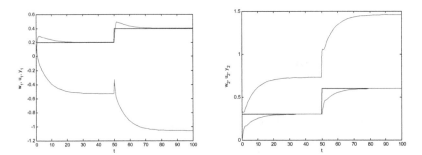

Fig. 8. Output of control – 1st subsystem (left), 2nd (right), for $k_1 = 0.2, T_1 = 1, k_2 = 1.4, T_2 = 1$

3.4 Increase of Both Interaction Poles

See Fig. 9.

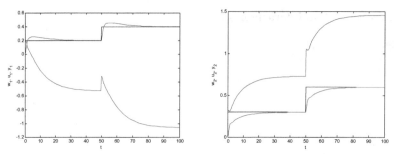

Fig. 9. Output of control – 1st subsystem (left), 2nd (right), for $k_1 = 0.2$, $T_1 = 1$, $k_2 = 1.4$, $T_2 = 5$

4 Conclusion

During the experimental verification it was found that the change of one value of the time constant and the gain will be reflected in both output control processes, the same it is for the change of quantities in both interactions. If the time constant is increased, this will be reflected in a long time to reach the set-point of the 2nd output. Increasing the gain value increases the overshoot value of the output signal at output 2.

References

1. Zhou, L., Zhao, J., Ma, L., Yang, C.: Decentralized composite suboptimal control for a class of two-time-scale interconnected networks with unknown slow dynamics. Neurocomputing **382**, 71–79 (2020)
2. Baigzadehnoe, B., Rahmani, Z., Khosravi, A., Rezaie, B.: Adaptive decentralized fuzzy dynamic surface control scheme for a class of nonlinear large-scale systems with input and interconnection delays. Eur. J. Control **54**, 33–48 (2020)
3. Ziras, C., Prostejovsky, A.M., Bindner, H.W., Marinelli, M.: Decentralized and discretized control for storage systems offering primary frequency control. Eng. Procedia **157**, 1253–1265 (2020)
4. Sun, H., Zong, G., Chen, P.: Adaptive decentralized output feedback PI tracking control design for uncertain interconnected nonlinear systems with input quantization. Inf. Sci. **512**, 186–206 (2020)
5. Hu, Q., Guo, C., Zhang, Y., Zhang, J.: Recursive decentralized control for robotic manipulators. Aerosp. Sci. Technol. **76**, 374–385 (2018)
6. Dong, B., Zhou, F., Liu, K., Li, Y.: Torque sensorless decentralized neuro-optimal control for modular and reconfigurable robots with uncertain environments. Neurocomputing **282**, 60–73 (2018)
7. Li, X.-J., Ren, X.-X., Yang, G.-H.: Backstepping-based decentralized tracking control for a class of interconnected stochastic nonlinear systems coupled via a directed graph. Inf. Sci. **477**, 302–320 (2019)
8. Pan, M., Chao, L., Zhou, W., Huang, J., Chen, Y.-H.: Robust decentralized control design for aircraft engines: a fractional type. Chin. J. Aeronaut. **32**(2), 347–360 (2019)
9. Zhou, Y., Cheng, M., Wu, J., Long, C.: Decentralized control of industrial heating loads for providing multiple levels and types of primary frequency control service. Eng. Procedia **158**, 3138–3143 (2019)

10. Xu, S., Yan, Z., Feng, D., Zhao, X.: Decentralized charging control strategy of the electric vehicle aggregator based on augmented Lagrangian method. Int. J. Elect. Power Eng. Syst. **104**, 673–679 (2019)
11. Dai, Y., Jiang, Z., Wang, S.: Decentralized control of parallel-connected chillers. Eng. Procedia **122**, 86–91 (2017)
12. Zhou, H., Zhai, J.: Adaptive output feedback control for a class of nonlinear time-varying delay systems. Appl. Math. Comput. **365**, 124692 (2020)
13. Nevaranta, N., Jaatinen, P., Vuojolainen, J., Sillanpää, T., Pyrhönen, O.: Adaptive MIMO pole placement control for commissioning of a rotor system with active magnetic bearings. Mechatronics **65**, 102313 (2020)
14. Haseltalab, A., Negenborn, R.R.: Adaptive control of autonomous ships with uncertain model and unknown propeller dynamics. Control Eng. Practice **91**, 104116 (2019)
15. Hu, H., Liu, L., Wang, Y., Cheng, Z., Luo, Q.: Active fault-tolerant attitude tracking control with adaptive gain for spacecrafts. Aerosp. Sci. Technol. **98**, 105706 (2020)
16. Maity, A., Höcht, L., Holzapfel, F.: Time-varying parameter model reference adaptive control and its applications to aircraft. Eur. J. Control **50**, 161–175 (2019)
17. Perutka, K.: Decentralized adaptive suboptimal LQ control in microsoft excel VBA. In: Machado, J., Soares, F., Veiga, G. (eds.) HELIX 2018. LNEE, vol. 505, pp. 116–123. Springer, Cham (2019). https://doi.org/10.1007/978-3-319-91334-6_17
18. Dostál, P., Bobál, V.: The suboptimal tracking problem in linear systems. In: Proceedings of the 7th Conference on Control and Automation, Haifa, Israel, pp. 667–673 (1999)
19. Bobal, V., Böhm, J., Fessl, J., Machacek, J.: Digital Self-tuning Controllers. Springer, London (2005)
20. Perutka, K.: Adaptive LQ control with pre-identification of two tanks laboratory model. In: Annals of DAAAM and Proceedings of the International DAAAM Symposium, Vienna, Austria, pp. 439–440 (2009)

Author Index

A
Amorim, Ana Rita, 129
Andrade, Marina A. P., 164
Arantes, Carlos, 449
Arfeen, M. Irfanullah, 433
Aștilean, Adina, 119, 141
Avram, Camelia, 119, 141
Avram, Mihai, 90

B
Bakircioğlu, Veli, 81
Bezerra, Karolina, 406
Bezerra, Karolina Celi Tavares, 422
Bonilla, Silvia Helena, 236
Borucka, Anna, 44
Braga, Ana Cristina, 1, 129

C
Çabuk, Nihat, 81
Calhamonas, Gabriel, 164
Cartal, Adrian, 189
Carvalho, Vítor, 119, 334, 377
Castelo Branco, Rodolfo Ramos, 422
Castillo-García, Fernando J., 313
Castro-Ribeiro, M. Leonor, 406
Cavacece, Massimo, 460
Císar, Miroslav, 112
Coandă, Philip, 90
Constantin, Victor, 90
Cunha, Pedro, 119

D
da Silva, António Ferreira, 301
de Morais, Misael Elias, 422
de Sousa, Mário, 301
Delgado, Pedro, 1
Dionísio, Rogério, 213
Dostatni, Ewa, 14

F
Ferreira, Marco, 154
Ferreira, Pedro José Gabriel, 236
Figliolini, Giorgio, 460
Filgueira, Anna Kellssya Leite, 422
Freitas, Luis, 221

G
Gallardo, Isabella Diniz, 422
Gazdos, Frantisek, 99
Gheorghe, Gheorghe, 270
Goliński, Marek, 351
Gonçalves, A. Manuela, 221
Gonçalves, Gil, 213, 322
Gonzalez-Rodríguez, Antonio, 313
Graczyk-Kucharska, Magdalena, 257
Grămescu, Bogdan, 90, 189
Gütmen, Selma, 257

H
Hashim, Ahmed Sachit, 189
Hrybiuk, Olena, 55

I
Ilie, Iulian, 270
Ivanov, Vitalii, 366

K
Kandera, Matej, 112
Karpov, Vadym, 201
Klarák, Jaromír, 112
Kotlarz, Piotr, 14
Kozłowski, Edward, 44
Kunz, Guilherme, 69
Kuric, Ivan, 112

L
Leiras, Valdemar, 129
Liu, Yiliu, 44
Lopes, Hugo Baptista, 334
Lopes, Isabel, 1
Lopes, Sérgio, 449

M
Machado, José, 285
Magalhães, António, 301
Malhão, Sérgio, 213
Malheiro, Teresa, 221
Marques, Mário J. Simões, 164
Martins, Cristina, 1
Martins, Ketinlly Yasmyne Nascimento, 422
Martins, Nuno, 178
Martynova, Nataliia, 366
Mazurkiewicz, Dariusz, 44
Miądowicz, Marek, 351
Modoranu, Mihai, 141
Monteiro, Caetano, 154
Monteiro, Paula, 221

N
Neto, Luís, 213
Nițu, Constantin, 189
Nunes, Isabel, 164

O
Olival, Ana, 406
Oliveira, Rui, 221
Ottaviano, Erika, 313, 460

P
Pavlenko, Ivan, 366
Pereira, Ana Paula, 393
Pereira, Eliseu, 322
Pereira, Filipe, 377
Perondi, Eduardo A., 25
Perutka, Karel, 468
Pires, Mariana A., 406
Providência, Bernardo, 245

R
Radu, Dan, 141
Ralha, Sónia, 178

Ramos, Jorge, 449
Rea, Pierluigi, 313, 460
Reis, João, 322
Reis, Luís Paulo, 322
Ribeiro, Pedro, 221
Rijo, Marcos G. Q., 25
Rocha, Ana Paula, 322
Rodrigues, Miguel, 154
Rojek, Izabela, 14

S
Sacomano, José Benedito, 236
Samokhvalov, Dmytro, 366
Santos, Adriano A., 301
Santos, Pedro, 393
Sarantis, Demetrios, 433
Sarmanho Jr., Carlos A. C., 25
Sarmanova, Lenka, 99
Seabra, Eurico, 129
Sena Esteves, João, 393, 449
Shah, Adil Ali, 433
Silva, Bárbara, 129
Silva, João, 285
Silva, Luís F., 129
Silva, Rute, 245
Silva, Vinícius, 393
Simion, Ionel, 37
Simoes, Ricardo, 178
Sivykh, Dmytro, 201
Soares, Filomena, 119, 377, 393
Sobczyk S., Mário R., 25
Sorin-Ionut, Badea, 270
Sousa, João, 285
Spacek, Lubos, 293
Spychała, Małgorzata, 351
Strimovskyi, Sergii, 201
Sylla, Cristina, 334
Szafrański, Maciej, 257

T
Tanasie, Marian, 37
Teixeira, André, 449
Teixeira, Humberto Nuno, 1
Teodoro, M. Filomena, 164
Tîrnovan, Tudor Claudiu, 119
Torres, Pedro, 213

V
Vasconcelos, Rosa, 377
Veretennikov, Ievgenii, 201
Veronica, Despa, 270
Viana, Rui, 129
Vicente, José, 221
Vilaça, A. A., 406
Vilarinho, Cândida, 406

Author Index

Vojtesek, Jiri, 293
Volontsevich, Dmitriy, 201

W
Weber, Gerhard Wilhelm, 257

Y
Yildirim, Şahin, 81

Z
Zajačko, Ivan, 112
Zhylenko, Tetiana, 366
Zuban, Yurii, 366

CPSIA information can be obtained
at www.ICGtesting.com
Printed in the USA
LVHW080845210621
690747LV00001B/20